Metareasoning: Theoretical and Methodological Developments

Metareasoning: Theoretical and Methodological Developments

Guest Editors

Linden Ball
Beth Richardson

MDPI

Basel • Beijing • Wuhan • Barcelona • Belgrade • Novi Sad • Cluj • Manchester

Guest Editors
Linden Ball
School of Engineering & Computing
University of Central Lancashire
Preston
United Kingdom

Beth Richardson
School of Engineering & Computing
University of Central Lancashire
Preston
United Kingdom

Editorial Office
MDPI AG
Grosspeteranlage 5
4052 Basel, Switzerland

This is a reprint of the Special Issue, published open access by the journal *Journal of Intelligence* (ISSN 2079-3200), freely accessible at: www.mdpi.com/journal/jintelligence/special_issues/metareasoning.

For citation purposes, cite each article independently as indicated on the article page online and using the guide below:

Lastname, Firstname, Firstname Lastname, and Firstname Lastname. Article Title. *Journal Name* **Year**, *Volume Number*, Page Range.

ISBN 978-3-7258-4052-6 (Hbk)
ISBN 978-3-7258-4051-9 (PDF)
https://doi.org/10.3390/books978-3-7258-4051-9

Contents

About the Editors

Linden Ball

Linden J. Ball is a Professor of Cognitive Psychology at the University of Central Lancashire, Preston, UK, where he is an Associate Dean (Research and Knowledge Exchange) in the School of Engineering and Computing. Linden conducts research on reasoning, problem-solving, and decision-making in both laboratory and real-world contexts, with his primary interests focused on metareasoning. Linden is the Editor-in-Chief of the *Journal of Cognitive Psychology*, an Associate Editor of *Thinking & Reasoning*, and an Editorial Board Member for several other journals. He is the Editor of Routledge's *Current Issues in the Thinking & Reasoning* book series, and he has recently co-authored a textbook entitled "Cognitive Psychology in a Changing World", published by Routledge as a volume in the British Psychological Society's "Core Textbooks" series. Linden has additionally co-edited two volumes in the Routledge *International Handbook* series, one on thinking and reasoning (2017) and the other on creative cognition (2024).

Beth Richardson

Dr Beth H. Richardson is a fully tenured Senior Research Fellow in the School of Engineering and Computing at the University of Central Lancashire, Preston, UK. Beth is broadly interested in the metacognitive aspects of group reasoning, problem-solving, and decision-making, including processes of alignment, dialogue, and intersubjectivity. Beth's research extends to address issues of real-world significance with a specific focus on advancing technological approaches to support decision-making. Beth is an Associate Editor of the *Journal of Cognitive Psychology*.

Journal of
Intelligence

MDPI

Editorial

Metareasoning: Theoretical and Methodological Developments

Linden J. Ball * and Beth H. Richardson

School of Engineering & Computing, University of Central Lancashire, Preston PR1 2HE, UK; bhrichardson@uclan.ac.uk
* Correspondence: lball@uclan.ac.uk

1. Introduction

This Special Issue aims to capture current theoretical and methodological developments in the field of metareasoning, which is concerned with the metacognitive processes that monitor and control our ongoing thinking and reasoning. Monitoring processes evaluate the efficacy of the "object-level" cognition that underpins task performance (Nelson and Narens 1990), including processes that are involved in understanding task information, drawing inferences and making decisions. In contrast, control processes allocate cognitive resources (e.g., attention and working memory), initiate new strategies and also terminate failing ones.

Progress in understanding metareasoning was given significant impetus several years ago, when Ackerman and Thompson (2017) published an important article in which they reviewed the state of the art in research and theorising relating to the metacognitive monitoring and control that arises in situations involving high-level cognition. A key outcome of this review was Ackerman and Thompson's presentation of a "metareasoning framework" used to integrate existing findings relating to metacognition in situations involving thinking and reasoning. Central to this framework is the idea that metacognitive monitoring takes places continually during task performance and is sensitive to fluctuating feelings of certainty and uncertainty that are experienced by a reasoner regarding how well an object-level process or strategy is unfolding. Likewise, metacognitive control continually responds to these shifting levels of experienced certainty or uncertainty—maintaining ongoing object-level processing if it is going well or else triggering strategy change if it is floundering. Metacognitive control processes are also needed to cease object-level processing if a satisfactory outcome is achieved or if object-level processing is failing to deliver an outcome.

At the time of proposing their metareasoning framework, Ackerman and Thompson (2017) fully acknowledged that it represented only a starting point to help catalyse further research and conceptual development in this area. However, since the inception of the framework, there has been little doubt that the concepts that it articulates have been highly influential in sparking a great deal of subsequent research and attendant theorising (e.g., Ackerman 2023; Ackerman and Thompson 2018; Mata 2020; Husselman et al. 2024; Morsanyi et al. 2019; Richardson et al. 2024; Undorf et al. 2021). In the present Special Issue, we showcase 13 articles that each contribute in important ways to the advancement of an understanding of metareasoning on the back of Ackerman and Thompson's (2017) initial foundation and that also contribute towards the associated methodological developments in this area of enquiry.

In organising the articles in this Special Issue, we have identified three key themes around which to showcase theoretical and methodological progress. The first theme relates to the interplay between metareasoning and object-level reasoning when the latter is conceptualised from a "dual-process" perspective, whereby relatively rapid, intuitive, and heuristic processes give rise to default responses that can be intervened upon by relatively slow, reflective, and analytic processes (e.g., Evans and Stanovich 2013a, 2013b). The second theme concerns the nature of individual differences in metareasoning, which is a topic that is being subjected to increasing scrutiny. The third theme concerns the way in which metareasoning research is rapidly extending its reach to address challenging questions in new domains of enquiry beyond those that have traditionally been the preserve of investigations. As will be demonstrated, such research is giving rise both to valuable methodological developments as well as important theoretical advancements.

2. Metareasoning and Dual-Process Theories

The study of metareasoning has featured extensively in many recent investigations that adopt a dual-process perspective, whereby object-level reasoning is viewed as involving two qualitatively distinct types of processes, which Evans and Stanovich (2013a, 2013b) refer to as Type 1 and Type 2 processes. Type 1 processes are described as intuitive, heuristic, and associative in nature, and are defined in terms of being relatively undemanding of working-memory resources as well as autonomous (i.e., they run to completion once they have been cued). Furthermore, correlated features of Type 1 processes suggest that they are typically high capacity, rapid, non-conscious, and capable of running in parallel. In contrast, Type 2 processes are described as reflective, deliberate, analytic, and controlled, and are defined in terms of requiring working-memory resources and as having a focus on hypothetical thinking. In addition, the correlated features of Type 2 processes suggest that they tend to be slow, capacity limited, serial, and conscious. Type 2 processes are additionally considered to be less prone to biases than Type 1 processes, although they are not immune from such bases, which can arise from the application of inappropriate or inadequate analytic operations (e.g., Evans 2018; Stanovich 2018).

The interplay between Type 1 and Type 2 processes in reasoning can readily be appreciated when considering the typical responding that is observed when people tackle items that make up the cognitive reflection test (CRT; Frederick 2005; Kahneman and Frederick 2002). A good example of a CRT item is the well-known bat-and-ball problem: A bat and a ball cost $1.10 in total. The bat costs $1.00 more than the ball. How much does the ball cost? The answer of 10 cents comes to mind quickly and intuitively because of Type 1 processing but it is in fact incorrect. Deriving the correct answer, which is 5 cents, seems to depend on the reasoner resisting the default, intuitive response and instead effortfully applying reflective and analytic Type 2 processing. The Type 2 response arises for only a minority of reasoners, with the majority committing to the erroneous Type 1 answer.

Ackerman and Thompson's (2017, 2018) metareasoning framework embraces the kind of "default-interventionist" dual-process perspective on object-level reasoning (e.g., Evans 2018; Howarth and Handley 2016) that is exemplified by people's performance on the CRT. According to the metareasoning framework, it is only those reasoners who experience a low "feeling of rightness" in the Type 1 response who are triggered to engage in analytic Type 2 processing that can then potentially overturn the incorrect Type 1 answer. Even though Ackerman and Thompson's metareasoning framework adopts dual-process principles, they are nonetheless careful to acknowledge that the requirement for metacognitive monitoring and control during reasoning also pertains to single-process theories, which eschew the idea that reasoning draws upon two distinct types of processes (e.g., see Kruglanski and Gigerenzer 2011). Regardless of the mechanisms that are envisaged to underpin object-level reasoning, this latter point underscores the importance of understanding the way in which people monitor rapidly generated initial answers and control the engagement of more in-depth processing. That said, the first set of articles in this Special

Issue all address the interplay between metareasoning and object-level reasoning when the latter is conceptualised from a dual-process viewpoint.

In their article, Byrd et al. (2023) examine evidence for a dual-process account of reasoning by focusing on people's performance with a purely *verbal* version of the CRT, referred to as the vCRT (originally developed and validated by Sirota et al. 2021). The vCRT is distinct from standard mathematical versions (e.g., ones that include the bat-and-ball problem) in that it is less reliant on numerical calculations. At the same time, it is conceptually equivalent to such versions in terms of being assumed to induce rapid, intuitive responses that are, in fact, incorrect, whereas correct answers presumably require reflective thinking. As an example of an item from the vCRT, consider the following: If you have only one match and you walk into a dark room where there is an oil lamp, a newspaper and wood, which thing would you light first? The intuitive "lured" answer that comes to mind quickly and effortlessly is the oil lamp. However, upon reflection, one may realise that the correct answer is the match. The vCRT has high internal consistency, high test–retest reliability, and benefits from less association with general numerical ability than the standard CRT, making it an important testbed for exploring theories of reasoning and metareasoning.

Byrd et al. (2023) set out to investigate whether performance on the vCRT aligns with a default-interventionist account. As they point out, however, this and other dual-process accounts of CRT performance have been contested in recent years based on results arising from several studies. For example, Bago and De Neys (2019) showed that most participants who answered a CRT item correctly following a period of reflection had already answered this item correctly when required to provide an initial response under extreme time pressure or under cognitive load. In addition, Stupple et al. (2017) found only a weak correlation between CRT response times and accuracy, whereas it would be expected that longer response times (indicative of more reflective processing) should be associated with increased accuracy. Item-level analyses by Stupple et al. (2017) also failed to demonstrate predicted response-time differences between correct analytic and incorrect intuitive answers for two of the three CRT items. Stupple et al.'s findings also resonate with evidence that derives from process-tracing research, which involves participants thinking aloud during the CRT. For example, Szaszi et al.'s (2017) think-aloud study indicated that most correct responses on the CRT were associated with participants articulating the correct answer from the very outset, whereas many incorrect lured responses persisted even after participants reflected on them. Such results appear to cast further doubt on a dual-process interpretation of CRT performance.

Byrd et al. (2023) report two studies using the think-aloud method, with an analysis of people's verbalised thoughts revealing that most (but not all) correct responses involved reflection and that most (but not all) lured responses lacked reflection. These data suggest that people's reasoning on the vCRT broadly aligns with a default-interventionist viewpoint—albeit not without some exceptions, as mentioned, which need to be accommodated in more advanced dual-process theorising. In addition, there was no evidence in Byrd et al.'s (2023) studies that thinking aloud had a disruptive effect on CRT performance when compared to a control group. This finding underscores the methodological advantages of using think-aloud protocol analysis to reveal insights regarding reasoning performance and associated metareasoning processes in cognitive reflection tests.

The next article in the Special Issue by Białek (2023) also takes a dual-process stance, but this time with a focus on the role of metacognition in debiasing reasoning and decision-making. Białek's contribution takes the form of a review and position piece in which he explores debiasing through the lens of the "foreign language effect", which arises when the presentation of information or instructions in a non-native language to either bilingual or multilingual participants serves to benefit their reasoning and decision-making (e.g., Keysar et al. 2012). In his article, Białek examines various proposals that have been advanced to explain the foreign language effect but finds these proposals to be inadequate given the inconsistent support that they have received. He therefore proposes an alternative working hypothesis, which is that the use of a foreign language *distorts* people's metacognitive processing by making them less sensitive to internal cues (e.g., affective experiences), which are typically suggestive of the correctness of intuitive responses. The lack of sensitivity to such cues means that people will sometimes reflect more on problems that have intuitively appealing but incorrect solutions. This reflection facilitates the discovery of less obvious yet normatively correct solutions, resulting in improved performance. According to this model, observed improvements in judgement and decision-making are essentially a fortuitous side-effect of distorted metacognition. The author acknowledges that this working hypothesis is currently supported only by limited evidence (e.g., Białek et al. 2019; Hayakawa et al. 2017; Hennig and Hütter 2021; Muda et al. 2018). Nevertheless, if corroborated, this explanation has important implications, not only for understanding the metacognitive cues that can trigger increased analytic thinking, but also for pointing toward approaches to debias reasoning and decision-making.

In the next article, Mata (2023) adopts the CRT as a paradigm with which to examine the metacognitive awareness that people have about their reasoning performance. His particular focus is on whether people have some awareness that their thinking is flawed on those occasions where they make errors. The first two studies reported by Mata involve a methodologically innovative comparison between people's confidence in their answers to general knowledge (GK) questions, which only require reflective thinking, versus their confidence in their answers to CRT problems, which have the potential to engender both intuitive and reflective thinking. The results showed that people are generally able to discriminate between their correct and incorrect answers, although this ability is far from perfect, and it is also greater for GK questions than for CRT problems. Interestingly, too, incorrect responses to CRT problems were found to be produced with approximately the same level of confidence, as were correct responses to GK questions. Furthermore, even though confidence was inappropriately high for incorrect responses to CRT problems, it was even higher for correct responses.

The results of these two studies are valuable in demonstrating differences in people's confidence across tasks and across correct and incorrect responses. However, the findings also raise questions about the underlying source of these confidence differences. This issue was addressed by Mata (2023) in two further studies. In one, participants were presented with standard "conflict" versions of the CRT (where intuition and reflection are at odds), as well as "no-conflict" versions of the CRT (where there is no intuitive response and only reflective thinking can give rise to an answer). The final study was a conceptual replication using a wider range of conflict and no-conflict problems. These two studies provided clear evidence that the key factor underlying confidence differences across tasks and across correct and incorrect responses was the *conflict* that problems pose between intuition (Type 1 processing) and reflection (Type 2 processing). More specifically, overconfidence among incorrect responders was greater when problems trigger an intuitive but incorrect response. This is because, for these tricky conflict items, responders generally do not realise that an alternative, correct response is available. On the other hand, correct responders for these tricky items realise that there are alternative solutions that, whilst tempting, are incorrect.

The next article, by Ferreira et al. (2022), likewise addresses the conflict that can arise between Type 1 intuitive responses and Type 2 reflective responses of the kind that require the application of logical rules or probabilistic principles. One classic example of such a task is the so-called "lawyer–engineer problem" (Kahneman and Tversky 1972), which reads as follows:

In a study, several psychologists interviewed a group of people. The group included 5 engineers and 995 lawyers. The psychologists prepared a brief summary of their impression of each interviewee. The following description was drawn randomly from the set of descriptions: Dan is 45. He is conservative, careful, and ambitious. He shows no interest in political issues and spends most of his free time on his many hobbies, which include carpentry, sailing, and mathematical puzzles. Which of the following is more likely?

(a) Dan is an engineer;
(b) Dan is a lawyer.

When tackling this problem, people often appear to rely on the similarity between Dan's description and the stereotype of an engineer, and thence infer that Dan is an engineer. People do this without seeming to consider the prior odds of someone being an engineer or a lawyer in this group of people (i.e., 5:995), which would make it more likely that Dan is a lawyer. What Kahneman and Tversky (1972) therefore demonstrated with this problem is that people's judgments often seem to rely on fast and intuitive Type 1 processes (i.e., a judgement by means of so-called "representativeness"), with little evidence for this default response being overridden by slower and more effortful reflective and analytic Type 2 processes that draw upon rules of probability (i.e., a judgement that involves considering the initial base rates in the group).

In their article, Ferreira et al. (2022) suggest that the dual-process research tradition (e.g., Evans and Stanovich 2013a, 2013b; Kahneman 2003), which has been applied to explain reasoning on tasks such as the lawyer–engineer problem, may have been overly focused on Type 1 processes as the source of decision-making errors, largely ignoring the potential contribution of Type 2 processes to such errors. The authors therefore set out to uncover judgement errors specifically associated with Type 2 processing in the form of an overgeneralisation bias, whereby a reasoner continues deliberatively to apply a probabilistic rule in circumstances in which it is no longer appropriate to do so. Such an effect would be suggestive of people's relatively superficial understanding of normative principles.

To investigate this idea experimentally, Ferreira et al. (2022) developed *rule-inadequate* versions of standard base-rate problems, in which base rates are made irrelevant. As an example, imagine that the lawyer–engineer problem included an additional premise stating that *on the first day of the study, the psychologists had randomly chosen two lawyers and two engineers to interview (from the total sample of lawyers and engineers).* In this situation, the same number of lawyers and engineers would have been interviewed, and therefore the base rates corresponding to the initial group composition (995 lawyers and 5 engineers) are not relevant, such that considering these base rates would be a mistake. Ferreira et al. (2022) conducted four experiments investigating metacognitive aspects of reasoning using both rule-adequate and rule-inadequate versions of base-rate problems. Across these studies, they observed *conflict sensitivity* (measured in terms of response latencies and response confidence) in people's responses both to standard, rule-adequate base-rate problems and to rule-inadequate versions of these problems. People's failure to discriminate between real and spurious conflict suggests that they often misuse statistical information and draw conclusions based on the Type 2 processing of irrelevant base rates.

Ferreira et al.'s (2022) findings have important implications for current debates regarding dual-process models of reasoning and their attendant assumptions about metareasoning. For example, with respect to the standard default-interventionist view (e.g., Evans 2018), the observation that many individuals who appear to respond normatively in standard base-rate problems are nevertheless also prone to relying on statistically irrelevant information calls into question the nature and role of Type 2 thinking in such problems. Further research is clearly needed to explore in more detail the timecourse and metareasoning components of intuitive and deliberate processing in rule-inadequate problems to clarify the theoretical implications of Ferreira et al.'s (2022) findings.

Base-rate tasks also feature in Valerjev and Dujmović's (2023) article, which reports research that again adopts a dual-process perspective to examine the nature of conflict and metacognition in reasoning. The authors' work builds upon a relatively recent idea in the reasoning literature, which is that people can rapidly access Type 1 "logical intuitions" when engaged in reasoning (e.g., De Neys 2012; Frey et al. 2018; Šrol and De Neys 2021). Such logical intuitions include ones that provide normatively correct solutions to mathematical, probabilistic, and logic problems. Moreover, these logical intuitions have the potential to compete with other Type 1 intuitions—more traditionally referred to as "heuristics" (e.g., matching, availability, and representativeness)—which tend to lead to non-normative and biased performance. Based on emerging evidence, the latest theorising (e.g., De Neys 2023) proposes that Type 1 processes can give rise to one or more initial intuitive responses, with the potential for a logical intuition to compete with a biased, heuristic intuition. This account provides a potential explanation of the phenomenon of base-rate neglect that is often seen in tasks such as the "lawyer–engineer problem" (Kahneman and Tversky 1972) discussed above, where people show a tendency to respond intuitively in accordance with the compelling stereotype that is presented rather than being sufficiently swayed by the conflicting base-rate information that points toward an alternative answer.

According to De Neys (2023), an interesting situation arises when two Type 1 intuitive responses are of similar strength and are in conflict, which is when a metacognitive monitoring process may trigger Type 2 activation to resolve the conflict. Indeed, many studies have shown that the presence of such conflict (together with the resulting metacognitive uncertainty that it arouses) increases the likelihood of Type 2 processing being triggered (e.g., Dujmović and Valerjev 2018; Pennycook et al. 2014, 2015; Thompson and Johnson 2014). As Valerjev and Dujmović (2023) explain, such Type 2 processing may involve the rationalisation of one of the competing responses, the consideration of additional evidence for one or both responses, or the attempt to try a new strategy and generate a novel response. In situations where metacognitive uncertainty is not at a sufficient level to trigger Type 2 processing, what is most likely is that the final response will simply be the stronger of the two Type 1 responses.

These theoretical proposals regarding the role of competing Type 1 responses in dual-process accounts of reasoning provided the inspiration for the two experiments reported by Valerjev and Dujmović (2023). These experiments involved a modification of the base-rate neglect task to enable the creation of task variants in which a belief-based Type 1 heuristic operated either in conflict or concert with another belief-based Type 1 heuristic. Moreover, Valerjev and Dujmović (2023) were able to design experiments to investigate accuracy, response time, and confidence for items where the strength of one heuristic drove responding in the opposite direction to a weaker heuristic (e.g., conflict items), as well for items where the two heuristics drove responding in the same direction (i.e., congruent items). Through sophisticated experimental manipulations, Valerjev and Dujmović demonstrate that the presence of conflict affects metareasoning, resulting in longer reasoning times and lower confidence ratings. These findings therefore favour a dual-process account of reasoning while also affirming the critical role played by metareasoning in this account.

3. Individual Differences in Metareasoning

The next set of papers in this Special Issue relate to the important topic of individual differences in people's metareasoning. In general terms, research on individual differences focuses on understanding the determinants of systematic variation between people. The questions posed in this field typically pivot on the potential existence of trait-like stability within and across individuals in relation to their personalities, dispositions, and cognitive capacities—but also in relation to their metacognitive abilities. The article by Law et al. (2022) addresses head on whether individual variation in "giving-up" behaviour during cognitive task performance exhibits such trait-like properties. According to Ackerman and Thompson's (2017, 2018) metareasoning framework, the process of giving up when a solution may not seem achievable reflects an adaptive metacognitive control decision, whereby individuals opt-out of responding to mitigate resource costs and to avoid making erroneous responses (Payne and Duggan 2011). However, as Law et al. (2022) note, research is needed to determine whether individuals systematically vary in this giving-up behaviour across different tasks. If this is indeed the case, then it is also vital to ascertain which variables are meaningfully associated with this behaviour.

In addressing these questions, Law et al. (2022) deployed three different uncertainty-monitoring paradigms that depend on perceptual decision-making, including one from research on animal metacognition, referred to as the sparse–uncertain–dense (SUD) task. In this task, participants are provided with boxes of varying degrees of pixel density and are required to decide whether a box is sparse, dense or whether they are uncertain, with the latter option reflecting an opt-out decision. Using these kinds of task-based methodological innovations, the authors report a study that not only established the factorial stability in people's giving-up tendencies but also supported the assumed adaptive nature of such tendencies by showing how the giving-up factor correlates positively with cognitive ability, rational decision-making, and academic performance. This research provides a valuable basis for further metareasoning studies to investigate the characteristics and underlying processes of this trait-like giving-up construct as well as the critical role that it plays in learning, cognitive processing, and decision-making.

The theme of individual differences in metareasoning is also central to the article by Morsanyi and Hamilton (2023), who report a study of the development of intuitive and analytic thinking in autism. When reasoning, it is often of critical importance for a person to be able to engage in metacognitive control to resist adhering to an initial, default intuitive response, and instead to apply more effortful and reflective analytic thinking that can potentially lead to the generation of a normatively correct response (e.g., Evans and Stanovich 2013a, 2013b). In their study, Morsanyi and Hamilton aimed to investigate whether closely matched autistic and neurotypical individuals share the same normatively incorrect intuitions that typically dominate responses to the cognitive reflection test (CRT; Frederick 2005; Kahneman and Frederick 2002), or whether individuals with autism show increased analytic processing, as suggested by some researchers (e.g., De Martino et al. 2008). The CRT represents a good testbed for investigating the balance of intuitive versus analytic processes that arise in autistic and neurotypical individuals, given that the problems that make up the test tend to elicit an intuitive but incorrect response, whereas the correct response requires effortful reflection.

Morsanyi and Hamilton's (2023) study included adolescents and young adults and involved methodological improvements on previous studies investigating CRT performance in autism. One such improvement was the use of the 6-item CRT-Long (Primi et al. 2016), which measures cognitive reflection more effectively in younger individuals, who often score zero on the original CRT. Like previous findings with the CRT, Morsanyi and Hamilton's (2023) results showed an age-related increase in analytic responding and a decrease in intuitive responding. Importantly, however, the proportion of both intuitive and analytic responses across autistic and neurotypical participants was identical in both age groups, suggesting similar object-level processing as well as similar levels of metacognitive control. As the authors acknowledge, finding no group differences in cognitive reflection does not necessarily imply that autistic and neurotypical samples would show an absence of differences on other reasoning or decision-making tasks. As such, more research is needed on this issue, including investigations that delve deeper into the metacognitive monitoring and control processes that accompany both intuitive and analytic responding in autism.

In the next article in this Special Issue, Stanovich and Toplak (2023) also deal with the topic of individual differences, this time in relation to the thinking disposition that is referred to as "actively open-minded thinking" (AOT). This construct is measured by means of items that tap reflective thought, the willingness to consider alternative opinions, sensitivity to evidence that is contradictory to current beliefs, and the willingness to postpone closure. AOT scales strongly predict performance on heuristics and biases tasks (e.g., Stanovich and West 2000) as well as the avoidance of reasoning traps, such as belief in conspiracy theories (e.g., Stanovich et al. 2016). As Stanovich and Toplak (2023) note, AOT has much conceptual overlap with metareasoning processes, both through its connection with notions of goal management and epistemic self-regulation (e.g., by means of monitoring ongoing thinking, allocating cognitive resources, and attentional switching) and via the concept of cognitive decoupling (i.e., the inhibitory override of intuitive responses together with the engagement of hypothetical reasoning and cognitive simulation, which are necessary to compute normatively correct responses).

In their article, Stanovich and Toplak (2023) not only provide a valuable review of their 25-year history of studying AOT and developing AOT measurement scales, but they also present their new, brief 13-item AOT scale, which addresses prior criticisms while building upon previous refinements. In discussing why AOT scales are such good predictors of performance on heuristics and biases tasks, the authors conclude that it is because these scales tap important processes of cognitive decoupling and decontextualisation that modernity increasingly requires (see Stanovich 2004)—processes that appear to be inherently tied to effective metareasoning, such as checking the validity of intuitive responses (cf. De Neys and Pennycook 2019; Pennycook et al. 2015). However, one paradox remains, and is discussed at length by Stanovich and Toplak, which is why it is that AOT scales are potent predictors of performance on most rational thinking tasks, yet they do not predict the attenuation of "myside bias"—the tendency to generate and evaluate evidence in a manner favourable toward one's prior opinions and attitudes. This consistent observation is disconcerting, because of all the biases that one would expect to be negatively correlated with AOT, myside bias is the primary one. Stanovich and Toplak present several pointers toward future research that might unravel the mystery of this absent correlation between AOT and the avoidance of myside bias, such as failures of metacognitive monitoring (e.g., the recognition of alternative beliefs). This topic seems ripe for further research, which could not be more timely or important for effective human reasoning in the modern world.

4. Extending the Reach of Metareasoning Research

Metareasoning research has been expanding rapidly in recent years and is continually extending its reach to address challenging issues in domains that range beyond the traditional ones that informed the development of Ackerman and Thompson's (2017, 2018) original metareasoning framework. Such traditional domains of enquiry feature in all the aforementioned articles, with their focus on deductive and probabilistic tasks as well as the CRT. In this section of the Special Issue, however, the scope broadens to consider metareasoning research that is taking place in domains beyond this established remit. As will be shown, such research is engendering valuable methodological developments, in addition to spearheading important conceptual advancements.

As a case in point, Shen et al. (2022) undertook a spectral electroencephalography (EEG) analysis to investigate the neural correlates of people being in a tip-of-the-tongue (TOT) state, which is a spontaneously occurring metacognitive state that is indicative of an answer to a query being almost at hand but not quite forthcoming. This TOT state is akin to what Ackerman and Thompson (2017, 2018) refer to as an initial "judgment of solvability", when people are presented with a question or problem. This judgement can trigger metacognitive control processes that deploy resources as the individual strives toward achieving an answer to the question at hand or a solution to a given problem. Indeed, although such TOT states are imbued with a sense of frustration, they are also associated with intense curiosity and a strong desire to find a resolution to the query, which translates into increased goal-directed cognitive processing (e.g., Metcalfe and Jacobs 2024; Schwartz and Metcalfe 2011).

In their study, Shen et al. (2022) showed that if an individual was in a TOT state when presented with a verbal query, then alpha suppression was in evidence in the spectral EEG analysis across centro-parietal regions prior to the experimenter's presentation of resolving feedback (note that the alpha frequency band ranges between 8 Hz and 12 Hz). Additional analyses also indicated that the occurrence of alpha suppression was associated with participants' verbal affirmations of being in a TOT state. Interestingly, too, Shen et al. (2022) observed that the opposite of alpha suppression in the centro-parietal regions—that is, alpha expression in these same regions—was associated with the feeling of "not knowing" the answer to a query. An increase in alpha power has often been interpreted as indicating reduced information processing and of being associated with so-called "cortical idling" and corresponding mental idling (Pfurtscheller 1999) as well as mind wandering (e.g., Baldwin et al. 2017). Overall, Shen et al.'s (2022) findings are important, as they are highly suggestive of alpha suppression being a neural signature of a spontaneous, conscious, and goal-directed metacognitive state of awareness. In this respect, it is noteworthy that the suppression of alpha band activity has traditionally been related to enhanced attention and more engaged information processing (Neuper and Pfurtscheller 2001; Pfurtscheller and Lopes da Silva 1999). The identification of neural markers of metacognition clearly represents an exciting and fruitful line of research for the future.

In the aforementioned article, Shen et al. (2022) note close similarities between the situation where a person is in a TOT state when trying to answer a given query and the situation where a problem-solver is currently unable to solve a problem but nevertheless has a high degree of certainty that a solution is within reach (cf. Polanyi 1974). In the next article in the Special Issue, Graf et al. (2023) focus on problems of a different type, whereby an initial, incorrect mental representation is triggered that needs to be restructured to enable the problem solver to find the solution. When attempting these so-called "insight problems", people often reach a point of "impasse", such that they have no sense that they can progress toward a solution. A period of impasse will sometimes then be followed by a sudden restructuring process or "representational change", which invokes an immediate insight into how the problem can successfully be solved (e.g., see Bowden et al. 2005; Ohlsson 2011). Restructuring reflects a change in strategy, which is presumably initiated by a metacognitive control process that is itself triggered by the metacognitive monitoring system identifying that no progress is being made with a current strategy. As Graf et al. (2023) point out, however, despite the theoretical assumption that this restructuring process happens suddenly with insight problems (engendering a typical "Aha!" experience), the evidence for this assumption remains inconsistent and inconclusive. Arguably, one key reason for this lack of clarity relates to the fact that most measures of restructuring and insight rely on the problem-solver's subjective experience of aspects of their own solution process. As such, a more *objective* measure of metacognition and the occurrence of sudden restructuring would be welcome.

Graf et al. (2023) propose that one way to obtain such objective data relating to problem restructuring and the occurrence of insight is to trace problem-solving processes using eye-movement tracking. Previous research using eye tracking to investigate restructuring processes in insight problem solving has employed matchstick arithmetic problems, which use matchsticks to present false arithmetic statements (written using Roman numerals, arithmetic operators, and equal signs), with the participant's task being to transform the false arithmetic statement into a true statement by moving only one stick (Knoblich et al. 2001). Four types of matchstick arithmetic problems have been defined, which involve varying levels of difficulty, depending on the constraints that need to be relaxed and the tightness of the chunks that need to be decomposed. Knoblich et al.'s eye-tracking study (Knoblich et al. 2001) provided good evidence that people initially fixate on elements of problems that are irrelevant to a solution (suggestive of an initial incorrect problem representation based on an effective solution strategy), with solvers only shifting their visual attention to relevant elements of problems in the final third of the problem-solving period (suggestive of a change in strategy and the construction of a more effective representation).

As Graf et al. (2023) note in their article, although Knoblich et al.'s (2001) study provides compelling evidence for the view that a restructuring of the problem representation takes place in problems that require constraint relaxation, what this previous research fails to show is whether this change is sudden or gradual. In their reported research, Graf et al. (2023) re-analyse their own previous eye-tracking data (from Bilalić et al. 2019) and demonstrate convincingly that it is possible to identify the occurrence of abrupt representational changes in the form of sudden bursts of attention to key aspects of a problem—albeit only through the application of nonlinear statistical models in conjunction with fine-grained change-point analysis. Additionally, they show that the provision of explicit solution hints to participants who had not solved a problem after five minutes served to reorient the focus of attention, demonstrably changing the dynamics of restructuring in insight problem solving. In sum, although an insight problem may well require the sudden restructuring of an initial mental representation, driven by metacognitive cues, it is only sophisticated analytical and statistical approaches that can uncover such strategic restructuring events.

In the next article, Kenett et al. (2023) maintain Graf et al.'s focus on the under-researched role of metacognitive cues in creative cognition, albeit not in relation to insight problem solving, but rather with respect to creative idea generation using the alternative uses task (AUT, e.g., Guilford 1967), whereby people are asked to come up with as many alternative uses as they can for common objects such as a bucket. Kenett et al. were specifically concerned with the heuristic cues that underlie people's metacognitive judgments of the *originality* of their ideas, with their particular research interest being on the role played by the "semantic distance" between concepts in informing people's originality judgments. Semantic distance is a measure derived from the computational analysis of large textual corpora, and it quantifies the conceptual *dis-similarity* between the AUT object and the words that are present in the open-ended responses that participants generate when coming up with alternative uses for that object (e.g., see Beaty and Johnson 2021).

In their research, Kenett et al. (2023) examined the extent to which semantic distance contributes explanatory value in predicting objective originality scores (i.e., the frequency with which an idea is generated within a sample of participants) as well as in predicting subjective originality judgments (i.e., metacognitive self-assessments of originality). In two experiments (one of which re-analysed previous data from Sidi et al. 2020), Kenett et al. found that semantic distance was a strong predictor of both objective originality scores and subjective originality judgments, while also being a cue that consistently predicted originality scores more strongly than originality judgments (i.e., people tend to *underestimate* their originality). Systematic priming manipulations that Kenett et al. built into their second experiment also revealed that the extent to which semantic distance can influence originality judgments is highly malleable and context dependent. Overall, this study highlights the role of semantic distance as a previously unacknowledged metacognitive cue that has the power to influence originality judgments in creative idea generation.

The next article in the Special Issue by Caro et al. (2022) focuses on metareasoning in machines, such as artificially intelligent systems, rather than in humans, representing a seemingly marked departure from the preceding articles. What is abundantly clear in Caro et al.'s reported work, however, is that many of the conceptual challenges that arise when considering metareasoning in non-human intelligent systems have close parallels in the domain of human intelligence. As these authors note, metareasoning research suffers from the "heterogeneity problem", whereby different researchers across different domains (e.g., psychology, education, computer science, artificial intelligence, and engineering) build diverse metareasoning models for intelligent systems that have comparable functionality but that use ambiguous terminology and contradictory descriptions as well as non-standardised conventions.

To address the heterogeneity problem, Caro et al. (2022) present an ontology-driven knowledge representation for metareasoning in the domain of artificial intelligence that is extendible to all intelligent systems, which they refer to as IM-Onto. This proposed ontology provides a visual means of sharing a common understanding of the structure and relationships between metareasoning terms and concepts. In their research, the authors also adopted a rigorous research method to ensure that the two main requirements of as ontology were met: integrity based on relevant knowledge and acceptance by researchers and practitioners. The data-driven evaluation of the ontology revealed a high accuracy rate, indicating that many of the knowledge elements in the ontology provide useful information for understanding metareasoning in intelligent systems. This research is important in providing a valuable stepping-stone toward a consistent ontology that can be used for sharing a common understanding of the structure and relationships among terms and concepts related to metareasoning across all intelligent systems, whether artificial or human.

The final article in the Special Issue by Richardson and Ball (2024) addresses a significant challenge for current metareasoning research, which relates to the fact that it is almost exclusively focused on explaining the monitoring and control processes that arise at the level of *individual* reasoners, with little consideration being given to the metareasoning that occurs when two or more individuals work conjointly on a reasoning task. Richardson and Ball argue that this omission in the metareasoning literature is significant, given the many domains in which team-based reasoning is critical, such as those that relate to design, innovation, process control, and defence and security. In their article, Richardson and Ball (2024) tackle the absence of a conceptual framework head on to address the nature of team metareasoning by discussing how Ackerman and Thompson's (2017) metareasoning framework can be extended to situations where people reason collaboratively to reach an agreed solution to a problem or to decide on a common course of action.

In developing a collaborative metareasoning framework, Richardson and Ball (2024) expand upon an important tripartite distinction relating to monitoring processes, which they have previously advanced in the literature (Richardson et al. 2024; see also

Pickering and Garrod 2021, who independently drew an equivalent distinction). This distinction aims to capture the difference between "self monitoring" (an individual's monitoring of their own performance), "other monitoring" (an individual's monitoring of the performance of others) and "joint monitoring" (the unified monitoring of collective performance). The tripartite distinction gives rise to many important questions regarding the kinds of cues that team members draw upon when engaging in other monitoring and joint monitoring, which are questions that Richardson and Ball (2024) address extensively in their article. Importantly, too, they propose a parallel tripartite distinction for control processes that can arise in teams. In articulating this novel distinction, the authors differentiate between "self-focused control" (i.e., an individual's procedural decisions about how to progress or terminate their own reasoning), "other-focused control" (i.e., an individual's procedural decisions about how to control the performance of others) and "joint control" (i.e., the unified control of procedural decisions regarding how to advance or terminate collective performance).

Richardson and Ball (2024) discuss in detail the nature of the control processes that operate at these three aforementioned levels. Their notion of other-focused control seems particularly fascinating in the context of collaborative metareasoning, as it captures the way in which people engage in metareasoning to persuade others through argumentation and negotiation. These various forms of other-focused control—which may even involve deception—can all be highly profitable for team progress toward a good outcome, such as when progress is stalling because of conflicted viewpoints. Indeed, the capacity to exert effective other-focused control over a team's ongoing reasoning has been shown to be an important aspect of a team leader's metareasoning repertoire (e.g., Ball and Ormerod 2000). In elaborating upon the distinct types of metacognitive monitoring and control that seem to pervade team-based reasoning, Richardson and Ball (2024) also discuss the prospect for developing a comprehensive collaborative metareasoning framework that centres on the importance of changes in team members' *language* as a means to track metacognitive uncertainty and misalignment over time. Such language-focused metacognitive tracking has significant potential to inform both theoretical developments and applied interventions to enhance team-based reasoning and decision-making performance.

5. Conclusions

The 13 articles in this Special Issue provide an exciting snapshot of current theoretical and methodological developments in the rapidly expanding field of metareasoning research. Many of the articles address the close links that exist between metareasoning concepts and dual-process accounts of reasoning, with new developments in dual-process theorising giving fresh impetus to a consideration of the role of metacognitive monitoring and control processes in many established reasoning, decision-making and problem-solving paradigms. We have also showcased articles that address the increasingly important topic of individual differences in people's metareasoning skills. There is clearly much value to be had in gaining a deeper understanding of the ways in which metareasoning varies in its functioning and effectiveness across individuals. In particular, such an understanding can inform the design and development of tailored approaches and interventions to support the training of metareasoning skills in ways that can enable the generation of sound inferences, improved decisions, and enhanced solutions. Finally, we have included a set of articles in the Special Issue that demonstrate how metareasoning research is addressing challenging questions in new domains beyond those that have traditionally been the focus of enquiry, such as in studying metareasoning in the tip-of-the-tongue phenomenon, in creative cognition, in artificial systems, and in team collaboration. Overall, we are in no doubt that metareasoning research is extremely vibrant and is constantly engendering valuable methodological developments and theoretical advancements.

Author Contributions: Conceptualization, L.J.B. and B.H.R.; writing—original draft preparation, L.J.B. and B.H.R.; writing—review and editing, L.J.B. and B.H.R.; project administration, L.J.B. and B.H.R. All authors have read and agreed to the published version of the manuscript.

Funding: This research received no external funding.

Conflicts of Interest: The authors declare no conflicts of interest.

References

Ackerman, Rakefet. 2023. Bird's-eye view of cue integration: Exposing instructional and task design factors which bias problem solvers. *Educational Psychology Review* 35: 55. [CrossRef]

Ackerman, Rakefet, and Valerie A. Thompson. 2017. Meta-reasoning: Monitoring and control of thinking and reasoning. *Trends in Cognitive Sciences* 21: 607–17. [CrossRef] [PubMed]

Ackerman, Rakefet, and Valerie A. Thompson. 2018. Meta-reasoning: Shedding meta-cognitive light on reasoning research. In *The Routledge International Handbook of Thinking and Reasoning*. Edited by Linden J. Ball and Valerie A. Thompson. Abingdon: Routledge, pp. 164–82.

Bago, Bence, and Wim De Neys. 2019. The smart System 1: Evidence for the intuitive nature of correct responding on the bat-and-ball problem. *Thinking & Reasoning* 25: 257–99.

Baldwin, Carryl L., Daniel M. Roberts, Daniela Barragan, John D. Lee, Neil Lerner, and James S. Higgins. 2017. Detecting and quantifying mind wandering during simulated driving. *Frontiers in Human Neuroscience* 11: 406. [CrossRef] [PubMed]

Ball, Linden J., and Thomas C. Ormerod. 2000. Putting ethnography to work: The case for a cognitive ethnography of design. *International Journal of Human-Computer Studies* 53: 147–68. [CrossRef]

Beaty, Roger E., and Dan R. Johnson. 2021. Automating creativity assessment with SemDis: An open platform for computing semantic distance. *Behavior Research Methods* 53: 757–80. [CrossRef] [PubMed]

Białek, Michał. 2023. Why should we study the foreign language effect: Debiasing through affecting metacognition? *Journal of Intelligence* 11: 103. [CrossRef]

Białek, Michał, Mariola Paruzel-Czachura, and Bertram Gawronski. 2019. Foreign language effects on moral dilemma judgments: An analysis using the CNI model. *Journal of Experimental Social Psychology* 85: 103855. [CrossRef]

Bilalić, Merim, Mario Graf, Nemanja Vaci, and Amory H. Danek. 2019. The temporal dynamics of insight problem solving: Restructuring might not always be sudden. *Thinking & Reasoning* 27: 1–37.

Bowden, Edward M., Mark Jung-Beeman, Jessica Fleck, and John Kounios. 2005. New approaches to demystifying insight. *Trends in Cognitive Sciences* 9: 322–28. [CrossRef] [PubMed]

Byrd, Nick, Brianna Joseph, Gabriela Gongora, and Miroslav Sirota. 2023. Tell us what you really think: A think aloud protocol analysis of the verbal Cognitive Reflection Test. *Journal of Intelligence* 11: 76. [CrossRef] [PubMed]

Caro, Manuel F., Michael T. Cox, and Raúl E. Toscano-Miranda. 2022. A validated ontology for metareasoning in intelligent systems. *Journal of Intelligence* 10: 113. [CrossRef]

De Martino, Benedetto, Neil A. Harrison, Steven Knafo, Geoff Bird, and Raymond J. Dolan. 2008. Explaining enhanced logical consistency during decision making in autism. *Journal of Neuroscience* 28: 10746–50. [CrossRef] [PubMed]

De Neys, Wim. 2012. Bias and conflict: A case for logical intuitions. *Perspectives on Psychological Science* 7: 28–38. [CrossRef] [PubMed]

De Neys, Wim. 2023. Advancing theorizing about fast-and-slow thinking. *Behavioral and Brain Sciences* 46: e111. [CrossRef]

De Neys, Wim, and Gordon Pennycook. 2019. Logic, fast and slow: Advances in dual-process theorizing. *Current Directions in Psychological Science* 28: 503–9. [CrossRef]

Dujmović, Marin, and Pavle Valerjev. 2018. The influence of conflict monitoring on meta-reasoning and response times in a base rate task. *Quarterly Journal of Experimental Psychology* 71: 2548–61. [CrossRef]

Evans, Jonathan St. B. T. 2018. Dual-process theories. In *The Routledge International Handbook of Thinking and Reasoning*. Edited by Linden J. Ball and Valerie A. Thompson. Abingdon: Routledge, pp. 151–66.

Evans, Jonathan St. B. T., and Keith E. Stanovich. 2013a. Dual-process theories of higher cognition: Advancing the debate. *Perspectives on Psychological Science* 8: 223–41. [CrossRef] [PubMed]

Evans, Jonathan St. B. T., and Keith E. Stanovich. 2013b. Theory and metatheory in the study of dual processing: Reply to comments. *Perspectives on Psychological Science* 8: 263–71. [CrossRef] [PubMed]

Ferreira, Mário B., Jerônimo C. Soro, Joana Reis, André Mata, and Valerie A. Thompson. 2022. When Type 2 processing misfires: The indiscriminate use of statistical thinking about reasoning problems. *Journal of Intelligence* 10: 109. [CrossRef] [PubMed]

Frederick, Shane. 2005. Cognitive reflection and decision making. *Journal of Economic Perspectives* 19: 25–42. [CrossRef]

Frey, Darren, Eric D. Johnson, and Wim De Neys. 2018. Individual differences in conflict detection during reasoning. *Quarterly Journal of Experimental Psychology* 71: 1–52. [CrossRef]

Graf, Mario, Amory H. Danek, Nemanja Vaci, and Merim Bilalić. 2023. Tracing cognitive processes in insight problem solving: Using GAMs and change point analysis to uncover restructuring. *Journal of Intelligence* 11: 86. [CrossRef] [PubMed]

Guilford, Joy P. 1967. Creativity: Yesterday, today and tomorrow. *The Journal of Creative Behavior* 1: 3–14. [CrossRef]

Hayakawa, Sayuri, David Tannenbaum, Albert Costa, Joanna D. Corey, and Boaz Keysar. 2017. Thinking more or feeling less? Explaining the foreign-language effect on moral judgment. *Psychological Science* 28: 1387–97. [CrossRef]

Hennig, Max, and Mandy Hütter. 2021. Consequences, norms, or willingness to interfere: A ProCNI model analysis of the foreignlanguage effect in moral dilemma judgment. *Journal of Experimental Social Psychology* 95: 104148. [CrossRef]

Howarth, Stephanie, and Simon Handley. 2016. Belief bias, base rates and moral judgment: Re-evaluating the default interventionist dual process account. In *Thinking Mind*. Edited by Niall Galbraith, Erica Lucas and David Over. New York: Taylor & Francis, pp. 97–111.

Husselman, Tammy-Ann, Edson Filho, Luca W. Zugic, Emma Threadgold, and Linden J. Ball. 2024. Stimulus complexity can enhance art appreciation: Phenomenological and psychophysiological evidence for the Pleasure-Interest Model of Aesthetic Liking. *Journal of Intelligence* 12: 42. [CrossRef]

Kahneman, Daniel. 2003. A perspective on judgment and choice: Mapping bounded rationality. *American Psychologist* 58: 697–720. [CrossRef]

Kahneman, Daniel, and Amos Tversky. 1972. Subjective probability: A judgment of representativeness. *Cognitive Psychology* 3: 430–54. [CrossRef]

Kahneman, Daniel, and Shane Frederick. 2002. Representativeness revisited: Attribute substitution in intuitive judgment. In *Heuristics and Biases: The Psychology of Intuitive Judgment*. Edited by Thomas Gilovich, Dale Griffin and Daniel Kahneman. Cambridge, MA: Cambridge University Press, pp. 49–81.

Kenett, Yoed N., Noam Gooz, and Rakefet Ackerman. 2023. The role of semantic associations as a metacognitive cue in creative idea generation. *Journal of Intelligence* 11: 59. [CrossRef]

Keysar, Boaz, Sayuri L. Hayakawa, and Sun Gyu An. 2012. The foreign-language effect: Thinking in a foreign tongue reduces decision biases. *Psychological Science* 23: 661–68. [CrossRef] [PubMed]

Knoblich, Günther, Stellan Ohlsson, and Gary E. Raney. 2001. An eye movement study of insight problem solving. *Memory & Cognition* 29: 1000–9.

Kruglanski, Arie W., and Gerd Gigerenzer. 2011. Intuitive and deliberate judgments are based on common principles. *Psychological Review* 118: 97–109. [CrossRef]

Law, Marvin K. H., Lazar Stankov, and Sabina Kleitman. 2022. I choose to opt-out of answering: Individual differences in giving up behaviour on cognitive tests. *Journal of Intelligence* 10: 86. [CrossRef]

Mata, André. 2020. Metacognition and social perception: Bringing meta-reasoning and social cognition together. *Thinking and Reasoning* 26: 140–49. [CrossRef]

Mata, André. 2023. Overconfidence in the Cognitive Reflection Test: Comparing confidence resolution for reasoning vs. General knowledge. *Journal of Intelligence* 11: 81. [CrossRef] [PubMed]

Metcalfe, Janet, and William J. Jacobs. 2024. The two faces of curiosity in creative cognition: Curiosity1, Curiosity2 (and their interaction). In *The Routledge International Handbook of Creative Cognition*. Edited by Linden J. Ball and Frederic Vallee-Tourangeau. Abingdon: Routledge.

Morsanyi, Kinga, and Jayne Hamilton. 2023. The development of intuitive and analytic thinking in autism: The case of Cognitive Reflection. *Journal of Intelligence* 11: 124. [CrossRef]

Morsanyi, Kinga, Niamh N. Cheallaigh, and Rakefet Ackerman. 2019. Mathematics anxiety and metacognitive processes: Proposal for a new line of inquiry. *Psihologijske Teme* 28: 147–69. [CrossRef]

Muda, Rafał, Pawel Niszczota, Michał Białek, and Paul Conway. 2018. Reading dilemmas in a foreign language reduces both deontological and utilitarian response tendencies. *Journal of Experimental Psychology: Learning, Memory, and Cognition* 44: 321–26. [CrossRef] [PubMed]

Nelson, Thomas O., and Louis Narens. 1990. Metamemory: A theoretical framework and new findings. In *The Psychology of Learning and Motivation: Advances in Research and Theory*. Edited by Gordon Bower. Cambridge, MA: Academic Press, pp. 125–73.

Neuper, Christa, and Gert Pfurtscheller. 2001. Event-related dynamics of cortical rhythms: Frequency-specific features and functional correlates. *International Journal of Psychophysiology* 43: 41–58. [CrossRef] [PubMed]

Ohlsson, Stellan. 2011. *Deep Learning: How the Mind Overrides Experience*. New York: Cambridge University Press.

Payne, Stephen J., and Geoffrey B. Duggan. 2011. Giving up problem solving. *Memory & Cognition* 39: 902–13.

Pennycook, Gordon, Dries Trippas, Simon J. Handley, and Valerie A. Thompson. 2014. Base rates: Both neglected and intuitive. *Journal of Experimental Psychology: Learning, Memory, and Cognition* 40: 544–54. [CrossRef] [PubMed]

Pennycook, Gordon, Jonathan A. Fugelsang, and Derek J. Koehler. 2015. What makes us think? A three-stage dual-process model of analytic engagement. *Cognitive Psychology* 80: 34–72. [CrossRef]

Pfurtscheller, Gert. 1999. EEG Event-Related Desynchronization (ERD) and Event-Related Synchronization (ERS). In *Electroencephalography: Basic Principles, Clinical Applications and Related Fields*, 4th ed. Edited by Ernst Niedermeyer and F. H. Lopes da Silva. Baltimore: Williams and Wilkins, pp. 958–67.

Pfurtscheller, Gert, and F. H. Lopes da Silva. 1999. Event-related EEG/MEG synchronization and desynchronization: Basic principles. *Clinical Neurophysiology* 110: 1842–57. [CrossRef] [PubMed]

Pickering, Martin J., and Simon Garrod. 2021. *Understanding Dialogue: Language Use and Social Interaction*. Cambridge, MA: Cambridge University Press.

Polanyi, Michael. 1974. Genius in Science. In *Methodological and Historical Essays in the Natural and Social Sciences*. Boston Studies in the Philosophy of Science. Edited by Robert S. Cohen and Marx W. Wartofsky. Dordrecht: Springer, vol. 14.

Primi, Caterina, Kinga Morsanyi, Francesca Chiesi, Maria Anna Donati, and Jayne Hamilton. 2016. The development and testing of a new version of the cognitive reflection test applying item response theory (IRT). *Journal of Behavioral Decision Making* 29: 453–69. [CrossRef]

Richardson, Beth H., and Linden J. Ball. 2024. Progressing the development of a collaborative metareasoning framework: Prospects and challenges. *Journal of Intelligence* 12: 28. [CrossRef] [PubMed]

Richardson, Beth H., Linden J. Ball, Bo T. Christensen, and John E. Marsh. 2024. Collaborative meta-reasoning in creative contexts: Advancing an understanding of collaborative monitoring and control in creative teams. In *The Routledge International Handbook of Creative Cognition*. Edited by Linden J. Ball and Frédéric Vallée-Tourangeau. Abingdon: Routledge, pp. 709–27.

Schwartz, Bennett L., and Janet Metcalfe. 2011. Tip-of-the-tongue (TOT) states: Retrieval, behavior, and experience. *Memory & Cognition* 39: 737–49.

Shen, Edmund Qian-Long, David Friedman, Paul Alexander Bloom, and Janet Metcalfe. 2022. Alpha suppression is associated with the Tip-of-the-Tongue (TOT) state whereas alpha expression is associated with knowing that one does not know. *Journal of Intelligence* 10: 121. [CrossRef]

Sidi, Yael, Ilan Torgovitsky, Daniela Soibelman, Ella Miron-Spektor, and Rakefet Ackerman. 2020. You may be more original than you think: Predictable biases in self-assessment of originality. *Acta Psychologica* 203: 103002. [CrossRef] [PubMed]

Sirota, Miroslav, Chris Dewberry, Marie Juanchich, Lenka Valuš, and Amanda Claire Marshall. 2021. Measuring cognitive reflection without maths: Development and validation of the verbal Cognitive Reflection Test. *Journal of Behavioral Decision Making* 34: 322–43. [CrossRef]

Stanovich, Keith E. 2004. *The Robot's Rebellion: Finding Meaning in the Age of Darwin*. Chicago: University of Chicago Press.

Stanovich, Keith E. 2018. Miserliness in human cognition: The interaction of detection, override and mindware. *Thinking & Reasoning* 24: 423–44.

Stanovich, Keith E., and Maggie E. Toplak. 2023. Actively open-minded thinking and its measurement. *Journal of Intelligence* 11: 27. [CrossRef] [PubMed]

Stanovich, Keith E., and Richard F. West. 2000. Individual differences in reasoning: Implications for the rationality debate? *Behavioral and Brain Sciences* 23: 645–726. [CrossRef] [PubMed]

Stanovich, Keith E., Richard F. West, and Maggie E. Toplak. 2016. *The Rationality Quotient: Toward a Test of Rational Thinking*. Cambridge, MA: MIT Press.

Stupple, Edward J. N., Melanie Pitchford, Linden J. Ball, Thomas E. Hunt, and Richard Steel. 2017. Slower is not always better: Response-time evidence clarifies the limited role of miserly information processing in the Cognitive Reflection Test. *PLoS ONE* 12: e0186404. [CrossRef] [PubMed]

Szaszi, Barnabas, Aba Szollosi, Bence Palfi, and Balazs Aczel. 2017. The Cognitive Reflection Test revisited: Exploring the ways individuals solve the test. *Thinking & Reasoning* 23: 207–34.

Šrol, Jakub, and Wim De Neys. 2021. Predicting individual differences in conflict detection and bias susceptibility during reasoning. *Thinking & Reasoning* 27: 38–68.

Thompson, Valerie A., and Stephen C. Johnson. 2014. Conflict, metacognition, and analytic thinking. *Thinking & Reasoning* 20: 215–44.

Undorf, Monika, Iris Livneh, and Rakefet Ackerman. 2021. Metacognitive control processes in question answering: Help seeking and withholding answers. *Metacognition & Learning* 16: 431–58.

Valerjev, Pavle, and Marin Dujmović. 2023. Single-heuristic reasoning: Is it still dual-process? *Journal of Intelligence* 11: 33. [CrossRef] [PubMed]

Journal of
Intelligence

MDPI

Article

Tell Us What You Really Think: A Think Aloud Protocol Analysis of the Verbal Cognitive Reflection Test

Nick Byrd [1,*], Brianna Joseph [2], Gabriela Gongora [3] and Miroslav Sirota [4]

[1] Intelligence Community Postdoctoral Research Fellowship Program, Stevens Institute of Technology, Hoboken, NJ 07030, USA
[2] AIOps, IBM, Atlanta, GA 30319, USA
[3] College of Business, Florida State University, Tallahassee, FL 32306, USA
[4] Department of Psychology, University of Essex, Colchester CO4 3SQ, UK; msirota@essex.ac.uk
* Correspondence: nick.a.byrd@gmail.com

Abstract: The standard interpretation of cognitive reflection tests assumes that correct responses are reflective and lured responses are unreflective. However, prior process-tracing of *mathematical* reflection tests has cast doubt on this interpretation. In two studies (N = 201), we deployed a validated think-aloud protocol in-person and online to test how this assumption is satisfied by the new, validated, less familiar, and non-mathematical verbal Cognitive Reflection Test (vCRT). Verbalized thoughts in both studies revealed that most (but not all) correct responses involved reflection and that most (but not all) lured responses lacked reflection. The think-aloud protocols seemed to reflect business-as-usual performance: thinking aloud did not disrupt test performance compared to a control group. These data suggest that the vCRT usually satisfies the standard interpretation of the reflection tests (albeit not without exceptions) and that the vCRT can be a good measure of the construct theorized by the two-factor explication of 'reflection' (as deliberate and conscious).

Keywords: cognitive reflection test; think-aloud protocol analysis; psychometrics; judgment- and decision-making; heuristics and biases

1. Introduction

If you were running a race and you passed the person in second place, what place would you be in now? The standard interpretation of a problem like this assumes that the answer that comes quickly and effortlessly to many people's mind is "first place". However, upon reflection, many people realize that the correct answer is "second place". This problem is considered a test of reflection because it is designed to lure us toward a particular response that, upon reflection, we can realize is incorrect (Byrd 2022b, 2022c). Thus, the standard interpretation of reflection labels lured responses with 'unreflective' and correct responses with 'reflective' (Pennycook et al. 2015a).

Since the introduction of the Cognitive Reflection Test (Kahneman and Frederick 2002), theories of reflection have advanced (Evans and Stanovich 2013). In the midst of this progress, some theorists distilled dozens of reflective-unreflective distinctions (see Frankish 2010, Table 1) down to just two somewhat orthogonal distinctions: automatic versus deliberate processing and conscious versus unconscious representations (Shea and Frith 2016). According to this two-factor explication of 'reflection', *reflective* thinking involves more consciousness representation (or awareness) of the relevant reasoning and more deliberate inhibition (as opposed to immediate acceptance of one's initial impulse) and *unreflective* thinking is less consciously represented and more automatic (Byrd 2022c). In other words, the *contents* of reflective reasoning are more accessible (e.g., to explain verbally) and the *process* of reflective thinking involves more reconsideration (e.g., doubting or double checking) than unreflective thinking (Byrd 2019). Theorists have also posited that the need for reflection may depend on context: in familiar reasoning domains, unreflective reasoning may be able to achieve desirable results, but in unfamiliar domains overcoming mistakes or biases might require some reflection (Pennycook et al. 2015a).

As theories of reflection progressed, so did our understanding of reflection tests. Some researchers have challenged the standard interpretation of mathematical reflection tests (mCRTs for short) and the two-factor explication of 'reflection' (Stanovich 2018). Indeed, some have found that most (67%) of those who answered a reflection test correctly after deliberation had already answered it correctly under time pressure or cognitive load before deliberation (Bago and De Neys 2019). While that two-response paradigm has helped test the default interventionist account of reflection—which posits that correct responses involve intervening on a default (a.k.a., lured) response (Evans 2007)—the paradigm overlooks plenty of useful information about the process of solving reflection test problems. So, others have listened to every word test-takers utter while thinking aloud during the reflection test. One such think-aloud study also "raise[d] doubts" about the standard interpretation of mCRTs: *most* (77%) correct responses on the mCRT "started . . . with the correct answer" and *many* (39%) lured responses on the mCRT involved "reflect[ing] on the . . . first response" (Szaszi et al. 2017). Some psychometric investigations of reflection tests suggest that these correct-but-unreflective and lured-yet-reflective responses might be explained by domain familiarity (Purcell et al. 2021), intelligence (Thompson and Johnson 2014), or strategy (Markovits et al. 2021). Although the predictive value of the mCRT remains after retaking the test (Białek and Pennycook 2018; Stagnaro et al. 2018), the best predictor of mCRT performance is often general math test performance (Attali and Bar-Hillel 2020; Erceg et al. 2020). So mCRTs may track not only reflection, but also mathematical competence (a.k.a., numeracy) and other factors.

1.1. The Verbal Cognitive Reflection Test

Sirota and colleagues (2021) developed and validated a new, 10-item, *non-mathematical* variant of Shane Frederick and colleagues' (Frederick 2005; Kahneman and Frederick 2002) well-known (mathematical) cognitive reflection test (mCRT) to address familiarity and numeracy problems (Byrd 2022d). One item from Sirota and colleagues' verbal cognitive reflection test (or vCRT for short) is the opening example: "If you were running a race, and you passed the person in 2nd place, what place would you be in now?" Multiple studies found that the vCRT enjoys high internal consistency, high test-retest reliability, and less association with general mathematical ability than the mCRT—even when translated to other languages and contexts (Sobkow et al. 2023). This suggests that the vCRT is a promising supplement or replacement for the mCRT in many research contexts.

1.2. Think-Aloud Protocol Analysis

Researchers have long called for investigation into the content and process of reflection rather than just the outcome (Stromer-Galley 2007). Fortunately, Ericsson and colleagues have developed and validated concurrent think-aloud protocols (Ericsson 2003; Ericsson and Simon 1993) that have been shown to overcome some well-known problems of early verbal report protocols such as confabulation (e.g., Wilson and Nisbett 1978) or performance interference on problems that seem unsolvable without some sort of insight (e.g., Schooler et al. 1993). For example, asking participants to *verbalize* or *recall* (rather than explain or justify) their thinking does not necessarily impair task performance or produce verbal reports inconsistent with their observed performance (Fox et al. 2011; Petitmengin et al. 2013). Concurrent verbalization can even *help* people think of a word that forms common compound words with three other words—e.g., thinking 'tree' when presented with 'apple', 'family', and 'house' (Ball and Stevens 2009; see also Blech et al. 2020). So, asking participants to think aloud while solving a problem (i.e., concurrently) may help participants tell us what they were really thinking better than retrospective requests to "explain", which can imply that participants should verbalize a justification (rather than a recollection) of what they were thinking (Ericsson and Simon 1980). Evans and colleagues fruitfully applied this realization about concurrent verbal reports to logical reflection tests to arbitrate between "rationalist" and "two-factor" or "dual process" theories of reasoning (Evans et al. 1983). Szaszi and colleagues also used such think-aloud methods to investigate the cognitive processes involved in solving the original CRT (2017) (Szaszi et al. 2017). So, concurrent think-aloud protocols may also be useful for tracing vCRT test-takers' thought processes and determining whether thinking aloud has a "reactive" effect on test performance (Ericsson 2003).

1.3. The Current Research

Our primary goals were (a) to test whether thinking aloud changed reflection test performance, (b) beta test an online think-aloud platform, (c) quantify the deviation between the standard interpretation of reflection tests and the two-factor explication of reflection, (d) to assess how vCRT performance depends on vCRT familiarity, and (e) assess the default interventionist account of reflection test responses. We pre-registered two hypotheses. First, thinking aloud during the vCRT will provide evidence of correct-but-unreflective responses and lured-yet-reflective responses. Second, thinking aloud will not significantly hinder vCRT performance—i.e., it will either not impact or improve reflection test performance. The results of an experiment and a follow-up study produced the hypothesized outcomes. They also detected that two-factor explication of 'reflection' strongly, albeit imperfectly, correlated with the standard interpretation of reflection test performance. All manipulations, measures, and exclusions are reported. All APA and IRB ethical guidelines were followed. Pre-registered hypotheses, methods, analytic strategy, data, and R scripts are on the Open Science Framework: https://osf.io/rk3jq.

2. Study 1

The first study primarily aimed to test the effect of thinking aloud on final responses to vCRT questions. The secondary aims were to test the difference in vCRT performance between familiar and naïve participants, the correlations between the standard interpretation of reflection tests and more recent explications of 'reflection', as well as the rate of correct-but-unreflective and lured-yet-reflective responses.

2.1. Method

Participants. People were recruited from public spaces on a university campus in the Southeastern United States. We pre-registered a target sample size of 100 participants—50 participants per condition (Simmons et al. 2013). After months of recruitment, reaching the pre-registered sample size with the in-person protocol became ethically and practically untenable because the World Health Organization announced a global pandemic (Ghebreyesus 2020), the university campus closed, and the university IRB announced that all in-person data collection must cease until further notice (Office for Human Subjects Protection 2020). Since the protocol could not be replicated online, we had to halt data collection after recruiting only 99 participants (mean age = 23.41 years; 48 identified as women, 50 as men, and 1 did not select a gender; 51 identified as White, 15 as Black, 16 as Hispanic or Latino, 1 as Pacific Islander, and 16 as other ethnicity).

2.2. Procedure and Materials

Manipulation. After consenting to participate, participants navigated to a Qualtrics survey using a QR code where they were randomly assigned to either a think-aloud condition or a control condition. To ensure that participants in both conditions completed the survey in front of a researcher, they were asked to remain at the research table until the end of the survey to receive their compensation: entry to win the smart speaker, water bottle, or books that were on the table.

Think-aloud protocol. Participants randomly assigned to the think-aloud condition were prompted to request instructions from a researcher. After the researcher explained the think-aloud protocol to participants, participants had a chance to ask for clarification and consent by selecting a button labeled "I received and understand the instructions from the researcher". Then a researcher began an audio recording on a smartphone and the participant practiced thinking aloud on a pre-survey task, "To practice thinking aloud, please say this sentence aloud, followed by the following number . . . ." The number each participant read aloud was generated randomly and used to anonymously pair survey responses with each corresponding think-aloud recording. Participants were reminded to think aloud as needed throughout the survey.

Verbal Cognitive Reflection Test. Participants completed the 10-item verbal Cognitive Reflection Test or vCRT (Sirota et al. 2021). Item order was fixed for all participants—no practice effects were observed (Appendix A Figure A1)—and responses were typed into text boxes. Following the standard interpretation of reflection tests, reflective scores were computed by summing correct responses (e.g., 2nd place) and unreflective scores were computed by summing lured responses (e.g., 1st place) on these verbal reasoning items. Non-lured incorrect responses were only 1.67% of all responses, lower than what has been observed for a widely used mCRT (Stupple et al. 2017, Table 5).

Questions about lures. To test whether correct responses followed lured responses—a la the default-interventionist account of reflection testing (Howarth and Handley 2016)—participants reported whether the lured response occurred to them after they submitted each reflection test answer. For instance, after answering the aforementioned question about passing the racer in 2nd place, participants were asked, "Have you thought at any point that '1st place' could be the answer?"

Deliberateness and consciousness in think-aloud recordings. The two-factors explication of 'reflection' holds that reasoning is reflective when if it is more deliberate and more consciously represented (Shea and Frith 2016). Reasoning is said to be deliberate when it does not merely accept the initial, automatic response and is said to be conscious when participants can articulate parts of their reasoning (Byrd 2019, 2022c). So, two student raters (who were not aware of the project's hypotheses), and then the first author, rated each response. Before raters became aware of participants' final answers, raters rated each response's deliberateness—i.e., whether the participant verbally reconsidered their initial response (Light's K_{avg} = 0.59)—and conscious representation—i.e., whether the participant verbalized a reason for or against any response (Light's K_{avg} = 0.43). Rating options included "yes", "no", and "indeterminate", and the "moderate" agreement between raters allowed each participant's "yes" ratings to be averaged (Landis and Koch 1977, p. 165).

Test familiarity. Prior work found that many participants were already familiar with reflection test questions and that such familiarity may be the best predictor of reflection test performance (Byrd 2022d; Stieger and Reips 2016). So the raters also rated whether each participant mentioned being familiar with any of the vCRT items (Light's K = 0.82). Determinations were labeled "yes", "no", and "indeterminate". The familiarity parameter was the average number of "yes" ratings per participant.

2.3. Results

We tested the rate of correct-but-unreflective responses and lured-yet-reflective responses on the vCRT, the effect of thinking aloud on reflection test performance, the correlation between standard reflection test scoring and recent explications of 'reflection', and the correlation between test familiarity and test performance.

Correct-but-unreflective and lured-yet-reflective responses. Think aloud verbal reports sometimes deviate from the standard interpretation of reflection tests. We expected some people to arrive at correct answers prior to reflection and without first thinking of the lured answer, and to be lured into accepting particular incorrect answers despite sustained reflection. Table 1 confirms this pre-registered expectation: the standard interpretation of correct and lured responses usually but imperfectly agrees with the two-factor explication of 'reflection'.

Table 1. Standard and two-factor categorizations of reflection test responses based on think-aloud protocol analysis of responses to the verbal reflection test in Study 1. Example verbalization based on the following reflection test question: "If you were running a race, and you passed the person in 2nd place, what place would you be in now?".

Category	Example Verbalization	Answer	Standard	Two-Factor	Rate
Correct-and-reflective	"1st obviously. No actually ... 2nd."	Correct	Reflective	Reflective	80.2%
Correct-but-unreflective	"2nd"	Correct	Reflective	Unreflective	19.8%
Lured-and-unreflective	"1st"	Lured	Unreflective	Unreflective	71.5%
Lured-yet-reflective	"I want to say 1st but, umm, yeah, 1st."	Lured	Unreflective	Reflective	28.5%

Thinking aloud did not impact performance. Figure 1 affirms our pre-registered hypothesis and prior meta-analytic work (Fox et al. 2011): we did not detect an interference effect of thinking aloud on the number of lured or correct responses on the vCRT.

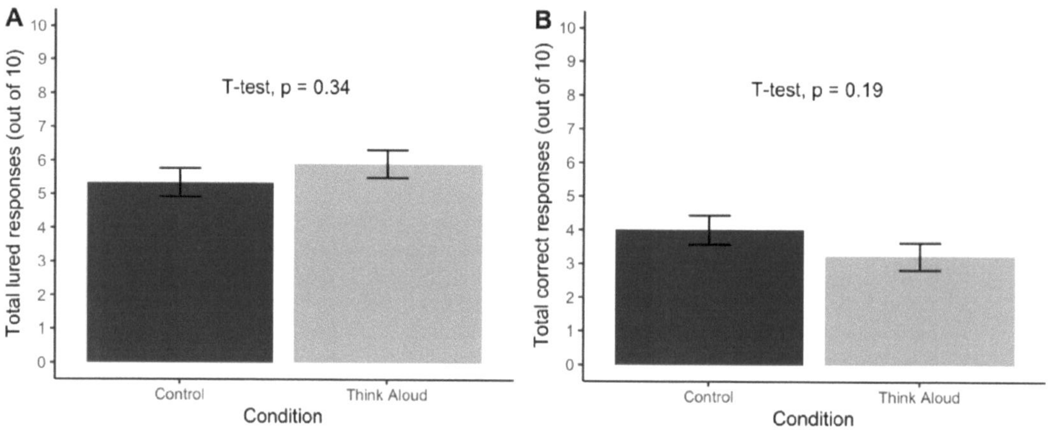

Figure 1. The effect of thinking aloud on (**A**) the number of lured responses and (**B**) the number of correct responses on the verbal cognitive reflection test (vCRT) in Study 1 (N = 99). Error bars represent a standard error.

Two factor interpretation predicts the standard interpretation. Regression analysis was employed to understand how well the standard interpretation of reflection tests aligns with dual process theorists' two-factor explication of 'reflection'. Figure 2 shows that they align well: the more that participants' verbalizations involved deliberate reconsideration of *any* response or conscious articulation of reasons for *any* response, the less likely participants were to accept lured responses and the more likely participants were to *accept* correct responses. In other words, the degree to which participants' *thinking* exhibited the two theoretical factors of reflection corresponded strongly with the degree to which participants' *final answers* were deemed reflective or unreflective by the standard interpretation of test responses.

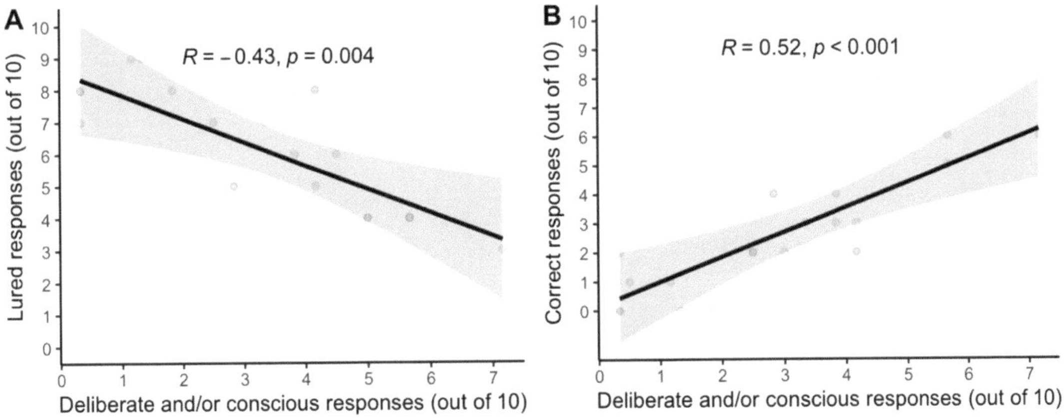

Figure 2. Correlations between the "two-factor" coding of reflective responses (deliberate and/or conscious) and (**A**) the number of lured—or so-called "unreflective"—responses and (**B**) the number of correct—or so-called "reflective"—responses on the verbal reflection test (vCRT) in the think-aloud condition of Study 1 (N = 47) with gray standard error bands.

Consideration of lured responses. Frederick (2005) observed that correct mCRT responses often involved consideration of the lured response. Some hypothesize that lures are appealing because they are more likely to feel correct (Thompson et al. 2013). If that is right, then people should not only be likely to *consider* lured responses, but those who consider lures should also be very likely to *accept* lures as their final answer. Figure 3 confirms this: consideration of lured responses *at some point* in each decision was relatively high (mean = 6.13, range = 0–10, S.D. = 2.59), and merely *considering* a lure almost perfectly predicted whether one *accepted* the lure as one's final answer.

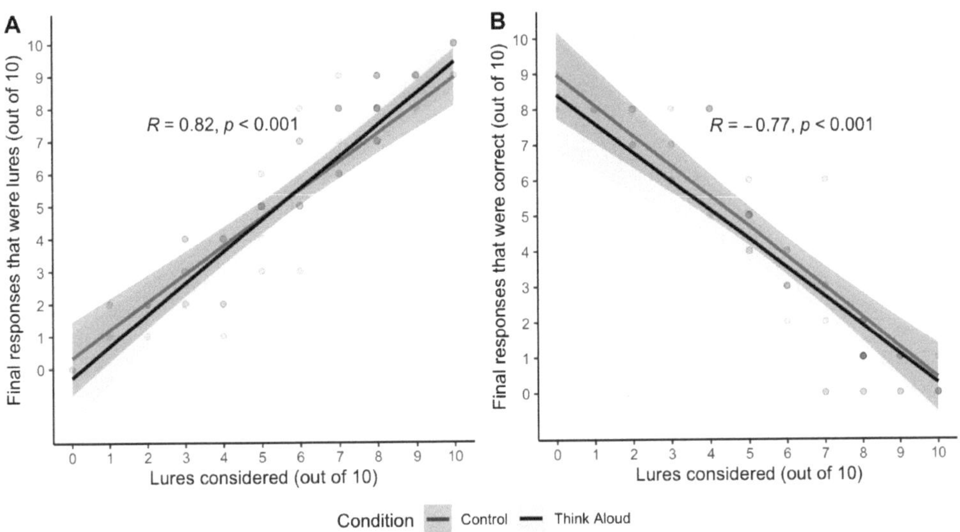

Figure 3. Correlations between *consideration* of lures and (**A**) *acceptance* of lures and (**B**) acceptance of correct responses on the verbal reflection test (vCRT) in Study 1 (N = 99). Gray bands represent a standard error.

Test familiarity predicted test performance. In about 27% of think-aloud recordings, participants mentioned prior familiarity with at least one item on the vCRT—e.g., "I've seen these questions on TikTok". Figure 4 shows a large difference in vCRT performance between familiar and naïve participants on both lured responses ($d = -0.87$) and correct responses ($d = 1.13$).

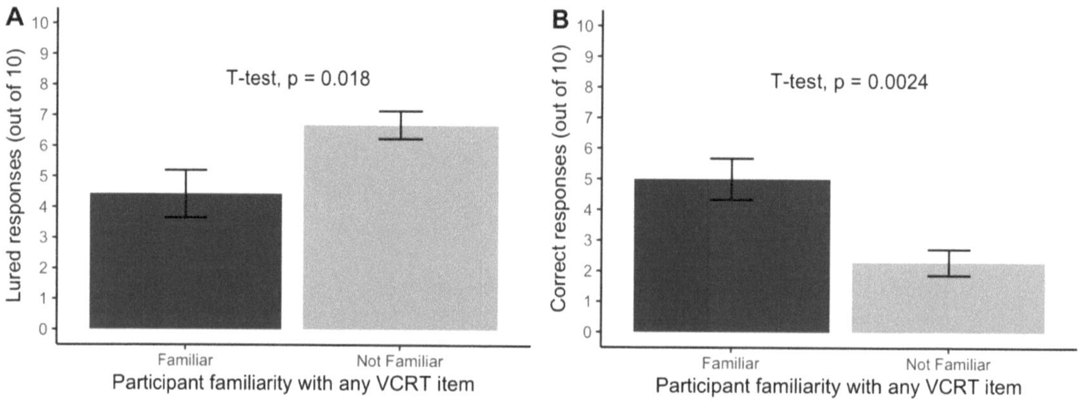

Figure 4. The number of (**A**) lured responses and (**B**) correct responses on the verbal reflection test (vCRT) among participants in the think-aloud condition of Study 1 (N = 47) depending on their unsolicited self-report of familiarity with the vCRT with standard error bars.

2.4. Discussion

These data suggest significant alignment between the dual process theorists' two-factor explication of 'reflection' and the standard interpretation of reflection test answers. They also suggest that the university participants were largely naïve to the vCRT even though self-reported familiarity remained a strong predictor of the standard interpretation of vCRT performance.

One might wonder whether these results will replicate in a larger think-aloud validation of the vCRT. Fortunately, our initial results suggest that the think-aloud protocol will not significantly influence vCRT performance. So, a larger replication is methodologically possible. Unfortunately, large-scale, in-person think-aloud protocols are prohibitively time-consuming, tedious, or—during a pandemic—unethical. To overcome these challenges, we partnered with a startup to develop a platform for large-scale, online, think-aloud surveys.

3. Study 2

Study 2 aimed to replicate the findings of Study 1 in a new sample of participants and test the feasibility of online think-aloud survey methodology. To do this, we reproduced all of the instructions and measures in the think-aloud condition of Study 1 in an online audio survey platform, Phonic (Perrodin and Todd 2021; Phonic Inc. 2020).

3.1. Method

Participants. English-speaking monolingual participants were recruited from Prolific (Palan and Schitter 2018; Peer et al. 2017) for an expected $9.85/hour based on average completion time of the think-aloud condition of Study 1. To ensure data quality, Prolific alerted candidate participants that compensation would depend on their consent and ability to provide usable recordings of their thoughts throughout the survey. We aimed to double the pre-registered sample size of the think-aloud condition of Study 1 (N = 47), recruiting 102 participants (mean age = 30.38; 57 identified as women, 38 as men, and 7 did not select a gender; 85 identified as White, 3 as Black, 3 as Hispanic or Latino, and 11 as other ethnicity).

3.2. Procedure and Materials

Phonic audio survey platform. We used an online audio survey platform (Phonic) to record concurrent verbalizations; this service was provided in exchange for beta testing the new survey platform.

Materials from Study 1. All materials from the think-aloud condition of Study 1 were included in Study 2. Participants practiced thinking aloud before the survey and then thought aloud while completing the same 10-item vCRT with reminders to verbalize all their thoughts throughout the test—again, no practice effect was observed (Appendix A Figure A2). After each answer was submitted, participants answered the same follow-up question about whether they considered the lured response. The deliberateness, consciousness, and test familiarity of participants' think-aloud recording for each question were rated by the principal investigator using the same rating options as Study 1. Replicating a result from Study 1, the rate of non-lured incorrect responses was very low (1.72%).

3.3. Results

We tested the correlation between test familiarity and test performance, the correlation between standard reflection test scoring and recent explications of 'reflection', as well as the rate of correct-but-unreflective responses and lured-yet-reflective responses on the vCRT.

Correct-but-unreflective and lured-yet-reflective responses. To test the agreement between the standard interpretation of reflection tests and more recent two-factor explications of 'reflection', the rates of correct-but-unreflective and lured-yet-reflective responses were determined by Prolific participants' think-aloud recordings. Table 2 shows a replication of the preponderant yet imperfect agreement between the standard interpretation and the two-factor explication of reflection test answers.

Table 2. Standard and two-factor categorizations of reflection test responses based on think-aloud protocol analysis of responses to the verbal reflection test in Study 2. Example verbalization based on the following reflection test question: "If you were running a race, and you passed the person in 2nd place, what place would you be in now?".

Categorization	Example Verbalization	Answer	Standard	Two-Factor	Rate
Correct-and-reflective	"1st. No ... 2nd. Whoopsie"	Correct	Reflective	Reflective	68.5%
Correct-but-unreflective	"2nd"	Correct	Reflective	*Unreflective*	31.5%
Lured-and-unreflective	"1st"	Lured	Unreflective	Unreflective	75.8%
Lured-yet-reflective	"1st ... or is that a trick? ... I'd say 1st."	Lured	Unreflective	*Reflective*	24.2%

Two-factor interpretation predicts the standard interpretation. Another regression analysis was employed to test how the standard interpretation of reflection tests aligns with dual process theorists' two-factor explication of 'reflection'. Figure 5 shows a replication of their correlation: increases in the number of participants' responses that involved deliberate or conscious thinking—as determined by think-aloud recordings—correlated with significant decreases in the number of lured responses and significant increases in the number of correct responses.

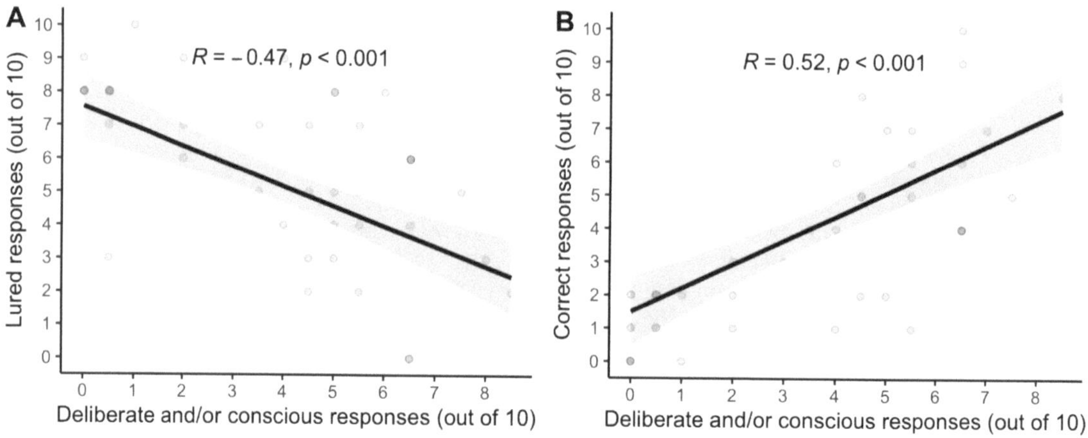

Figure 5. Correlations between the "two-factor" coding of reflective responses (deliberate and/or conscious) and (**A**) the number of lured—or so-called "unreflective"—responses and (**B**) the number of correct—or so-called "reflective"—responses on the verbal reflection test (vCRT) performance in Study 2 (N = 102). Gray bands represent a standard error.

Consideration of lured responses. Figure 6 shows a replication of the strong feeling of rightness of lured responses. Indeed, lure *consideration* was not only relatively high (mean = 5.37, range = 0–10, S.D. = 2.38), it remained the best predictor of *accepting* both lured and correct responses on the vCRT.

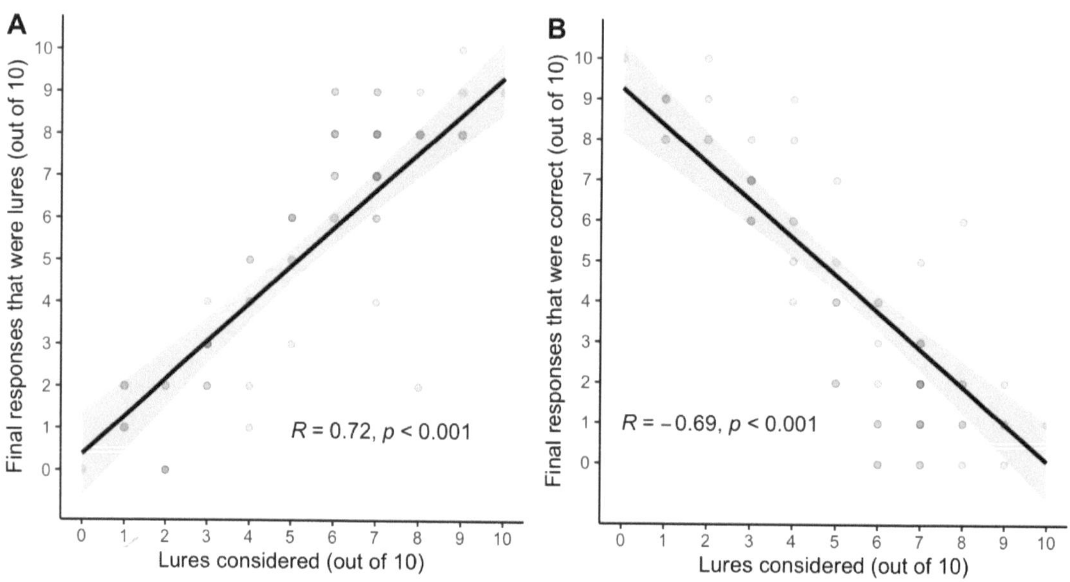

Figure 6. Correlations between consideration of lures and (**A**) *acceptance* of lures and (**B**) acceptance of correct responses on the verbal reflection test (vCRT) performance Study 2 (N = 102). Gray bands represent a standard error.

Test familiarity predicted test performance. In about 17% of think-aloud recordings, Prolific participants mentioned prior familiarity with at least one item on the vCRT—significantly less familiarity than the 27% familiarity among our university participants, $t = -2.9$, 95% CI [0.09, 0.24], $p = 0.005$. Figure 7 shows a replication of the large difference in vCRT performance between familiar and naïve participants for both lured responses ($d = -0.91$) and correct responses ($d = 0.90$).

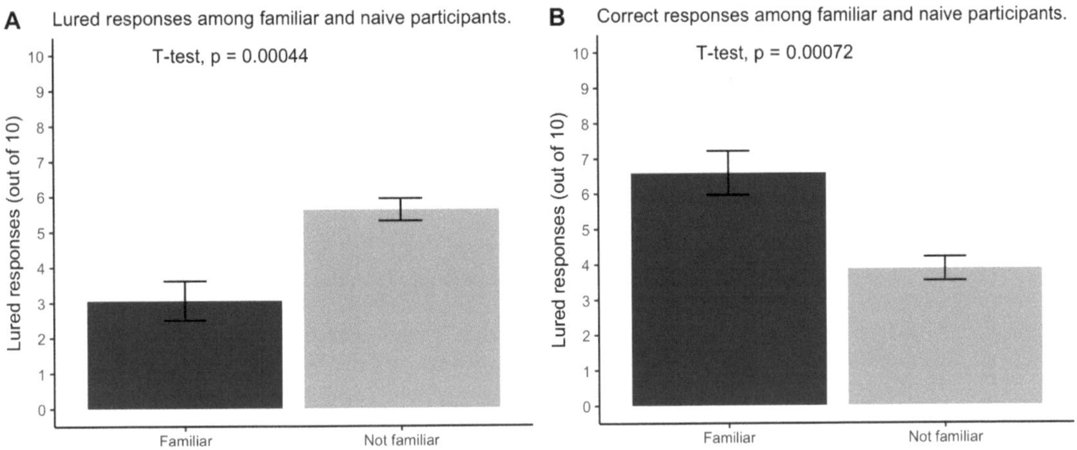

Figure 7. The number of (**A**) lured responses and (**B**) correct responses on the verbal reflection test (vCRT) among participants in Study 2 (N = 102) depending on their unsolicited self-report of familiarity with the vCRT. Error bars represent a standard error.

4. General Discussion

In two think-aloud protocol studies, we found that our participants' correct responses on the verbal cognitive reflection test occurred after thinking of the intuitively appealing but wrong initial responses (i.e., lures). When accepting lures as their *final* answers, our participants' thinking typically lacked deliberative attempts to correct the initial, appealing, and wrong (lured) response. Exceptions to these two expected thinking trajectories on the vCRT comprised only 19–31% of responses—notably fewer than the 39–77% of exceptional responses detected for a widely used mCRT (Szaszi et al. 2017). Thus, our findings largely aligned with the standard interpretation of vCRT responses (Pennycook et al. 2015a) and more recent two-factor explications of reflective reasoning (Byrd 2019, 2022a, 2022c; Shea and Frith 2016). Nonetheless, the substantial minority of correct-but-unreflective and lured-yet-reflective responses indicate opportunities to reduce the measurement error of reflection tests (Machery 2021) to avoid falsely labeling some correct responses as reflective and some lured responses as unreflective or miserly (Toplak et al. 2014).

Further, we did not find that thinking aloud interfered with performance on the verbal reflection test. These studies also confirmed that most university participants and Prolific participants were naïve to the vCRT. Nonetheless, unsolicited think-aloud self-reports of familiarity with the vCRT were a strong predictor of the standard interpretation of vCRT performance in both studies. This evidence replicates and extends some of the promising features of the vCRT (Sirota et al. 2021).

4.1. Methodological Implications

The present studies also suggest that think-aloud protocols can reveal valuable and otherwise undetected nuance in cognitive reflection test performance. For instance, think-aloud recordings revealed that the standard interpretation of reflection test responses mislabeled 19–31% of responses as either reflective or unreflective.

This insight seems to increase the justificatory burden of employing the standard interpretation of reflection tests or of *not* employing the think-aloud protocols (Byrd 2022b) that more accurately detected the deliberate and conscious features of reflective thinking (Shea and Frith 2016). Even if researchers do not rethink their interpretation of or reliance on reflection tests, they may nonetheless need to justify the status quo.

In addition to implications for *measuring* reflection, there may also be implications for *manipulating* reflection. Recall the two-response paradigm of measuring reflection (e.g., Bago and De Neys 2019). It may not provide insight into the process that changes participants' initial response, but it does indicate whether this initial response was already correct. So the two-response paradigm could be a compromise between the ease of the standard interpretation of reflection tests and the tedium of coding reflection test transcripts (e.g., Burič and Šrol 2020). Ideally, however, this second response would come after some sort of reflective task—as opposed to an imposed waiting period in which participants can do whatever they want—to ensure that observed changes are the result of reflection and not some other factor (Isler et al. 2020).

4.2. Theoretical Implications

Another result of the present studies was reliable support for the "feeling of rightness" explanation of reflection test performance (Pennycook et al. 2015b; Thompson et al. 2011). Most responses involved *consideration* of the lure and *consideration* of lures was the best predictor of *accepting* both lured and correct responses on the vCRT. If lures were not significantly more appealing than other possible responses, then it would be difficult to explain this preponderance of lure *consideration*, lure *acceptance*, and their strong correlation.

This may have implications for the debate between default interventionist accounts of reflection and their alternatives (Howarth and Handley 2016). Those who arrived at the correct response were very unlikely to have considered the lured response. In other words, the so-called reflective (i.e., correct) response on reflection tests may not usually involve intervening on a default (lured) response after all. Of course, these data confirm that there are *some* cases of reflective default intervention. So the current evidence may not falsify the default interventionist account so much as show that it is not an exhaustive explanation of reflection test performance.

4.3. Limitations

The current studies were limited by resources for listening to and coding think-aloud verbal reports. This resulted in minimal sample sizes for the research questions addressed in this paper (Simmons et al. 2013). Although the expected effects were detected—some in multiple populations, both in-person and online—there remains an opportunity for researchers with more resources to conduct larger-scale replications and extensions of the existing work. For example, researchers could annotate additional aspects of participants' transcripts from reflection tests—e.g., whether overcoming faulty intuitions varies depending on the *kinds* of reasons participants consider: reasons for their initial response, reasons for subsequent responses, reasons against their initial response, or reasons against subsequent responses (Cullen et al. 2022). We look forward to seeing what other researchers discover with our and others' (open) datasets. There are also opportunities for online think-aloud survey platforms to improve their speech transcription, sentiment analysis, and other features enough to automate and therefore scale up think-aloud protocol research.

5. Conclusions

The present studies partially replicate and clarify existing validations of the verbal cognitive reflection test, thanks in part to novel online audio survey technology. Most participants are naïve to the test and the standard interpretation of reflection testing largely aligns with more advanced explications of reflective reasoning. Taken together with existing work showing that verbal reflection tests can have high internal consistency, high test-retest reliability, and less association with mathematical ability or gender, the present evidence suggests that the vCRT could be a promising supplement or replacement for widely used reflection tests. Nonetheless, there may still be opportunities to improve our understanding of reflection by redeploying online think-aloud protocols for larger-scale research. Thus, both verbal reflection tests and online think-aloud protocols are promising tools for advancing our understanding of reflective reasoning and its alternatives.

Author Contributions: Conceptualization, N.B. and M.S.; Methodology, N.B.; Formal analysis, N.B.; Investigation, N.B.; Data curation, N.B., B.J. and G.G.; Writing—original draft, N.B.; Writing—review & editing, N.B. and M.S.; Visualization, N.B.; Supervision, N.B.; Project administration, N.B.; Funding acquisition, N.B. All authors have read and agreed to the published version of the manuscript.

Funding: This research benefitted from $250 and two research assistants (B. Joseph & G. Gongora) via the Undergraduate Research Opportunity Program at Florida State University. The APC for this article was waived.

Institutional Review Board Statement: This research was deemed "exempt" by the Office for Human Subjects Protection at Florida State University (STUDY00000697, 2019-12-13: https://osf.io/z7bua).

Informed Consent Statement: An "Information Sheet" was recommended instead of a consent form by the Office for Human Subjects Protection at Florida State University (STUDY00000697, 2019-12-13: https://osf.io/ax85v).

Data Availability Statement: All data and analysis files available at https://osf.io/rk3jq/.

Acknowledgments: Thanks to the late Anders Ericsson for advising the development of participant instructions as well as students in Ericsson's final *Expertise & Protocol Analysis* seminar for feedback on the initial presentation of this research. Thanks also to Jeff Larsen, Sunny Khemlani, Megan Peters, the Cognitive Research Group at the University of Surrey, the Center for Research on Self and Identity at University of Southampton, the Department of Social and Decision Sciences at Carnegie Mellon University, and participants of the 2022 Society for Personality and Social Psychology, 2022 Society for Judgment and Decision-making, 2021 Behavioral Research Online, and 2021 Association For Psychological Science conferences for feedback on presentations of this project.

Conflicts of Interest: Phonic audio survey service was provided in exchange for beta testing its platform and features (Study 2).

Appendix A

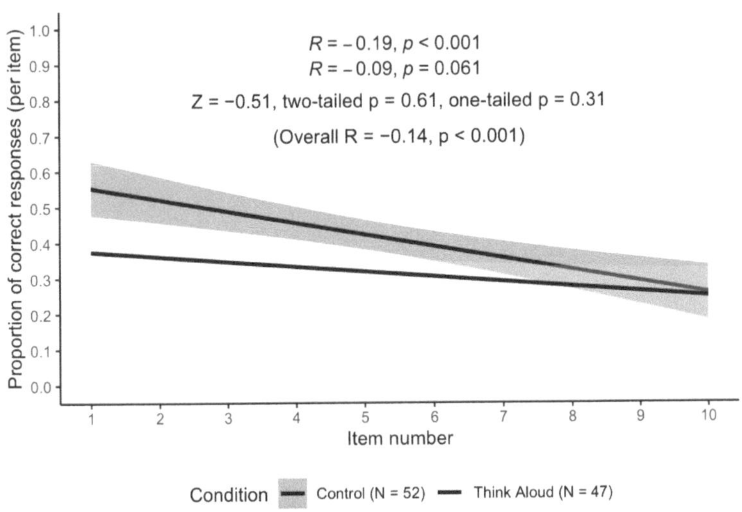

Figure A1. Practice effect tests in Study 1 (N = 99). Performance did not increase but decreased during the survey. The magnitude of this decrease was not dramatically different between conditions (Z-score). Error bands indicate standard error.

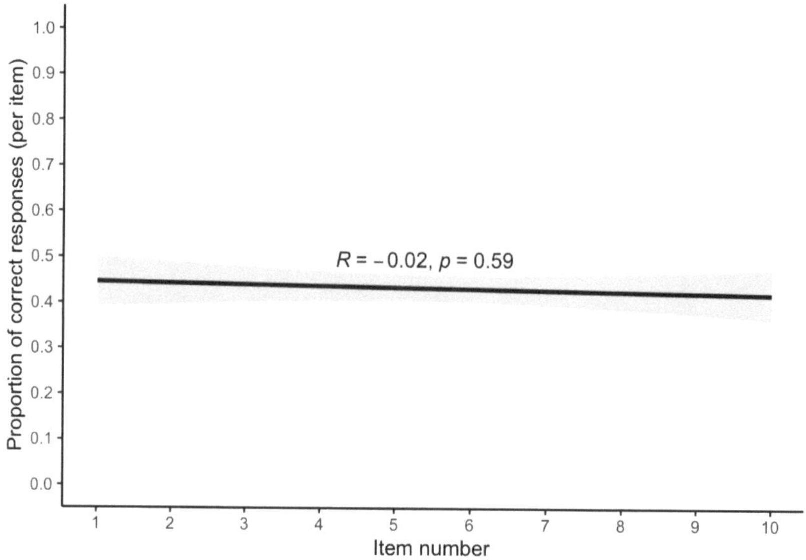

Figure A2. Practice effect test in Study 2 (N = 102). Performance did not increase or decrease during the survey. Error bands indicate standard error.

Think Aloud Instructions

In this experiment we are interested in what you think about when you find answers to some questions that I am going to ask you to answer. In order to do this we ask you to THINK ALOUD as you work on the problem given.

What I mean by 'think aloud' is that I want you to tell me EVERYTHING you are thinking from the time you first see the question until you give an answer. I would like you to talk aloud CONSTANTLY from the time you start the survey to the time that you finish.

Please don't plan out what you say or try to explain to me what you are saying. Just act as if you are alone in the room speaking to yourself.

It is most important that you keep talking. If you are silent for any long period of time I will ask you to talk.

Do you understand what we need you to do?

Good, now you can practice thinking aloud on this sample question.

[Ensure they read aloud (including the code) and continue to think aloud while practicing]

Ok. Now you will begin the survey.

[Use phrase like 'remember to say your thoughts aloud' if they are silent for a few seconds.]
[Thank them when they are done.]

Modified Verbal Cognitive Reflection Test (Sirota et al. 2021)

(1) Mary's father has 5 daughters but no sons—Nana, Nene, Nini, Nono. What is the fifth daughter's name probably?
[correct answer: Mary, lured answer: Nunu]
[Page break]
"Do you remember thinking at any point that 'Nunu' could be the answer?"
Yes No

(2) If you were running a race, and you passed the person in 2nd place, what place would you be in now?
[correct answer: 2nd, lured answer: 1st]
[Page break]
"Do you remember thinking at any point that '1st' could be the answer?"

Yes No

(3) It's a stormy night and a plane takes off from JFK airport in New York. The storm worsens, and the plane crashes-half lands in the United States, the other half lands in Canada. In which country do you bury the survivors?
[correct answer: don't bury survivors, lured answers: answers about burial location]
[Page break]
"Do you remember thinking at any point that survivor burial was an option?"
Yes No

(4) A monkey, a squirrel, and a bird are racing to the top of a coconut tree. Who will get the banana first, the monkey, the squirrel, or the bird?
[correct answer: no banana on coconut tree, lured answer: any of the animals]
[Page break]
"Do you remember thinking at any point that 'bird', 'squirrel', or 'monkey' could be the answer?"
Yes No

(5) In a one-storey pink house, there was a pink person, a pink cat, a pink fish, a pink computer, a pink chair, a pink table, a pink telephone, a pink shower—everything was pink! What colour were the stairs probably?
[correct answer: a one-storey house probably doesn't have stairs, lured answer: pink]
[Page break]
"Do you remember thinking at any point that 'pink' could be the answer?"
Yes No

(6) How many of each animal did Moses put on the ark?
[correct answer: none; lured answer: two]
[Page break]
"Do you remember thinking at any point that 'two' could be the answer?"
Yes No

(7) The wind blows west. An electric train runs east. In which cardinal direction does the smoke from the locomotive blow?
[correct answer: no smoke from an electric train, lured answer: west]
[Page break]
"Do you remember thinking at any point the locomotive will produce smoke?"
Yes No

(8) If you have only one match and you walk into a dark room where there is an oil lamp, a newspaper and wood—which thing would you light first?
[correct answer: match, lured answer: oil lamp]
[Page break]
"Do you remember thinking at any point that 'oil lamp', 'newspaper', or 'wood' could be the answer?"
Yes No

(9) Would it be ethical for a man to marry the sister of his widow?
[correct answer: not possible, lured answer: yes, no]
[Page break]
"Do you remember thinking at any point that it is possible for a man to marry the sister of his widow?"
Yes No

(10) Which sentence is correct: (a) "the yolk of the egg are white" or (b) "the yolk of the egg is white"?
[correct answer: the yolk is yellow, lured answer: b]
[Page break]
"Do you remember thinking at any point that 'a' or 'b' could be the answer?"
Yes No

Scoring and Coding

- Conditions: Control condition = 0. Think Aloud condition = 1.
- Standard coding: Sum correct and lured answers for reflective and unreflective parameters, respectively.
- From transcripts of think-aloud condition, create variables for (a) whether each participant reconsidered each their initial responses, (b) whether each participant verbalized any reason(s) for or against any response(s), and (c) whether a participant mentioned being familiar with any verbal reflection test question.

References

Attali, Yigal, and Maya Bar-Hillel. 2020. The False Allure of Fast Lures. *Judgment and Decision Making* 15: 93–111. [CrossRef]

Bago, Bence, and Wim De Neys. 2019. The Smart System 1: Evidence for the Intuitive Nature of Correct Responding on the Bat-and-Ball Problem. *Thinking & Reasoning* 25: 257–99. [CrossRef]

Ball, Linden J., and Alexandra Stevens. 2009. Evidence for a Verbally-Based Analytic Component to Insight Problem Solving. *Proceedings of the Thirty-First Annual Conference of the Cognitive Science Society* 31: 1060–65.

Białek, Michał, and Gordon Pennycook. 2018. The Cognitive Reflection Test Is Robust to Multiple Exposures. *Behavior Research Methods* 50: 1953–59. [CrossRef]

Blech, Christine, Robert Gaschler, and Merim Bilalić. 2020. Why Do People Fail to See Simple Solutions? Using Think-Aloud Protocols to Uncover the Mechanism behind the Einstellung (Mental Set) Effect. *Thinking & Reasoning* 26: 552–80. [CrossRef]

Burič, Roman, and Jakub Šrol. 2020. Individual Differences in Logical Intuitions on Reasoning Problems Presented under Two-Response Paradigm. *Journal of Cognitive Psychology* 32: 460–77. [CrossRef]

Byrd, Nick. 2019. What We Can (and Can't) Infer about Implicit Bias from Debiasing Experiments. *Synthese* 198: 1427–55. [CrossRef]

Byrd, Nick. 2022a. A Two-Factor Explication of "Reflection": Unifying, Making Sense of, and Guiding the Philosophy and Science of Reflective Reasoning. pp. 1–21. Available online: https://researchgate.net/publication/370131881 (accessed on 21 March 2023).

Byrd, Nick. 2022b. All Measures Are Not Created Equal: Reflection Test, Think Aloud, and Process Dissociation Protocols. Available online: https://researchgate.net/publication/344207716 (accessed on 21 March 2023).

Byrd, Nick. 2022c. Bounded Reflectivism & Epistemic Identity. *Metaphilosophy* 53: 53–69. [CrossRef]

Byrd, Nick. 2022d. Great Minds Do Not Think Alike: Philosophers' Views Predicted by Reflection, Education, Personality, and Other Demographic Differences. *Review of Philosophy and Psychology*. [CrossRef]

Cullen, Simon, Nick Byrd, and Philipp Chapkovski. 2022. Experiments in Reflective Equilibrium Using the Socrates Platform. Paper presented at Remotely to Reflection on Intelligent Systems: Towards a Cross-Disciplinary Definition, Stuttgart, Germany, October 20–21; Available online: https://researchgate.net/publication/370132037 (accessed on 21 March 2023).

Erceg, Nikola, Zvonimir Galic, and Mitja Ružojčić. 2020. A Reflection on Cognitive Reflection—Testing Convergent Validity of Two Versions of the Cognitive Reflection Test. *Judgment & Decision Making* 15: 741–55. [CrossRef]

Ericsson, K. Anders. 2003. Valid and Non-Reactive Verbalization of Thoughts During Performance of Tasks Towards a Solution to the Central Problems of Introspection as a Source of Scientific Data. *Journal of Consciousness Studies* 10: 1–18.

Ericsson, K. Anders, and Herbert A. Simon. 1980. Verbal Reports as Data. *Psychological Review* 87: 215–51. [CrossRef]

Ericsson, K. Anders, and Herbert A. Simon. 1993. *Protocol Analysis: Verbal Reports as Data*, revised ed. Cambridge: Bradford Books/MIT Press.

Evans, Jonathan. 2007. On the Resolution of Conflict in Dual Process Theories of Reasoning. *Thinking & Reasoning* 13: 321–39. [CrossRef]

Evans, Jonathan, and Keith E. Stanovich. 2013. Dual-Process Theories of Higher Cognition Advancing the Debate. *Perspectives on Psychological Science* 8: 223–41. [CrossRef] [PubMed]

Evans, Jonathan, Julie L. Barston, and Paul Pollard. 1983. On the Conflict between Logic and Belief in Syllogistic Reasoning. *Memory & Cognition* 11: 295–306. [CrossRef]

Fox, Mark C., K. Anders Ericsson, and Ryan Best. 2011. Do Procedures for Verbal Reporting of Thinking Have to Be Reactive? A Meta-Analysis and Recommendations for Best Reporting Methods. *Psychological Bulletin* 137: 316–44. [CrossRef]

Frankish, Keith. 2010. Dual-Process and Dual-System Theories of Reasoning. *Philosophy Compass* 5: 914–26. [CrossRef]

Frederick, Shane. 2005. Cognitive Reflection and Decision Making. *Journal of Economic Perspectives* 19: 25–42. [CrossRef]

Ghebreyesus, Tedros Adhanom. 2020. WHO Director-General's Opening Remarks at the Media Briefing on COVID-19 on March 11. Available online: who.int/director-general/speeches/detail/who-director-general-s-opening-remarks-at-the-media-briefing-on-covid-19---11-march-2020 (accessed on 21 March 2023).

Howarth, Stephanie, and Simon Handley. 2016. Belief Bias, Base Rates and Moral Judgment: Re-Evaluating the Default Interventionist Dual Process Account. In *Thinking Mind*. Edited by Niall Galbraith, Erica Lucas and David Over. New York: Taylor & Francis, pp. 97–111.

Isler, Ozan, Onurcan Yilmaz, and Burak Dogruyol. 2020. Activating Reflective Thinking with Decision Justification and Debiasing Training. *Judgment and Decision Making* 15: 926–38. [CrossRef]

Kahneman, Daniel, and Shane Frederick. 2002. Representativeness Revisited: Attribute Substitution in Intuitive Judgment. In *Heuristics and Biases: The Psychology of Intuitive Judgment*. Edited by Thomas Gilovich, Dale W. Griffin and Daniel Kahneman. New York: Cambridge University Press, pp. 49–81.
Landis, J. Richard, and Gary G. Koch. 1977. The Measurement of Observer Agreement for Categorical Data. *Biometrics* 33: 159–74. [CrossRef]
Machery, Edouard. 2021. A Mistaken Confidence in Data. *European Journal for Philosophy of Science* 11: 34. [CrossRef]
Markovits, Henry, Pier-Luc de Chantal, Janie Brisson, Éloise Dubé, Valerie Thompson, and Ian Newman. 2021. Reasoning Strategies Predict Use of Very Fast Logical Reasoning. *Memory & Cognition* 49: 532–43. [CrossRef]
Office for Human Subjects Protection. 2020. Temporary Cessation to Some FSU Human Subjects Research. *Florida State University News (blog)*. March 23. Available online: https://news.fsu.edu/announcements/covid-19/2020/03/23/temporary-cessation-to-some-fsu-human-subjects-research/ (accessed on 23 March 2020).
Palan, Stefan, and Christian Schitter. 2018. Prolific.Ac—A Subject Pool for Online Experiments. *Journal of Behavioral and Experimental Finance* 17: 22–27. [CrossRef]
Peer, Eyal, Laura Brandimarte, Sonam Samat, and Alessandro Acquisti. 2017. Beyond the Turk: Alternative Platforms for Crowdsourcing Behavioral Research. *Journal of Experimental Social Psychology* 70: 153–63. [CrossRef]
Pennycook, Gordon, James Allan Cheyne, Derek J. Koehler, and Jonathan A. Fugelsang. 2015a. Is the Cognitive Reflection Test a Measure of Both Reflection and Intuition? *Behavior Research Methods* 48: 341–48. [CrossRef]
Pennycook, Gordon, Jonathan A. Fugelsang, and Derek J. Koehler. 2015b. What Makes Us Think? A Three-Stage Dual-Process Model of Analytic Engagement. *Cognitive Psychology* 80: 34–72. [CrossRef]
Perrodin, David D, and Richard Watson Todd. 2021. Choices in Asynchronously Collecting Qualitative Data: Moving from Written Responses to Spoken Responses for Open-Ended Queries. *DRAL4* 2021: 11. Available online: https://sola.pr.kmutt.ac.th/dral2021/wp-content/uploads/2022/06/3.pdf (accessed on 21 March 2023).
Petitmengin, Claire, Anne Remillieux, Béatrice Cahour, and Shirley Carter-Thomas. 2013. A Gap in Nisbett and Wilson's Findings? A First-Person Access to Our Cognitive Processes. *Consciousness and Cognition* 22: 654–69. [CrossRef]
Phonic Inc. 2020. Surveys You Can Answer with Your Voice. Available online: phonic.ai (accessed on 21 February 2020).
Purcell, Zoë A., Colin A. Wastell, and Naomi Sweller. 2021. Domain-Specific Experience and Dual-Process Thinking. *Thinking & Reasoning* 27: 239–67. [CrossRef]
Schooler, Jonathan W., Stellan Ohlsson, and Kevin Brooks. 1993. Thoughts beyond Words: When Language Overshadows Insight. *Journal of Experimental Psychology: General* 122: 166–83. [CrossRef]
Shea, Nicholas, and Chris D. Frith. 2016. Dual-Process Theories and Consciousness: The Case for "Type Zero" Cognition. *Neuroscience of Consciousness* 2016: niw005. [CrossRef] [PubMed]
Simmons, Joseph P., Leif D. Nelson, and Uri Simonsohn. 2013. Life after P-Hacking. Paper presented at Meeting of the Society for Personality and Social Psychology, New Orleans, LA, USA, January 17–19; p. 38. Available online: https://papers.ssrn.com/abstract=2205186 (accessed on 21 March 2023).
Sirota, Miroslav, Lenka Kostovičová, Marie Juanchich, Chris Dewberry, and Amanda Claire Marshall. 2021. Measuring Cognitive Reflection without Maths: Developing and Validating the Verbal Cognitive Reflection Test. *Journal of Behavioral Decision Making* 34: 322–43. [CrossRef]
Sobkow, Agata, Angelika Olszewska, and Miroslav Sirota. 2023. The Factor Structure of Cognitive Reflection, Numeracy, and Fluid Intelligence: The Evidence from the Polish Adaptation of the Verbal CRT. *Journal of Behavioral Decision Making* 36: e2297. [CrossRef]
Stagnaro, Michael N., Gordon Pennycook, and David G. Rand. 2018. Performance on the Cognitive Reflection Test Is Stable across Time. *Judgment and Decision Making* 13: 260–67. [CrossRef]
Stanovich, Keith E. 2018. Miserliness in Human Cognition: The Interaction of Detection, Override and Mindware. *Thinking & Reasoning* 24: 423–44. [CrossRef]
Stieger, Stefan, and Ulf-Dietrich Reips. 2016. A Limitation of the Cognitive Reflection Test: Familiarity. *PeerJ* 4: e2395. [CrossRef]
Stromer-Galley, Jennifer. 2007. Measuring Deliberation's Content: A Coding Scheme. *Journal of Public Deliberation* 3. [CrossRef]
Stupple, Edward J. N., Melanie Pitchford, Linden J. Ball, Thomas E. Hunt, and Richard Steel. 2017. Slower Is Not Always Better: Response-Time Evidence Clarifies the Limited Role of Miserly Information Processing in the Cognitive Reflection Test. *PLoS ONE* 12: e0186404. [CrossRef]
Szaszi, Barnabas, Aba Szollosi, Bence Palfi, and Balazs Aczel. 2017. The Cognitive Reflection Test Revisited: Exploring the Ways Individuals Solve the Test. *Thinking & Reasoning* 23: 207–34. [CrossRef]
Thompson, Valerie A., and Stephen C. Johnson. 2014. Conflict, Metacognition, and Analytic Thinking. *Thinking & Reasoning* 20: 215–44. [CrossRef]
Thompson, Valerie A., Jamie A. Prowse Turner, and Gordon Pennycook. 2011. Intuition, Reason, and Metacognition. *Cognitive Psychology* 63: 107–40. [CrossRef] [PubMed]
Thompson, Valerie A., Jonathan Evans, and Jamie I. D. Campbell. 2013. Matching Bias on the Selection Task: It's Fast and Feels Good. *Thinking & Reasoning* 19: 431–52. [CrossRef]

Toplak, Maggie E., Richard F. West, and Keith E. Stanovich. 2014. Assessing Miserly Information Processing: An Expansion of the Cognitive Reflection Test. *Thinking & Reasoning* 20: 147–68. [CrossRef]

Wilson, Timothy, and Richard E. Nisbett. 1978. The Accuracy of Verbal Reports About the Effects of Stimuli on Evaluations and Behavior. *Social Psychology* 41: 118–31. [CrossRef]

Journal of
Intelligence

MDPI

Essay

Why Should We Study the Foreign Language Effect: Debiasing through Affecting Metacognition?

Michał Białek

Psychology of Management Unit, Institute of Psychology, University of Wroclaw, 50-527 Wrocław, Poland; michal.bialek3@uwr.edu.pl

Abstract: Debiasing is a method of improving people's decisions by reducing their reliance on salient intuitions causing them to behave suboptimally or biasedly. However, many of the known debiasing techniques have limited effectiveness or can only remedy a one-shot decision, rather than having a lasting impact. In this work, I focus on the role of metacognition in debiasing decision-making and discuss how it can be better understood through the lens of the foreign language effect. The foreign language effect suggests that using a foreign language can sometimes benefit people's decision-making without providing them with additional information or instructions regarding the task. However, we do not fully understand how the foreign language effect works and its limitations. I conclude by urging scientists to research this effect with the hope of having a lasting positive impact on society.

Keywords: intuition; reflection; metacognition; debiasing; foreign language effect

1. Introduction

Following the 9/11 World Trade Center attack, the United States established an "enhanced interrogation program" in detention camps such as Guantanamo Bay with the intention of preventing future terrorist attacks. However, as early as three months after the program was launched, reports suggested that techniques such as waterboarding and sleep deprivation were ineffective in preventing attacks (Senate Select Committee on Intelligence 2012). Despite this, the program persisted for over a decade. In hindsight, continuing (and even starting) the program was a bad decision.

There are many possible explanations for why people make bad decisions that, in hindsight, appear to be obviously incorrect. Some errors are simply random noise, as people decide by sampling from their internal distribution of values (Stewart et al. 2006), or by selectively assigning queries to internalized preferences (Johnson et al. 2007). In other words, people do not have a fixed set of preferences but rather decide ad hoc. Our preferences are not fixed parameters (e.g., 7 out of 10), but rather a distribution of parameters, in which some values are more probable than others, but many are possible nonetheless. To illustrate, assume you have developed a taste for lasagna, but you may wish to eat it more or less strongly on a given day. Decisions are being made similarly, and even experts make mistakes due to the natural "noisiness" of human cognition (Kahneman et al. 2022). However, some errors are predictable, to the extent that for some type of problems most people tend to commit the same error (Kahneman et al. 1982). Decision errors, such as starting the torture program by the US government, may occur because decision-makers lack the required knowledge or expertise to assess all arguments and potential consequences of their decision (Larrick and Feiler 2015). People may also lack the cognitive abilities to process all the available information (Maran et al. 2020). Finally, some people may be blinded by their beliefs and ignore inconsistent information (Kunda 1990).

However, there is something potentially more profound and often overlooked in the literature that contributes to imperfect decision-making—metacognition, that is thinking about thinking (Dunlosky and Metcalfe 2008). By allowing the decision-maker to take an outside perspective and reevaluate their thinking, metacognition can improve decision-making. It helps an individual to monitor their cognition, resulting in subjective feeling of its quality and accuracy—confidence or feeling for rightness (Ackerman and Thompson 2017; Thompson et al. 2011). If the confidence is low, an individual allocates more cognitive resources into the ongoing cognitive processing, which in turn is likely to improve its quality and accuracy. In this essay, I will explore the functions of metacognition and examine whether it can be facilitated or distorted.

Metacognition is quite different from cognitive abilities, such as intelligence, which are responsible for information processing. The best illustration of the difference between cognitive abilities and metacognition in the context of decision-making I can think of is chess. When you try to solve a chess puzzle, you know there is a winning move. All you have to do is to consider all the possible moves and their follow-ups and see which of these combinations leads to a win. This is a lot of information you must process, and the outcome of your thinking depends on your cognitive abilities. However, things are far more complicated in a real-life competitive chess game for which you have limited time. You can lose a game when having a worse position and ending up being checkmated, but you can also lose on time. Hence, you cannot go into such deep deliberation as in a chess puzzle for every move. To manage your time efficiently, you have to know when to reflect and search for this single good move, and when no deep thinking is needed because there are several equally good moves and you can select any of them without much impact on the game. Deciding when to think depends on metacognition. This may refer to thinking before deciding, but also to cases when one has to re-evaluate a decision that has already been made.

The torture program of the USA I described in this work illustrates a failure of metacognition—the error stems from decision-makers not realizing they must stop and think, reevaluating all available evidence. They may have been operating under the assumption that these techniques were necessary for obtaining valuable intelligence and protecting national security. However, they may not have taken the time to critically evaluate the effectiveness of these methods, or to consider the ethical and moral implications of using such methods. Additionally, they may not have reflected on the potential long-term consequences of using these techniques, such as damage to the reputation of the country, legal challenges, or creating more radicalized individuals. In this scenario, the lack of metacognitive insight—the ability to step back and reflect on one's own thought process—led to a decision that was not fully informed and not considering the broader context and implications. Even after learning the results of the program were not as expected, the decision-makers failed to stop and reevaluate the decision they once made. This example illustrates that by going forward and not reflecting on their intuitions and past decisions, even the most gifted person will fail at a task because they have not even started thinking. Without metacognitive insight, whatever made these certainly intelligent individuals decide in a particular way will never be corrected, in turn escalating commitment to a bad decision, and risking them repeating the same decision error in the future.

As mentioned earlier, a failure of metacognition can occur when an individual fails to reflect on their own thinking. However, there is another less obvious but equally problematic failure of metacognition: overthinking. For example, a person may be spending too much time analyzing and comparing options that are only slightly different from each other. This inefficient allocation can result in a waste of time and resources for both the individual and those involved in the decision-making process. Effective metacognition, on the other hand, helps individuals allocate their cognitive resources efficiently by directing their attention toward the most important and relevant tasks.

Hopefully, it is becoming clear that metacognition, or thinking about our own thinking, is crucial for good decision-making. It helps decision-makers become aware of their internal processes, the cues they use, and to monitor and calibrate their confidence in their decisions (Ackerman and Thompson 2017). In this work, we are going to look into metacognition more closely. We will learn about what it is and how it works, and we will see if we can find ways to make it better. Specifically, I will present research on the foreign language effect, which suggests that using a non-native language can influence metacognitive processes and enhance certain types of decision-making, such as risky choices (Circi et al. 2021; Stankovic et al. 2022). My ultimate goal is to determine if the insights gained from studying the foreign language effect and metacognition can be applied to improve everyday decision-making. However, while we can artificially boost metacognition in experimental settings, it is not always beneficial, and we still do not fully understand why it works and when it will be effective. Therefore, although the foreign language effect research may not provide an immediate solution to decision-making difficulties, I believe that the long-term potential of this research is promising.

This essay is intended as an introductory reading to help you familiarize yourself with the topics of debiasing, metacognition, and the foreign language effect. I will discuss basic concepts and current directions in debiasing research. In conclusion, after examining other existing explanations of the effect and finding inconsistent support, I propose a working hypothesis: the use of a foreign language distorts our metacognitive processing, leading us to reflect more on problems that have intuitively appealing but incorrect solutions. I suggest that people become less sensitive to internal cues indicating the correctness of their intuitive responses. Consequently, they allocate cognitive effort more randomly, occasionally reflecting on problems that elicit highly appealing intuitions. This approach helps individuals avoid relying on incorrect but intuitively appealing responses and instead discover less obvious yet normatively correct solutions, resulting in improved performance but potentially unnecessary reflection on simple problems. In this model, improved judgment and decision-making is merely a side-effect of distorted metacognition. However, this is only a working hypothesis, supported by limited evidence, mostly coming from my own research.

2. How to Know When to Think?

As I argued before, much of our thinking relies on mental shortcuts, preserving many of our cognitive resources at the cost of slightly reduced accuracy. These shortcuts, or heuristics, are typically well-calibrated and produce accurate and satisfying solutions (Gigerenzer 2000). The problem with thinking is that heuristics are not always accurate, but one cannot always verify them because that would be burdensome (De Neys 2022; Stanovich 2020). Hence, metacognition has the duty to monitor our thinking and intervene when needed by triggering reflective processes.

Some insight into the question of how people know when to stop and reflect can be gained from a line of research started by Wim De Neys. He presented people with tasks that have salient incorrect intuitions (which he called conflict tasks) and tasks in which such an intuition is correct (no-conflict tasks). He then compared intuition-based responses in these two types of tasks, finding that people are less confident, memorize more details of a task, and require a longer time to provide an intuition-based response in conflict tasks than in co-conflict tasks (De Neys and Glumicic 2008). Brain areas responsible for response inhibition (i.e., lateral prefrontal cortex) were activated in cases when a person responded correctly to a conflict task, but also when responding incorrectly, suggesting each time a person had to inhibit one of their considered responses (De Neys et al. 2008). Hence, even though a person responded incorrectly to a conflict task, they still suffered internal conflict between their gut feelings and a normatively correct response. This metacognitive experience of internal conflict is claimed to trigger reflection allowing one to decouple and verify the intuitive response (De Neys and Pennycook 2019; Pennycook et al. 2015)[1].

Individuals differ in their sensitivity to metacognitive cues. Typically, metacognition is inferred from confidence ratings and their relationship to the response accuracy (Fleming and Lau 2014). Here, one is expected to show higher confidence in own responses when they are correct compared to when they are wrong, or, to put this differently, confidence should be strongly positively correlated to response accuracy. Another instance of faulty metacognition is the Dunning-Kruger effect, in which poor performers overestimate their performance, while top performers sometimes underestimate their performance (Kruger and Dunning 1999). An alternative and extremely popular method used to measure metacognitive sensitivity is the Cognitive Reflection Test (Frederick 2005)[2]. In its original version, the test consists of just three items. One of them is the famous bat-and-ball problem:

A baseball bat and a ball cost $1.10 together, and the bat costs $1.00 more than the ball, how much does the ball cost?

Oftentimes, people reply the ball costs 10 cents, which is incorrect. If a ball would cost 10 cents, and the bat $1.00, then the two items would indeed cost together $1.10, but the difference between the two items would be only 90 cents rather than $1.00. What is most surprising, people seem to not improve in this task after multiple exposures—their performance is constant (Stagnaro et al. 2018), and it correlates similarly with other individual differences or decision performance in naïve and experienced participants (Białek and Pennycook 2018). In other words, people who provide an incorrect response are typically so satisfied that they never reinvestigate it (Raoelison and De Neys 2019).

Cognitive reflection is a crucial aspect of effective decision-making. Individuals who are more cognitively reflective tend to display a number of positive attributes such as less steep discounting (Frederick 2005), a reduced susceptibility to decision frames in the discounting problem (Białek and Sawicki 2018), a greater ability to differentiate between true and false news (Pennycook and Rand 2019), less overconfidence (Białek and Domurat 2018), and fewer dogmatic beliefs and paranormal beliefs (Pennycook et al. 2012). Surprisingly, the benefits of cognitive reflection are independent of their cognitive abilities (Stanovich and West 2008). Differently put, without cognitive reflection the benefits of high cognitive abilities are much smaller (but smarter people may have more accurate intuitions, Raoelison et al. 2020).

Finally, people's metacognitive sensitivity to a conflict seems stable over time. In a study by Voudouri et al. (2022), participants answered a set of heuristic-and-bias tasks quickly and under cognitive load, and then were given an opportunity to reflect and change their response (a so-called two-response paradigm, Thompson et al. 2011). This task was then administered to them again two weeks later, again using the same two-response procedure. The majority of people consistently provided the same response, i.e., they either responded correctly or incorrectly in the intuitive and reflective stages of the experiment. Those who were more conflicted in the intuitive stage tended to change their responses more eagerly when given an opportunity to further reflect on a problem in the same session and after repeating the same task after several weeks. Critically, people were more likely to change their responses in conflict rather than no-conflict problems, again suggesting that they detect the intuition-based response they provided may be incorrect. That said, even those who answer incorrectly tend to be subtly aware of the fact that their reasoning may be biased. They use this metacognitive information to engage in reflection given the opportunity, and in turn, are more likely to change their response.

3. Can Metacognition Be Improved?

Above, I argued that metacognitive sensitivity to cues (i.e., cognitive reflection) is critical for good decision-making because it can trigger reflection to (1) correct an intuitive response and (2) allow for a response change given the opportunity to make the same decision a second time. Since people differ in their metacognitive sensitivity, a question arises whether this trait is fixed or can be improved. I already reviewed some studies that showed that cognitive reflection seems to be relatively stable across time and situations. Some studies reported increased performance in the Cognitive Reflection Test (CRT) after prior exposure (Haigh 2016; Stieger and Reips 2016). However, such a result is likely an artifact because those who performed well on the CRT are simultaneously more likely (1) to improve and (2) to recall even taking the test (Stagnaro et al. 2018). Hence, when the prior experience with the test is self-reported, it biases the results of the analysis because those who performed well on it are also more likely to recall having taken it.

As I mentioned in the previous subsection, people seem not to gain better insight into a task by merely repeating it. What is more, even providing feedback on accuracy in a conflict task does not seem to help people improve. To illustrate, Janssen et al. (2020) observed that the few who did improve after receiving feedback on the bat-and-ball problem also showed improved conflict detection. That is, it appears that they learned to be sensitive to the logic-based cues and metacognitively realize that these cues conflict with the intuition-based cues. Hence, by learning how a task is designed and how to solve it, some people were able to increase their metacognitive sensitivity to logical intuitions.

Consistent findings are reported in the research on the Monty Hall problem (Selvin 1975). Here, as in the famous TV show "Let's Make a Deal", the setup is as follows: you are a contestant on a game show and you are presented with three doors. Behind one door is a valuable prize (e.g., a car), and behind the other two doors are goats. You choose one door, but before the prize behind that door is revealed, the host, Monty Hall, who knows what is behind the doors, opens one of the other two doors to reveal a goat. Monty then offers you the option to stick with your original choice or to switch to the remaining door. The question is "what should you do to maximize your chances of winning the prize?"

The solution to the Monty Hall problem is that the contestant should switch their choice. The reasoning behind this is that the probability of the prize being behind the original choice is 1/3 since there are three doors and one of them has the prize. However, after Monty opens one of the other doors to reveal a goat, the probability of the prize being behind the remaining closed door is 2/3. This is because the probability of the prize being in any of the unselected doors is 2/3. Since Monty would not open a door if the prize was behind it, and since there are now only two doors remaining, the odds of the prize being behind the unopened door are now 2/3, which is a significant improvement over the 1/3 chance of winning by sticking with the original choice.

The problem is hard and very abstract, and thus it is unsurprising that many people fail to respond correctly to it (Gilovich et al. 1995).

What is more surprising, people tend to stick with their initial choice (i.e., not to switch doors) even when given the opportunity to learn the true probability of winning for each option. This was shown in an experiment where participants repeatedly solved the problem over 80 trials. Although the decision to switch became more common in later trials, it still occurred in only slightly more than 50% of decisions. Additionally, even with various forms of feedback, participants did not improve their understanding of the task (Saenen et al. 2015). So, as with the Cognitive Reflection Test, people seem to learn very little from their own mistakes.

My own efforts in improving the Cognitive Reflection Test scores were somewhat more promising: in one line of research, I asked people additional questions about the Cognitive Reflection Test items before asking them the main question. For example, for the bat-and-ball problem, I asked them a question that has a correct response, "What is the most popular sport in the US?", or an opinion question that has no single correct response, "Name a sport, other than baseball, that can be played with bat and ball". My idea was that once people start deliberating on a problem, they will subsequently default to that processing mode, but only when reflection would be accuracy-oriented (as in questions but not in opinions). Consistently with my hypothesis, the Cognitive Reflection Test scores improved only when preceded by questions and not by opinions (Białek 2016). However, the effect was borderline significant, observed only in a subsample of naïve participants, and I subsequently failed to replicate it in larger samples from the Amazon Mechanical Turk.

The second line of research that showed improvement in the Cognitive Reflection Test score was serendipitous: we investigated moral judgments under alcohol intoxication (Paruzel-Czachura et al. 2021). One of the tested ideas was that maybe the effect of alcohol on moral decision-making would be mitigated by high cognitive reflection. The study had a control group that drank orange juice, a placebo group that drank orange juice sprayed with alcohol, and an experimental group that drank orange juice mixed with alcohol. We found that the performance in the CRT improved in the placebo and experimental groups. We believe that people thought that their cognition may have been impaired due to intoxication and were therefore more metacognitively aware of their internal thought processes. Having intensified their metacognitive awareness, they could have spotted the internal conflict between heuristic and logical intuitions and reflected on the problem more than participants in the control group. However, there is a flip side to the coin. It is possible that what we observed is not an improvement in cognitive reflection, but rather, an alteration in the performance on the Cognitive Reflection Test. That is, the results of a psychological test, such as an IQ test, depend not only on the true level of the tested trait but also on the motivation to perform well. To illustrate, people motivated by financial incentives can improve their performance on an intelligence test by 0.64 SD (Duckworth et al. 2011). Differently put, one can appear to gain almost 10 IQ points when motivated properly to perform the test well. Analogously to the effect of incentives on intelligence, believing to be drunk (or being slightly tipsy, as in our study, where BAC ranged from 0.65 to 0.73) could only motivate participants to try harder and thus improve their performance. In both cases (i.e., the intelligence test and the cognitive reflection test) the underlying psychological trait is hypothesized to be constant; what changes is the performance in a test that should measure it. Whichever route the manipulation affected our participants, motivating them to put maximum effort or boosting the cognitive reflection itself, it appears our participants could have reflected slightly more. Yet, it should be stressed again, the effect of alcohol on cognitive reflection is serendipitous, and we worked on replicating it in a preregistered set of experiments.

To conclude this section, it appears that individuals frequently overlook their thinking errors, even when repeating tasks or receiving performance feedback. Although experimental manipulations have shown some promise in enhancing metacognitive awareness, the evidence remains weak and requires further validation.

4. The Foreign Language Effect

As a researcher in the field of decision-making, I have observed there are limited psychological interventions that have been robustly shown to improve decision-making outcomes. This is particularly evident in the case of physical exercise, where a large-scale study involving over 60,000 US fitness chain customers found that only approximately half of the 53 interventions tested improved gym participation compared to placebo intervention, and only five interventions performed better than the baseline intervention of text-message reminders and small financial incentives for attendance (Milkman et al. 2021). However, a study by Keysar et al. (2012) caught my attention. The study found that bilingual participants who used their foreign language were less susceptible to cognitive biases compared to those who used their native language. This finding, if robust, has important implications as it suggests that the use of a foreign language can be used as a tool to improve decision-making[3]. This research on the "foreign language effect" on decision-making has been a rapidly growing field and many researchers are excited to see the potential practical applications that could arise from it (McFarlane et al. 2020).

Some promising research suggested people's risky choices are better when using a foreign language, with bilinguals showing to be less affected by the framing effect, that is, to prefer risk when the outcomes are described as losses or gains (Costa et al. 2014; Winskel et al. 2016). Moreover, their risky choices were more strategic, with people who decided in their foreign language taking more risks when a lottery was beneficial and taking less risk when the bet was unbeneficial (Hayakawa et al. 2019). What this means is that people using their foreign language take more risks when the expected value of the gamble is positive. Expected value is a multiplication of the potential payoff and its probability, e.g., a 50% chance to win \$10 is $EV = 0.5 \times 10 = 5$. A positive gamble is when the EV of a gamble is higher than a certain gain, e.g., of $4. Yet, given such a choice, most people tend to opt for certain smaller gains, displaying so-called risk aversion (Pratt 1964). Since people using their foreign language gambled more often in such positive-EV gambles, using a foreign language is therefore suggested to reduce this aversion (Circi et al. 2021).

Another risk-related benefit of using a foreign language is that participants were observed to be less affected by the hot hand fallacy when gambling (Gao et al. 2015). The hot hand fallacy is the belief that a person who has experienced success in a particular activity has a higher chance of continued success in the immediate future (Gilovich et al. 1985). In other words, the belief is that a player who has just made a few successful shots in basketball is more likely to make the next shot. However, statistical analysis has shown that this belief is actually a fallacy, as past success does not increase the probability of future success in subsequent activities.

As the research on the foreign language effect progresses, the effect appears more complex and malleable than initially hoped. For example, we found no foreign language effect on risky choice (Borkowska et al. 2023; Muda et al. 2020) and using a foreign language has been shown to distort the ability to delay gratifications (Białek et al. 2022), reason logically (Białek et al. 2020) or the ability to discern true from false news (Muda et al. 2023). Numerous studies from other labs are also reporting no effects of using a foreign language on decision-making. For example, the performance in various heuristic-and-biases tasks was unaffected by the language used (Mækelæ and Pfuhl 2019; Vives et al. 2018). The inconsistencies in the findings can be seen as a positive development, allowing us to better understand the boundary conditions of the foreign language effect, and explain when using a foreign language will affect decision-making, whether the effect will be beneficial, and in turn, whether we can adopt some of this knowledge to debiasing overall, not to bilinguals only.

5. The Mechanisms of the Foreign Language Effect

All of this is observed in proficient bilinguals, so it is hard to argue that the foreign language effect occurs because participants do not understand the problem enough to fall into a trap the task is designed to set. Therefore, there has to be a specific mechanism that explains the effect. Understanding how people using a foreign language decide better can lead researchers to propose debiasing techniques that would be much more effective than what is currently proposed. The most promising mechanism, given the current state of the knowledge, is that people who use their foreign language have different access to their metacognitive processes. However, let us discuss the three initial mechanisms through which using a foreign language was supposed to work first (Hayakawa et al. 2016).

The first speculation was that people using their foreign language experience lower affective arousal than those using their native language. Dedicated studies showed that although people using their foreign language are sometimes less aroused by the problem they face, this change in arousal was not predicting the decision change (Geipel et al. 2015b; Hadjichristidis et al. 2015; Muda et al. 2023). However, in one research this hypothesis was confirmed, showing that people using their foreign language when solving moral dilemmas used less affect-rich vocabulary, and this change predicted their shift to more outcome-based responding (Kyriakou et al. 2022). Notably, this hypothesis was also criticized for using proxies for emotions, lacking a generalizable theory of emotion, and unclear baseline levels of neutrality (McFarlane and Cipolletti Perez 2020).

The second explanation—greater willingness to reflect when using a foreign language—also failed to be consistently confirmed. For example, people performed equally well (or, rather, equally poor) in the CRT in their foreign as in their native language (Costa et al. 2014; Milczarski et al. 2022; Vega-Mendoza et al. 2021). Had they reflected more, they should have performed in this test better.

The third explanation, proposed by Hayakawa et al. (2016)—that people use more abstract construals to represent a problem—gained little attention and was never directly tested. The single study I am aware of showed the two effects—changes in the decision and more abstract construals—in two separate experiments, never evidencing a direct path between the two effects (Shin and Kim 2017).

So, how does using a foreign language affect people's decisions through metacognition? Some insight can be gained through our research on the bias blind spot (Niszczota et al. 2022). We asked people to rate how likely they (and regular people) are to fall into some of the decision traps. As the Bible warned, people tend to look at the speck of sawdust in someone's brother's eye and pay no attention to the plank in one's own eye (*English Standard Version Bible* 2001, Mat. 7:3–5). As in past research on this issue (Pronin et al. 2002), our results showed that people indeed see much more bias in others than they see in themselves. To avoid the bias blind spot trap, one must recognize that they are just as susceptible to bias as any other person and must be aware of their own potential biases. This requires metacognition. We observed a decrease in the bias blind spot in people using their foreign language. So, maybe they gained some sort of additional metacognitive awareness when using their foreign language? I discuss this hypothesis in the subsequent section.

6. Debiasing through Metacognition?

The answer to the question from the section above can be "Yes". It becomes more plausible when considered in the context of findings suggesting that being bilingual or multilingual can provide cognitive benefits, such as improved attention, memory, and problem-solving abilities, compared to being monolingual (Bialystok 1988). This (highly debated and criticized, see e.g., Paap 2019) bilingual advantage hypothesis suggests that the process of constantly switching between languages and adjusting to different linguistic contexts can lead to improved cognitive flexibility, which is the ability to adapt to new information and changing task demands[4]. Speculatively, the benefits of being bilingual may be more pronounced when using a foreign language, leading to greater cognitive control and flexibility, and resulting in more metacognitive awareness of one's mental states. Therefore, the foreign language effect would be a special case where the bilingual advantage is especially visible through more efficient metacognition. The reduced bias blind spot fits this hypothesis well.

Given this research, it appears that using a foreign language helps people to better use their metacognition and show less bias. If this is a true and replicable effect, this suggests that metacognition can be experimentally boosted, opening new avenues for the debiasing research. We dedicated some more resources to studying this hypothesis, i.e., that bilinguals, especially when using their foreign language, decide better when to think and when it is enough to rely on their gut feelings. We tested this hypothesis in experiments on reasoning, where people are asked to evaluate whether a syllogism (constructed of two premises and a conclusion) is logically valid. A logically valid syllogism is one in which its conclusion is always true given the premises are true. However, people tend to solve this task differently than asked—they first look at the believability of a conclusion, displaying so-called belief bias (Evans et al. 1983). When the conclusion is believable, people adopt a liberal responding strategy, preferring to accept the syllogism as valid. Differently put, they prefer to err by accepting an invalid syllogism as valid rather than by rejecting a valid syllogism as invalid. When the conclusion is unbelievable, they adopt a conservative response strategy, preferring to rather err by rejecting a valid syllogism (Dube et al. 2010). To reason correctly, one needs to reflect on the logical structure of the syllogism and ignore the intuitively available judgment of its conclusions' believability. To make this process efficient, rather than inspecting logical structure of all syllogisms, one should query their metacognition. More specifically, one has to detect that logic and intuition are in conflict, and dedicate additional cognitive resources to reflect only on those syllogisms where logic and intuition are incongruent (Ackerman and Thompson 2017).

Across three experiments, we asked participants to evaluate a handful of syllogisms, either in their native or foreign language (Białek et al. 2020). Contrary to what I expected, we found decreased performance in the foreign language condition. The critical aspect was the conflict detection, a metacognitive cue that the initial response may be incorrect and that further reflection is needed. We observed that confidence ratings were higher in correct no-conflict trials (e.g., a valid syllogism with a believable conclusion) compared to incorrect conflict trials (e.g., an invalid syllogism with a believable conclusion). This shows that despite that the decision to accept a syllogism as valid reflected its believability, our participants were more conflicted when the validity and believability clashed. This is evidence for them being (to some degree) metacognitively aware of the possible error. However, the critical insight of our research was that conflict detection was lower in the foreign language condition. In other words, bilinguals using their foreign language were less able to detect when they might be wrong. As a result, they failed to reflect when reflection was needed.

This finding was surprising: it showed that using a foreign language can indeed affect metacognition, but it distorts rather than boosts its quality. This made me formulate a model in which people differently track intuitions produced in their foreign vs native language. For example, foreign-language intuitions could be deprived of their affective component and thus less salient than native-language intuitions. Thus, people's decisions made in a foreign language could be less affected by these foreign-language intuitions, resulting in better performance in tasks where an intuitive response is always wrong. However, this does not mean the everyday decisions made in a foreign language will be any better. Intuitions are evolutionarily calibrated quite accurately and are very useful (Gigerenzer 2000; Heyes 2019). Having our intuitions attenuated can deprive us of a hunch about what to do and can result in reflecting on things that do not require it or failing to reflect when needed. The former type of metacognitive error—underthinking—can lead us to make little use of our cognitive capacities and likely lead to suboptimal choices. The problems with the latter type of metacognitive error—overthinking—can pose two risks: computational errors and inefficient use of cognitive resources.

While this hypothesis appears plausible, it has not yet been tested empirically. Some indirect support for this hypothesis comes from the moral foreign language effect. In moral psychology, we study how people weigh two moral inclinations against each other—deontological reliance on moral rules (e.g., do not kill) versus utilitarian reliance on the outcomes of a decision (e.g., kill one person to save five other people) (Conway and Gawronski 2013). Both of these considerations are intuitive (Bago and De Neys 2018; Białek and De Neys 2016). When a moral problem is considered in a foreign language, deontology and utilitarianism both lose their salience (Białek et al. 2019; Hayakawa et al. 2017; Hennig and Hütter 2021; Muda et al. 2018). Thus, a moral problem becomes less of a moral one and more of a regular problem where human lives transcend into mere numbers. This also aligns with the attenuated intuitions hypothesis. However, direct experimental evidence for the attenuated metacognition hypothesis, which is currently absent, is needed.

7. Summary and Conclusions

In the above, across almost 6.5 thousand words and about 90 citations, I argued how metacognition is critical to good decision-making and reviewed some of the recent research on debiasing through metacognition. This lack of thinking about their thinking and decision processes can lead people to bad decisions. If one focuses on the goal, even if it is an undoubtedly good goal such as national security and prevention of terrorist attacks, a lack of awareness of one's own biased thinking can lead to disastrous decisions, such as the Guantanamo Bay torture program by the US government. Oftentimes, to decide well is to stop and reconsider one's own decision process.

Much of the research on debiasing is unsuccessful in achieving its effects, the effects are short-lived and malleable, and their boundary conditions are not well understood (IJzerman et al. 2020). However, there is a light in the tunnel: some manipulations do work in some conditions. Yet, we are still far from understanding how to help people. It seems that neither repeatedly solving similar problems nor being provided with feedback on past performance has a lasting effect on human performance. Using a foreign language also does not robustly affect decision-making, and if it does, its results are not always beneficial.

Most of the manipulations that do work, such as the Nobel Prize-winning nudge (Thaler and Sunstein 2008), assist people in making a specific decision but do not necessarily lead to better decisions in the long term; but see (Van Rookhuijzen et al. 2021). A nudge is a type of choice architecture that reframes a problem to encourage a particular choice while still allowing decision-makers the freedom to decide. For instance, individuals may be automatically enrolled in a retirement savings plan but can choose to opt out. This way, more people sign in compared to when given a free choice whether to sign in or not. While leaving the debate about its moral permissibility aside (Bovens 2009; Schmidt and Engelen 2020), because of its incidental nature, the nudge direction of interventions may not be ideal. In the retirement savings example, people may save more money, but their overall attitude towards saving or their willingness to save actively in other ways may remain unchanged. Instead, the end goal of debiasing should be long-term, unsupervised, uncoerced, and free decision-making that results in better wealth, health, and happiness.

I believe that we can achieve this goal and help people make better decisions by encouraging them to reflect on their own thinking. We need more experimental work on metacognition, preferably through field studies that capture real behavioral data, to test whether people can be trained to be more aware of their own thought processes, have better calibrated confidence, and use their cognitive resources more efficiently. Potentially, learning a second language could also help, but we need longitudinal data to determine whether metacognition and foreign language proficiency are positively correlated.

As I argue, metacognition is to some extent malleable and can be changed when using a foreign language. Among other consequences, people using their foreign language are less attracted to intuitively appealing but incorrect responses. Had we dedicated more resources to understanding how the foreign language affects metacognition, we could identify the critical features of this process and find methods that have similar features but do not require knowledge of a foreign language.

Funding: This research received no external funding.

Institutional Review Board Statement: Not applicable.

Informed Consent Statement: Not applicable.

Data Availability Statement: Not applicable.

Conflicts of Interest: The author declare no conflict of interest.

Notes

1 People engage in reflection not necessarily to challenge their own decision, but sometimes they reflect to find support for their initial choice and to counter arguments for alternative solutions (Haidt 2012; Kunda 1990; Mercier and Sperber 2011). Hence, reflection is not a panacea to decision-making problems.

2 Altough the original is the most popular version of the CRT, there are several newer versions of the test, and all of them consist of just several items (Primi et al. 2016; Sirota et al. 2021; Thomson and Oppenheimer 2016; Toplak et al. 2014). Of these, the version by Sirota et al. (2021) is claimed to measure cognitive reflection with little to no interference of numeracy, which is a common issue in other versions of the test.

3 Much of the foreign language effect research investigates moral judgments (Białek et al. 2019; Geipel et al. 2015a; Hayakawa et al. 2017). However, as it is challenging to establish which morality is correct or erroneous, discussing "improved performance" in this context is not feasible. Thus, I will not delve extensively into this issue in this essay.

4 As with many psychological observations, the bilingual advantage is also not a definitive effect, with it being only conditionally observed, e.g., in some tasks, or mostly in older adults (Ware et al. 2020). Additionally, some of the bilingual advantages can be rather a result of higher socioeconomic status (Morton and Harper 2007) or a cultural effect (Samuel et al. 2018).

References

Ackerman, Rakefet, and Valerie A. Thompson. 2017. Meta-Reasoning: Monitoring and Control of Thinking and Reasoning. *Trends in Cognitive Sciences* 21: 607–17. [CrossRef]

Bago, Bence, and Wim De Neys. 2018. The Intuitive Greater Good: Testing the Corrective Dual Process Model of Moral Cognition. *Journal of Experimental Psychology: General* 33: 461. [CrossRef]

Białek, Michał. 2016. What Color Are the Lilies? Forced Reflection Boosts Performance in the Cognitive Reflection Test. *SSRN Electronic Journal*, 1–19. [CrossRef]

Białek, Michał, and Wim De Neys. 2016. Conflict Detection during Moral Decision-Making: Evidence for Deontic Reasoners' Utilitarian Sensitivity. *Journal of Cognitive Psychology* 28: 631–39. [CrossRef]

Białek, Michał, and Artur Domurat. 2018. Cognitive Abilities, Analytic Cognitive Style and Overconfidence: A Commentary on Duttle (2016). *Bulletin of Economic Research* 70: e119–25. [CrossRef]

Białek, Michał, and Gordon Pennycook. 2018. The Cognitive Reflection Test Is Robust to Multiple Exposures. *Behavior Research Methods* 50: 1953–59. [CrossRef]

Białek, Michał, and Przemysław Sawicki. 2018. Cognitive Reflection Effects on Time Discounting. *Journal of Individual Differences* 39: 99–106. [CrossRef]

Białek, Michał, Artur Domurat, Mariola Paruzel-Czachura, and Rafał Muda. 2022. Limits of the Foreign Language Effect: Intertemporal Choice. *Thinking and Reasoning* 28: 97–124. [CrossRef]

Białek, Michał, Mariola Paruzel-Czachura, and Bertram Gawronski. 2019. Foreign Language Effects on Moral Dilemma Judgments: An Analysis Using the CNI Model. *Journal of Experimental Social Psychology* 85: 103855. [CrossRef]

Białek, Michał, Rafał Muda, Kaiden Stewart, Paweł Niszczota, and Damian Pieńkosz. 2020. Thinking in a Foreign Language Distorts Allocation of Cognitive Effort: Evidence from Reasoning. *Cognition* 205: 104420. [CrossRef]

Bialystok, Ellen. 1988. Levels of Bilingualism and Levels of Linguistic Awareness. *Developmental Psychology* 24: 560–67. [CrossRef]

Borkowska, Anna, Wojciech Milczarski, and Michal Bialek. 2023. No Foreign Language Effect in Incentivized Gambles with Verbal Probability Expressions. *PsyArXiv*, 1–22. [CrossRef]

Bovens, Luc. 2009. The ethics of nudge. In *Preference Change: Approaches from Philosophy, Economics and Psychology*. Edited by Till Grüne-Yanoff and Sven Ove Hansson. Berlin: Springer, pp. 207–19. [CrossRef]

Circi, Riccardo, Daniele Gatti, Vincenzo Russo, and Tomaso Vecchi. 2021. The Foreign Language Effect on Decision-Making: A Meta-Analysis. *Psychonomic Bulletin and Review* 28: 1131–41. [CrossRef] [PubMed]

Conway, Paul, and Bertram Gawronski. 2013. Deontological and Utilitarian Inclinations in Moral Decision Making: A Process Dissociation Approach. *Journal of Personality and Social Psychology* 104: 216–35. [CrossRef] [PubMed]

Costa, Albert, Alice Foucart, Inbal Arnon, Melina Aparici, and Jose Apesteguia. 2014. 'Piensa' Twice: On the Foreign Language Effect in Decision Making. *Cognition* 130: 236–54. [CrossRef]

De Neys, Wim. 2022. Advancing Theorizing about Fast-and-Slow Thinking. *Behavioral and Brain Sciences*, 1–68. [CrossRef]

De Neys, Wim, and Gordon Pennycook. 2019. Logic, Fast and Slow: Advances in Dual-Process Theorizing. *Current Directions in Psychological Science* 28: 503–9. [CrossRef]

De Neys, Wim, and Tamara Glumicic. 2008. Conflict Monitoring in Dual Process Theories of Thinking. *Cognition* 106: 1248–99. [CrossRef]

De Neys, Wim, Oshin Vartanian, and Vinod Goel. 2008. Smarter than We Think: When Our Brains Detect That We Are Biased. *Psychological Science* 19: 483–89. [CrossRef]

Dube, Chad, Caren M. Rotello, and Evan Heit. 2010. Assessing the Belief Bias Effect with ROCs: It's a Response Bias Effect. *Psychological Review* 117: 831–63. [CrossRef]

Duckworth, Angela Lee, Patrick D. Quinn, Donald R. Lynam, Rolf Loeber, and Magda Stouthamer-Loeber. 2011. Role of Test Motivation in Intelligence Testing. *Proceedings of the National Academy of Sciences* 108: 7716–20. [CrossRef] [PubMed]

Dunlosky, John, and Janet Metcalfe. 2008. *Metacognition*. Thousand Oaks: SAGE Publications.

English Standard Version Bible. 2001. ESV Online. Available online: https://esv.literalword.com/ (accessed on 12 March 2023).

Evans, Jonathan S. B. T., Julie L. Barston, and Paul Pollard. 1983. On the Conflict between Logic and Belief in Syllogistic Reasoning. *Memory and Cognition* 11: 295–306. [CrossRef] [PubMed]

Fleming, Stephen M., and Hakwan C. Lau. 2014. How to Measure Metacognition. *Frontiers in Human Neuroscience* 8: 443. [CrossRef]

Frederick, Shane. 2005. Cognitive Reflection and Decision Making. *Journal of Economic Perspectives* 19: 25–42. [CrossRef]

Gao, Shan, Ondrej Zika, Robert D. Rogers, and Guillaume Thierry. 2015. Second Language Feedback Abolishes the 'Hot Hand' Effect during Even-Probability Gambling. *The Journal of Neuroscience* 35: 5983–89. [CrossRef] [PubMed]

Geipel, Janet, Constantinos Hadjichristidis, and Luca Surian. 2015a. How Foreign Language Shapes Moral Judgment. *Journal of Experimental Social Psychology* 59: 8–17. [CrossRef]

Geipel, Janet, Constantinos Hadjichristidis, and Luca Surian. 2015b. The Foreign Language Effect on Moral Judgment: The Role of Emotions and Norms. *PLoS ONE* 10: e0131529. [CrossRef]

Gigerenzer, Geerd. 2000. *Adaptive Thinking: Rationality in the Real World*. Oxford: Oxford University Press.

Gilovich, Thomas, Robert Vallone, and Amos Tversky. 1985. The Hot Hand in Basketball: On the Misperception of Random Sequences. *Cognitive Psychology* 17: 295–314. [CrossRef]

Gilovich, Thomas, Victoria Husted Medvec, and Serena Chen. 1995. Commission, Omission, and Dissonance Reduction: Coping with Regret in the 'Monty Hall' Problem. *Personality and Social Psychology Bulletin* 21: 182–90. [CrossRef]

Hadjichristidis, Constantinos, Janet Geipel, and Lucia Savadori. 2015. The Effect of Foreign Language in Judgments of Risk and Benefit: The Role of Affect. *Journal of Experimental Psychology: Applied* 21: 117–29. [CrossRef]

Haidt, Jonathan. 2012. *The Righteous Mind: Why Good People Are Divided by Politics and Religion*. New York: Vintage.

Haigh, Matthew. 2016. Has the Standard Cognitive Reflection Test Become a Victim of Its Own Success? *Advances in Cognitive Psychology* 12: 145. [CrossRef]

Hayakawa, Sayuri, Albert Costa, Alice Foucart, and Boaz Keysar. 2016. Using a Foreign Language Changes Our Choices. *Trends in Cognitive Sciences* 20: 791–93. [CrossRef] [PubMed]

Hayakawa, Sayuri, Becky Ka Ying Lau, Sophie Holtzmann, Albert Costa, and Boaz Keysar. 2019. On the Reliability of the Foreign Language Effect on Risk-Taking. *Quarterly Journal of Experimental Psychology* 72: 29–40. [CrossRef]

Hayakawa, Sayuri, David Tannenbaum, Albert Costa, Joanna D. Corey, and Boaz Keysar. 2017. Thinking More or Feeling Less? Explaining the Foreign-Language Effect on Moral Judgment. *Psychological Science* 28: 1387–97. [CrossRef]

Hennig, Max, and Mandy Hütter. 2021. Consequences, Norms, or Willingness to Interfere: A ProCNI Model Analysis of the Foreign Language Effect in Moral Dilemma Judgment. *Journal of Experimental Social Psychology* 95: 104148. [CrossRef]

Heyes, Cecilia. 2019. Précis of Cognitive Gadgets: The Cultural Evolution of Thinking. *Behavioral and Brain Sciences* 42: e169. [CrossRef]

IJzerman, Hans, Neil Anthony Lewis, Netta Weinstein, Lisa Marie DeBruine, Stuart James Ritchie, Simine Vazire, and Patrick S. Forscher. 2020. Use caution when applying behavioural science to policy. *Nature Human Behaviour* 4: 1092–94. [CrossRef] [PubMed]

Janssen, Eva M., Matthieu Raoelison, and Wim De Neys. 2020. 'You're Wrong!': The Impact of Accuracy Feedback on the Bat-and-Ball Problem. *Acta Psychologica* 206: 103042. [CrossRef]

Johnson, Eric J., Gerald Häubl, and Anat Keinan. 2007. Aspects of Endowment: A Query Theory of Value Construction. *Journal of Experimental Psychology: Learning Memory and Cognition* 33: 461–74. [CrossRef]

Kahneman, Daniel, Olivier Sibony, and Cass R. Sunstein. 2022. *Noise*. New York: HarperCollins UK.

Kahneman, Daniel, Paul Slovic, and Amos Tversky. 1982. *Judgment under Uncertainty: Heuristics and Biases*. Cambridge: Cambridge University Press.

Keysar, Boaz, Sayuri L. Hayakawa, and Sun Gyu An. 2012. The Foreign-Language Effect: Thinking in a Foreign Tongue Reduces Decision Biases. *Psychological Science* 23: 661–68. [CrossRef]

Kruger, Justin, and David Dunning. 1999. Unskilled and Unaware of It: How Difficulties in Recognizing One's Own Incompetence Lead to Inflated Self-Assessments. *Journal of Personality and Social Psychology* 77: 1121. [CrossRef] [PubMed]

Kunda, Ziva. 1990. The Case for Motivated Reasoning. *Psychological Bulletin* 108: 480–98.

Kyriakou, Andreas, Alice Foucart, and Irini Mavrou. 2022. Moral Judgements in a Foreign Language: Expressing Emotions and Justifying Decisions. *International Journal of Bilingualism*, 136700692211341. [CrossRef]

Larrick, Richard P., and Daniel C. Feiler. 2015. Expertise in Decision Making. In *The Wiley Blackwell Handbook of Judgment and Decision Making*. Oxford: John Wiley and Sons, Ltd., pp. 696–721. [CrossRef]

Mækelæ, Martin, and Gerit Pfuhl. 2019. Deliberate Reasoning Is Not Affected by Language. *PLoS ONE* 14: e0211428. [CrossRef] [PubMed]

Maran, Thomas, Theo Ravet-Brown, Martin Angerer, Marco Furtner, and Stefan E. Huber. 2020. Intelligence Predicts Choice in Decision-Making Strategies. *Journal of Behavioral and Experimental Economics* 84: 101483. [CrossRef]

McFarlane, Steven, and Heather Cipolletti Perez. 2020. Some Challenges for Research on Emotion and Moral Judgment: The Moral Foreign-Language Effect as a Case Study. *Diametros* 17: 56–71. [CrossRef]

McFarlane, Steven, Heather Cipolletti Perez, and Christine Weissglass. 2020. Thinking in a Non-Native Language: A New Nudge? *Frontiers in Psychology* 11: 549083. [CrossRef]

Mercier, Hugo, and Dan Sperber. 2011. Why do humans reason? Arguments for an argumentative theory. *Behavioral and Brain Sciences* 34: 57–74. [CrossRef]

Milczarski, Wojciech, Anna Borkowska, Mariola Paruzel-Czachura, and Michał Białek. 2022. Using a Foreign Language Does Not Make You Think More: Null Effects of Using a Foreign Language on Cognitive Reflection and Numeracy, December. *PsyArxiv*, 1–14. [CrossRef]

Milkman, Katherine L., Dena Gromet, Hung Ho, Joseph S. Kay, Timothy W. Lee, Pepi Pandiloski, Yeji Park, Aneesh Rai, Max Bazerman, John Beshears, and et al. 2021. Megastudies Improve the Impact of Applied Behavioural Science. *Nature* 600: 478–83. [CrossRef]

Morton, J. Bruce, and Sarah N. Harper. 2007. What did Simon say? Revisiting the bilingual advantage. *Developmental Science* 10: 719–26. [CrossRef]

Muda, Rafał, Alexander C. Walker, Damian Pieńkosz, Jonathan A. Fugelsang, and Michał Białek. 2020. Foreign Language Does Not Affect Gambling-Related Judgments. *Journal of Gambling Studies* 36: 633–52. [CrossRef] [PubMed]

Muda, Rafał, Gordon Pennycook, Damian Pieńkosz, and Michał Białek. 2023. People Are Worse at Detecting Fake News in Their Foreign Language. *Journal of Experimental Psychology: Applied, in press*. [CrossRef]

Muda, Rafał, Pawel Niszczota, Michał Białek, and Paul Conway. 2018. Reading Dilemmas in a Foreign Language Reduces Both Deontological and Utilitarian Response Tendencies. *Journal of Experimental Psychology: Learning, Memory, and Cognition* 44: 321–26. [CrossRef] [PubMed]

Niszczota, Paweł, Magdalena Pawlak, and Michał Białek. 2022. Bilinguals Are Less Susceptible to the Bias Blind Spot in Their Second Language. *International Journal of Bilingualism* 7: 13670069221110384. [CrossRef]

Paap, Kenneth. 2019. The Bilingual Advantage Debate: Quantity and Quality of the Evidence. In *The Handbook of the Neuroscience of Multilingualism*. Edited by John W. Schwieter and Michel Paradis. New York: John Wiley & Sons Ltd., pp. 701–35.

Paruzel-Czachura, Mariola, Katarzyna Pypno, Jim A. C. Everett, Michal Bialek, and Bertram Gawronski. 2021. The Drunk Utilitarian Revisited: Does Alcohol Really Increase Utilitarianism in Moral Judgment? *Personality and Social Psychology Bulletin* 49: 20–31. [CrossRef]
Pennycook, Gordon, and David G. Rand. 2019. Lazy, Not Biased: Susceptibility to Partisan Fake News Is Better Explained by Lack of Reasoning than by Motivated Reasoning. *Cognition* 188: 39–50. [CrossRef]
Pennycook, Gordon, James Allan Cheyne, Paul Seli, Derek J. Koehler, and Jonathan A. Fugelsang. 2012. Analytic Cognitive Style Predicts Religious and Paranormal Belief. *Cognition* 123: 335–46. [CrossRef]
Pennycook, Gordon, Jonathan A. Fugelsang, and Derek J. Koehler. 2015. What Makes Us Think? A Three-Stage Dual-Process Model of Analytic Engagement. *Cognitive Psychology* 80: 34–72. [CrossRef]
Pratt, John W. 1964. Risk Aversion in the Small and in the Large. *Econometrica* 33: 122–36. [CrossRef]
Primi, Caterina, Kinga Morsanyi, Francesca Chiesi, Maria Anna Donati, and Jayne Hamilton. 2016. The Development and Testing of a New Version of the Cognitive Reflection Test Applying Item Response Theory (IRT). *Journal of Behavioral Decision Making* 29: 453–69. [CrossRef]
Pronin, Emily, Daniel Y. Lin, and Lee Ross. 2002. The Bias Blind Spot: Perceptions of Bias in Self Versus Others. *Personality and Social Psychology Bulletin* 28: 369–81. [CrossRef]
Raoelison, Matthieu, and Wim De Neys. 2019. Do We De-Bias Ourselves?: The Impact of Repeated Presentation on the Bat-and-Ball Problem. *Judgment and Decision Making* 14: 170–78. [CrossRef]
Raoelison, Matthieu, Valerie A. Thompson, and Wim De Neys. 2020. The Smart Intuitor: Cognitive Capacity Predicts Intuitive Rather than Deliberate Thinking. *Cognition* 204: 104381. [CrossRef] [PubMed]
Saenen, Lore, Wim Van Dooren, and Patrick Onghena. 2015. A Randomised Monty Hall Experiment: The Positive Effect of Conditional Frequency Feedback. *Thinking and Reasoning* 21: 176–92. [CrossRef]
Samuel, Steven, Karen Roehr-Brackin, Hyensou Pak, and Hyunji Kim. 2018. Cultural effects rather than a bilingual advantage in cognition: A review and an empirical study. *Cognitive Science* 42: 2313–41. [CrossRef]
Schmidt, Andreas T., and Bart Engelen. 2020. The Ethics of Nudging: An Overview. *Philosophy Compass* 15: e12658. [CrossRef]
Selvin, Steve. 1975. A Problem in Probability. *The American Statistician* 29: 67–71. [CrossRef]
Senate Select Committee on Intelligence. 2012. Committee Study of the Central Intelligence Agency's Detention and Interrogation Program. Available online: http://www.intelligence.senate.gov/study2014/sscistudy1.pdf (accessed on 2 March 2023).
Shin, Hong Im, and Juyoung Kim. 2017. Foreign Language Effect and Psychological Distance. *Journal of Psycholinguistic Research* 46: 1339–52. [CrossRef]
Sirota, Miroslav, Chris Dewberry, Marie Juanchich, Lenka Valuš, and Amanda C. Marshall. 2021. Measuring Cognitive Reflection without Maths: Development and Validation of the Verbal Cognitive Reflection Test. *Journal of Behavioral Decision Making* 34: 322–43. [CrossRef]
Stagnaro, Michael N., Gordon Pennycook, and David G. Rand. 2018. Performance on the Cognitive Reflection Test Is Stable across Time. *Judgment and Decision Making* 13: 260–67. [CrossRef]
Stankovic, Michelle, Britta Biedermann, and Takeshi Hamamura. 2022. Not All Bilinguals Are the Same: A Meta-Analysis of the Moral Foreign Language Effect. *Brain and Language* 227: 105082. [CrossRef]
Stanovich, Keith E. 2020. Why Humans Are Cognitive Misers and What It Means for the Great Rationality Debate. In *Routledge Handbook of Bounded Rationality*, 1st ed. Edited by Riccardo Viale. Routledge: Taylor and Francis Group, pp. 196–206.
Stanovich, Keith E., and Richard F. West. 2008. On the Relative Independence of Thinking Biases and Cognitive Ability. *Journal of Personality and Social Psychology* 94: 672–95. [CrossRef] [PubMed]
Stewart, Neil, Nick Chater, and Gordon D. A. Brown. 2006. Decision by Sampling. *Cognitive Psychology* 53: 1–26. [CrossRef] [PubMed]
Stieger, Stefan, and Ulf-Dietrich Reips. 2016. A Limitation of the Cognitive Reflection Test: Familiarity. *PeerJ* 4: e2395. [CrossRef]
Thaler, Richard H., and Cass Sunstein. 2008. *Nudge: Improving Decisions about Health, Wealth, and Happiness*. Yale: Yale University Press.
Thompson, Valerie A., Jamie A. Prowse Turner, and Gordon Pennycook. 2011. Intuition, Reason, and Metacognition. *Cognitive Psychology* 63: 107–40. [CrossRef]
Thomson, Keela S., and Daniel M. Oppenheimer. 2016. Investigating an Alternate Form of the Cognitive Reflection Test. *Judgment and Decision Making* 11: 99–113. [CrossRef]
Toplak, Maggie E., Richard F. West, and Keith E. Stanovich. 2014. Assessing Miserly Information Processing: An Expansion of the Cognitive Reflection Test. *Thinking and Reasoning* 20: 147–68. [CrossRef]
Van Rookhuijzen, Merije, Emely De Vet, and Marieke A. Adriaanse. 2021. The Effects of Nudges: One-Shot Only? Exploring the Temporal Spillover Effects of a Default Nudge. *Frontiers in Psychology* 12: 683262. [CrossRef] [PubMed]
Vega-Mendoza, Mariana, Patrik Hansson, Daniel Eriksson Sörman, and Jessica K. Ljungberg. 2021. Testing the Foreign Language Effect on Cognitive Reflection in Older Adults. *Brain Sciences* 11: 1527. [CrossRef]
Vives, Marc-Lluís, Melina Aparici, and Albert Costa. 2018. The Limits of the Foreign Language Effect on Decision-Making: The Case of the Outcome Bias and the Representativeness Heuristic. Edited by Mariano Sigman. *PLoS ONE* 13: e0203528. [CrossRef]
Voudouri, Aikaterini, Michał Białek, Artur Domurat, Marta Kowal, and Wim De Neys. 2022. Conflict detection predicts the temporal stability of intuitive and deliberate reasoning. *Thinking & Reasoning*, 1–29. [CrossRef]

Ware, Anna T., Melissa Kirkovski, and Jarrad A. G. Lum. 2020. Meta-analysis reveals a bilingual advantage that is dependent on task and age. *Frontiers in Psychology* 11: 1458. [CrossRef] [PubMed]

Winskel, Heather, Theeraporn Ratitamkul, Victoria Brambley, Tulaya Nagarachinda, and Sutheemar Tiencharoen. 2016. Decision-Making and the Framing Effect in a Foreign and Native Language. *Journal of Cognitive Psychology* 28: 427–36. [CrossRef]

Journal of *Intelligence*

MDPI

Article

Overconfidence in the Cognitive Reflection Test: Comparing Confidence Resolution for Reasoning vs. General Knowledge

André Mata

CICPSI, Faculdade de Psicologia, Universidade de Lisboa, 1649-013 Lisboa, Portugal; aomata@psicologia.ulisboa.pt

Abstract: This research examines the metacognitive awareness that people have about their reasoning performance in the Cognitive Reflection Test (CRT). The first two studies compare confidence judgments about the CRT vs. general knowledge (GK) questions. Results show that (1) people are generally able to discriminate between correct and incorrect answers, but this ability is far from perfect, and it is greater for GK questions than for CRT problems. Indeed, and strikingly, (2) incorrect responses to CRT problems are produced with approximately the same level of confidence as *correct* responses to GK questions. However, (3) even though confidence is high for incorrect responses to CRT problems, it is even higher for correct responses. The results of two additional studies show that these differences in confidence are ultimately related to the conflict that CRT problems pose between intuition and deliberation. These findings have implications for the possibility of implicit error monitoring and dual-process models of overconfidence.

Keywords: cognitive reflection; overconfidence; error monitoring; conflict detection

The Cognitive Reflection Test (CRT; Frederick 2005) has become quite popular in research on reasoning and behavioral economics (Bialek and Pennycook 2018; Branas-Garza et al. 2019). The problems in the test pose a conflict between an intuitive but incorrect response and a correct response that requires effortful deliberation. A good example is the well-known bat-and-ball problem: A bat and a ball together cost 1 dollar and 10 cents. The bat costs 1 dollar more than the ball. How much does the ball cost? The answer "10 cents" comes to mind immediately, but it is incorrect (the correct answer is 5 cents). The fact that the incorrect answer is so intuitive makes it likely (1) that many people fall for it, and (2) that they are highly confident in those answers, that is, that they are overconfident (e.g., Mata et al. 2013b; Pennycook et al. 2017).

A key question in research on reasoning and metacognition is whether people who make reasoning errors are aware that their thinking is flawed. Incorrect responses to the CRT are often associated with great confidence. For instance, incorrect responders are willing to bet on the accuracy of their responses, they believe that most people would give the same responses as they did, and they tend to make high confidence ratings (e.g., De Neys et al. 2013; Johnson et al. 2016; Mata 2019a, 2020; Mata et al. 2013b). This would suggest that incorrect responders are completely unaware of the biased nature of their responses.

Other studies, however, suggest that incorrect responders are aware, at least to some extent, of the biased nature of their reasoning. In particular, conflict detection studies (e.g., De Neys et al. 2011, 2013) show that reasoners react differently to conflict problems and no-conflict control versions of those problems (e.g., they take longer or are less confident for the former than the latter). Once again using the bat-and-ball problem as an example, it is easy to modify it (by eliminating the "more than the ball" part of the second premise) in order to remove the conflict: A bat and a ball together cost 1 dollar and 10 cents. The bat costs 1 dollar. How much does the ball cost? Now the intuitive response "10 cents" is correct. Because of this, these control problems produce virtually no errors, and the few participants who make those errors are removed from the analysis (e.g., De Neys et al. 2013).

The present research takes a different approach to exploring overconfidence and error awareness in reasoning: Instead of (or in addition to—see Study 4) using control problems where intuition is correct, the studies presented here use a new type of control problems where there is no intuitive response, and for which participants must rely solely on deliberation. This type of control has a key advantage: It produces errors. That is, responses to these problems can be hard, just as for conflict problems, but unlike conflict problems, they are not tricky, as intuition is mute. This enables a novel approach to assessing the role of intuition and deliberation in overconfidence. Moreover, it is important to measure confidence for errors on such non-tricky questions, as they provide a relevant baseline against which to gauge error awareness in reasoning. Research on error monitoring for conflict reasoning problems (e.g., De Neys et al. 2013; Mata et al. 2013b) has thus far only examined confidence for three types of responses: Correct responses to tricky problems, incorrect responses to tricky problems, and correct responses to non-tricky problems. The missing cell—incorrect responses to non-tricky problems—enables two things: (1) A relevant baseline against which to measure the confidence in errors for conflict reasoning problems; and (2) a measure of resolution, that is, the degree to which responders discriminate between correct and incorrect responses, and whether that differs for conflict problems such as those included in the CRT vs. the no-conflict control problems introduced in these studies.

Participants in the present studies responded to tricky conflict questions such as the bat-and-ball problem (i.e., questions for which intuition and deliberation suggest different responses), as well as non-tricky no-conflict questions (i.e., questions for which there is no intuition, and a person must rely on deliberation alone). Whereas Studies 3–4 manipulate conflict within the same domain (i.e., control versions of conflict problems were created), Studies 1–2 manipulate conflict across domains: Confidence was compared for CRT problems (i.e., conflict) vs. typical (i.e., no-conflict) general knowledge (GK) questions, which also represents a new approach in comparison to previous research.

General knowledge is a domain where there has been extensive research on overconfidence. Here too people often show overconfidence (e.g., Atir et al. 2015; Fischhoff et al. 1977; Moore and Healy 2008). Critically, though, typical GK questions do not pose a conflict between intuition and deliberation. People either know the correct response (by deliberatively thinking about it), or they do not, but unlike CRT-type problems, GK questions are not designed to trigger (faulty) intuitions. As a consequence, even though in both domains people are known to be overconfident, confidence resolution should be higher for GK questions than for reasoning problems. That is, for GK questions, people should know better when they know the answer and when they do not, whereas, for CRT-type problems, they should be misled into trusting their intuitive but incorrect response. In other words, still, the double curse in the Dunning–Kruger effect (i.e., people who lack knowledge or competence in some area also lack awareness of those shortcomings; Dunning 2011; Kruger and Dunning 1999) should be greater when a problem triggers a misleading intuition than when that is not the case.

For correct responses, on the other hand, confidence should be greater for CRT-type problems than for GK questions. This is because arriving at a correct response in tricky reasoning problems presumably implies overcoming the intuitive incorrect response. Thus, there is a two-step reasoning process, whereby the reasoner first thinks of the intuitive response and only later comes to the correct response (though I should note that the logical intuition framework would not predict this two-step process—for a review, see De Neys 2012; for more on this topic, see Section 5). This awareness of having overcome the intuitive but incorrect response and having solved a tricky problem should give correct reasoners a confidence boost (Mata 2019a; Mata et al. 2013b).

For all studies, confidence ratings were collected for the four cells described above. Several comparisons are of theoretical relevance:

Comparing resolution (i.e., the difference in confidence for correct vs. incorrect responses) for CRT-type problems vs. GK questions shows whether the ability to tell correct responses from incorrect ones differs across domains. Conflict detection studies (e.g., De Neys et al. 2011, 2013) test whether responders react differently to conflict problems vs. control no-conflict problems, but they are not informative about how good their resolution is. The present studies test people's ability to distinguish between correct and incorrect responses in cognitive reflection, as compared to their ability to monitor their knowledge in other domains.

(a) Comparing confidence for correct responses in conflict vs. no-conflict problems provides a test of the metacognitive advantage of having awareness of alternative solutions. According to the metacognitive advantage model (Mata 2019a, 2019b, 2020; Mata and Almeida 2014; Mata et al. 2013b), correct responders should have a confidence boost for conflict problems, as they realize that they were able to overcome a tricky problem for which there is a compelling but misleading response. For no-conflict problems, on the other hand, it is less likely that there is such an awareness of alternative responses, and therefore confidence for correct responses should be high, but not so high as for conflict problems.
(b) Comparing confidence for incorrect responses in conflict vs. no-conflict problems provides a test of whether reasoners can detect their reasoning errors to the same extent when there is an intuitive (though incorrect) response that comes to mind vs. when intuition is absent and responders rely solely on deliberation.
(c) Finally, comparing confidence for incorrect responses in conflict problems vs. confidence for *correct* responses to no-conflict problems provides an extreme test of failure to detect errors. Because tricky conflict problems are designed to trigger an intuitive (though incorrect) response, the confidence that comes with that response should be greater than the confidence that comes with errors in the no-conflict problems. Still, if reasoners realize somehow that they made a mistake, they should rate their confidence lower than that for correct responses to the control problems. If they do not, this is revealing of their awareness (or lack thereof) of their errors.

1. Study 1

1.1. Method

1.1.1. Participants

Sixty participants, students at the University of Heidelberg, were recruited to take part in this study; fifty-nine completed the study.

1.1.2. Procedure

Participants completed the CRT, as well as a GK quiz (for all the studies, the task was computerized). The order in which participants completed the CRT and the quiz was counterbalanced across subjects. Within each block of problems (CRT or GK), the order of items was randomized.

The GK quiz was a questionnaire used in research on overconfidence (Moore and Healy 2008). It contained 10 questions on several topics (see Appendix A). Responses were open-ended.

After each CRT or GK question, participants rated how confident they were that their response was correct, on a scale from 1: "not at all confident" to 9: "very confident". After they responded to all the questions in a test (CRT or GK), they estimated how many of the X questions that they had just seen (3 for the CRT; 10 for GK) they answered correctly. I shall refer to these measures as local confidence and global confidence, respectively (Liberman 2004).

1.2. Results

1.2.1. Performance

On average, participants gave 1.86 (SD = 1.06) correct responses for the CRT, and 6.64 (SD = 2.42) for GK. In percentage, there is no difference in performance across tests,

$t < 1$. This eliminates a potential source of differences in overconfidence across domains: Difficulty (Kruger 1999; Pulford and Colman 1997; Schraw and Roedel 1994).

1.2.2. Global Confidence

Comparing the estimated number of correct responses against that actual number provides a measure of over- or under-estimation. In the quiz, participants estimated that they were correct for 5.92 of the 10 questions ($SD = 2.44$), which is close to the actual number of correct responses (6.64), but significantly lower (i.e., under-estimation), $t(58) = -2.92$, $p = .005$, $d = 0.38$. For CRT problems, there was over-estimation: Participants estimated that they were correct for 2.41 of the 3 problems ($SD = 0.65$), which is significantly higher than the actual number of correct responses (1.86), $t(58) = 4.65$, $p < .001$, $d = 0.61$. Comparing these estimation errors for CRT problems vs. GK questions, there is a significant difference, $t(58) = 5.85$, $p < .001$, $d = 0.76$. Not only are these errors in estimation different in quality (i.e., over- vs. under-estimation), but they are also different in quantity: Comparing the absolute difference from estimated to actual correct responses shows that the miscalibration was greater for the CRT than for GK, $t(58) = 2.05$, $p = .045$, $d = 0.27$.

1.2.3. Local Confidence

Mean confidence ratings were compared for correct and incorrect responses to each kind of question (see Table 1).

Table 1. Mean confidence ratings (and SD) by accuracy and type of problem (Study 1).

	CRT	GK
Correct	8.49 (1.32)	7.11 (1.45)
Incorrect	6.82 (2.23)	4.35 (2.23)

The analyses presented here include both an ANOVA comparing confidence for correct and incorrect responses for both types of problem, as well as t-tests for specific comparisons of theoretical relevance, which were laid out in the Introduction (moreover, the ANOVA sacrifices several data points, as it requires that participants give all four types of response, whereas t-tests comparing only two cells preserve more data).

Paired-sample t tests show that (1) correct CRT responses were given with greater confidence than incorrect CRT responses, $t(29) = 2.58$, $p = .015$, $d = 0.47$; (2) correct GK responses were also associated with greater confidence than incorrect GK responses, $t(45) = 8.38$, $p < .001$, $d = 1.24$; (3) correct CRT responses were given with greater confidence than correct GK responses, $t(50) = 4.95$, $p < .001$, $d = 0.69$; (4) incorrect CRT responses were associated with greater confidence than incorrect GK responses, $t(30) = 4.55$, $p < .001$, $d = 0.82$; (5) indeed, confidence for incorrect CRT responses was above the midpoint (5) of the scale (in a one-sample t test, $p < .001$; compare that with the confidence shown for incorrect GK responses, which is below the midpoint of the scale, $p = .056$); (6) in fact, incorrect CRT responses were associated with the same confidence levels as correct GK responses, $t < 1$ (note that degrees of freedom change for the different comparisons, as not all participants gave all four types of responses).

Considering those participants who provided all four types of response, an ANOVA reveals a main effect of type of problem, $F(1, 23) = 25.48$, $p < .001$, $\eta_p^2 = .53$, such that confidence was higher for the CRT than for the GK quiz; a main effect of accuracy, $F(1, 23) = 31.84$, $p < .001$, $\eta_p^2 = .58$, such that confidence was higher for correct responses than for incorrect ones; and an interaction effect, $F(1, 23) = 4.97$, $p = .036$, $\eta_p^2 = .18$, such that the difference in confidence for correct vs. incorrect responses was larger for the GK quiz than for the CRT.

2. Study 2

The second study seeks to replicate the findings of Study 1 with a larger sample and a different GK test that (1) contains more questions, (2) is pre-tested for the specific subject population, and (3) has a different format: Multiple choice instead of open-ended. In Study 1, there was no overconfidence for the GK quiz in the global confidence measure. However, even though this result is not unusual in global confidence measures (Liberman 2004), the recognition exercise involved in multiple-choice tests might lead to greater overconfidence than the recall exercise that is called for in open-ended questions. Still, it is important to note that there were no differences in difficulty (as measured by performance) across tests in the first study. The second study again controls for difficulty.

2.1. Method

2.1.1. Participants

Ninety students from the University of Heidelberg took part in the study.

2.1.2. Procedure

Participants completed the CRT and a GK quiz. The order in which participants completed the CRT and the quiz was counterbalanced across subjects. Within each block of problems (CRT or GK), the order of items was randomized.

The GK quiz was a questionnaire designed to study overconfidence in a German sample (Michailova and Katter 2014). It contained 18 questions from several domains (see Appendix A). For each question, three response options were presented for participants to choose from.

After each CRT or GK question, participants rated how confident they were that their response was correct, on a scale from 1: "not at all confident" to 9: "very confident" (local confidence). When all items in a test (CRT or GK) were completed, participants were asked to estimate how many of the X questions they had just seen (3 for the CRT; 18 for the GK quiz) they had answered correctly (global confidence).

2.2. Results

2.2.1. Performance

On average, participants had 1.69 (SD = 1.13) correct responses in the CRT and 11.39 (SD = 1.93) in the GK quiz. In percentage, there is no significant difference in performance across tests, $t(89) = 1.66$, $p = .101$, $d = 0.16$. This again suggests that any potential differences in overconfidence across domains are not due to differences in difficulty.

2.2.2. Global Confidence

On average, participants estimated that they were correct for 9.62 of the 18 GK questions (SD = 2.85), which is below the actual number of correct responses (M = 11.39), $t(89) = 5.61$, $p < .001$, $d = 0.59$. For the CRT, in contrast, there was over-estimation: Participants estimated that they were correct for 2.34 of the 3 problems (SD = 0.72), which is significantly higher than the actual number of correct responses (M = 1.69), $t(88) = -6.11$, $p < .001$, $d = 0.59$. Comparing these estimation errors for CRT problems vs. GK questions, there is a significant difference, $t(88) = 7.83$, $p < .001$, $d = 0.83$.

Once again, these estimation errors differ not only in quality (i.e., over- vs. under-estimation) but also in quantity: Comparing the absolute difference from estimated to actual correct responses in proportion to the total number of responses in a test shows that the miscalibration was greater for CRT (M = 25.66%, SD = 31.09) than for GK (M = 14.70%, SD = 12.62), $t(88) = 2.89$, $p = .005$.

2.2.3. Local Confidence

Mean confidence ratings were compared for correct and incorrect responses for each kind of question (see Table 2).

Table 2. Mean confidence ratings (and SD) by accuracy and type of problem (Study 2).

	CRT	GK
Correct	7.99 (1.74)	6.72 (1.05)
Incorrect	6.56 (2.04)	4.09 (1.58)

Paired-sample *t*-tests revealed that (1) correct CRT responses were given with greater confidence than incorrect CRT responses, $t(41) = 2.91$, $p = .006$, $d = 0.45$; (2) correct GK responses were also associated with greater confidence than incorrect GK responses, $t(89) = 14.94$, $p < .001$, $d = 0.83$; (3) correct CRT responses were associated with greater confidence than correct GK responses, $t(71) = 6.97$, $p < .001$, $d = 0.82$; (4) incorrect CRT responses were associated with greater confidence than incorrect GK responses, $t(59) = 7.52$, $p < .001$, $d = 0.97$; (5) indeed, confidence for incorrect CRT responses was above the midpoint (5) of the scale (in a one-sample *t* test, $p < .001$), whereas confidence for incorrect GK responses was below the midpoint of the scale ($p < .001$); (6) in fact, incorrect CRT responses were given with the same confidence as correct GK responses, $t < 1$.

An ANOVA comparing the confidence ratings of those participants who gave all four types of response revealed a main effect of type of problem, $F(1, 41) = 53.82$, $p < .001$, $\eta_p^2 = .57$, such that confidence was higher for the CRT than for the GK quiz; a main effect of type of responding, $F(1, 41) = 53.61$, $p < .001$, $\eta_p^2 = .57$, such that confidence in correct responses was higher than confidence in incorrect responses; and an interaction, $F(1, 41) = 9.00$, $p = .005$, $\eta_p^2 = .18$, such that the difference in confidence for correct vs. incorrect responses was greater for GK than for the CRT.

3. Study 3

Study 3 sought to provide further evidence about the source of the overconfidence of biased reasoners: Conflict. Whereas the previous studies manipulated conflict across domains (CRT problems were tricky, whereas GK questions were not), Study 3 compares confidence for conflict vs. no-conflict problems within the same domain: Cognitive reflection. Specifically, this study compares confidence for the CRT problems used in the previous studies (i.e., conflict problems where intuition and deliberation are at odds) with confidence for problems that are structurally similar, but which do not pose a conflict between intuition and deliberation, as there is no intuitive response (i.e., there is no response that comes immediately to mind, and only via deliberative reasoning can a person come to a response). The main prediction is that, if the reason for the overconfidence of incorrect CRT responders is that they are tricked by the intuitive lure, then resolution (i.e., the difference in confidence for correct vs. incorrect responses) should be higher for problems where there is no intuition competing for the response (i.e., deliberation-only problems) than for conflict problems where intuition and deliberation are at odds, and this is because confidence for incorrect responses should be higher in the latter case, as participants are tricked by their intuition.

To the best of my knowledge, this control (no lure) has not been used for CRT-type problems, which is also a contribution of this study. As explained in the Introduction, typical control problems in this research prompt a correct intuitive response. By eliminating the intuitive pull of the problems, the control used in the present study provides a test of how poor resolution in cognitive reflection relates to faulty intuition.

3.1. Method

3.1.1. Participants

One hundred participants were recruited from the Prolific Academic platform for online studies (a total of 101 completed the task; 49.5% female, 50.5% male; *M* age = 33.15, *SD* = 9.34). The language was English.

3.1.2. Procedure

Participants answered 3 CRT-type problems (the conflict problems) as in Studies 1–2, as well as 3 problems that were analogous to the conflict problems, but where there was no intuitive response (e.g., A cake and a piece of fruit cost $8.47 in total. The cake costs $6.71. How much does the piece of fruit cost?—see Appendix A). For each problem, participants made confidence ratings from 1: "not at all confident" to 9: "very confident" (global confidence was not assessed in this study). The order of the 6 problems was randomized.

3.2. Results

3.2.1. Performance

Even though the mathematics required to solve conflict problems is presumably easier than that required for the no-conflict/deliberation-only problems, the number of correct responses was lower for conflict vs. no-conflict problems, demonstrating the intuitive pull of conflict problems: $M = 1.36$, $SD = 1.18$ vs. $M = 2.44$, $SD = 0.73$, $t(100) = 8.93$, $p < .001$, $d = 0.89$.

3.2.2. Confidence

Confidence ratings are presented in Table 3 as a function of conflict and performance. Replicating the previous studies, (1) responders were more confident when they were correct vs. incorrect for conflict problems, $t(42) = 4.64$, $p < .001$, $d = 0.71$; (2) the same difference holds for no-conflict problems, $t(42) = 4.60$, $p < .001$, $d = 0.70$; (3) correct responses to conflict problems were produced with greater confidence than correct responses to no-conflict problems, $t(67) = 2.72$, $p = .008$, $d = 0.33$; (4) incorrect responses to conflict problems were presented with greater confidence than incorrect responses to no-conflict problems, though this difference was marginally significant, $t(34) = 1.88$, $p = .068$, $d = 0.32$, which most likely relates to the fact that this was the pairwise comparison (out of the 6 tests presented here) with fewer datapoints; indeed, the difference in overall values for these cells (not conditional on whether participants committed both types of error) is larger when one considers the values in Table 3 than when one considers the smaller range of cases where both types of error occurred; (5) confidence for incorrect responses to conflict problems was above the midpoint (5) of the scale (in a one-sample t test, $p < .001$, whereas confidence for incorrect responses to no-conflict problems was not, $p = .386$); and (6) finally, in this study, confidence for incorrect responses to conflict problems was lower than that expressed for correct responses to no-conflict problems, $t(73) = -4.27$, $p < .001$, $d = 0.50$. Still, confidence for incorrect responses to conflict problems was quite high ($M = 6.27$, on a 9-point scale).

Table 3. Mean confidence ratings (and SD) by accuracy and type of problem (Study 3).

	Conflict	No-Conflict
Correct	8.05 (1.49)	7.33 (1.59)
Incorrect	6.27 (2.17)	5.37 (2.81)

More importantly, and replicating the previous studies, resolution was lower for conflict vs. no-conflict problems: An ANOVA comparing the confidence ratings of those participants who gave all four responses revealed the predicted conflict-by-accuracy interaction, $F(1, 14) = 4.90$, $p = .044$, $\eta_p^2 = .23$, such that the difference in confidence for correct vs. incorrect responses was larger for no-conflict than for conflict problems (see Table 3); there was also a main effect of accuracy, $F(1, 14) = 9.51$, $p = .008$, $\eta_p^2 = .40$, such that confidence was higher for correct vs. incorrect responses; the main effect of conflict was not significant, $F < 1$.

4. Study 4

Study 4 seeks to replicate the results of the previous studies with a broader set of items that have been used in previous research on reasoning and general knowledge (adapted from Ackerman 2014; Boland 2013; Erickson and Mattson 1981; Fischhoff et al. 1977; Frederick 2005; Juslin et al. 2000; Koriat 1995, 2008; Mata et al. 2013a; Oldrati et al. 2016; Sirota et al. 2021; Thomson and Oppenheimer 2016; Van Dooren et al. 2005; Weber 2016). In particular, various problems tapping various thinking-and-reasoning domains were used in this study, including standard CRT-type items, verbal CRT problems (less dependent on mathematical reasoning), GK items, semantic illusions, and others (see Appendix A), so as to establish the robustness of the key finding of Studies 1–3 (the fact that resolution is lower for conflict vs- no-conflict problems) with a wide range of problems. Moreover, by using a larger set of problems as well as a larger sample, this study seeks to increase the number of participants who give all four responses (i.e., both correct and incorrect responses to conflict and no-conflict problems), which are necessary for the resolution analysis.

Critically, and as in Study 3, this study includes conflict and no-conflict versions of the same problems, in order to control for the critical factor presumed to have generated the difference in resolution for conflict vs. no-conflict items in the first studies: Conflict. Ultimately, there is nothing inherent to CRT problems and GK questions that makes the former necessarily trickier and therefore more prone to generate overconfidence than the latter. Indeed, CRT problems can be simple (e.g., the no-conflict control problems used in research on conflict detection and implicit error monitoring) and GK questions can be tricky (e.g., Erickson and Mattson 1981; Fischhoff et al. 1977; Juslin et al. 2000; Koriat 1995). To the extent that one domain is trickier than the other, it should produce greater overconfidence. The key hypothesis, again, is that the discrimination between good and bad performance (i.e., resolution) should be poorer for conflict tasks, which trigger an intuitive but incorrect response, than for no-conflict tasks, which do not trigger those faulty intuitions.

Finally, this study uses a different type of control problem: Whereas in Studies 1–3, no-conflict problems were such that deliberation was called for, as there was no intuitive response that was prompted by the problems, the problems used in this study trigger an intuitive response that is incorrect. This is the type of control problem that is used in research on cognitive reflection and conflict detection (e.g., De Neys et al. 2011, 2013; Mata 2020; Mata and Almeida 2014). The goal of using this type of control in this study is two-fold: First, it seeks to test whether the difference in resolution for conflict vs. no-conflict problems observed in the previous studies holds using the type of control problems typically used in this area of research. More importantly, by using no-conflict control problems that trigger an intuitive though incorrect response, this study enables an assessment of the automatic vs. controlled processes underlying overconfidence in cognitive reflection. Specifically, by using a large and balanced number of conflict and no-conflict problems, this study enables using the Process Dissociation Procedure (PDP; Jacoby 1991; Jacoby et al. 1993). This analytical method uses performance in conflict and no-conflict trials to disentangle the controlled (C) vs. automatic (A) components of the mental processing that goes into solving a cognitive task (see equations in the results section). This procedure has been successfully applied to both reasoning (Ferreira et al. 2006; Mata et al. 2013c) and general knowledge tasks (Mata et al. 2013a). The hypotheses are that the higher the A scores, the more compelling the intuitive responses will be, and therefore the more overconfident participants will be in their incorrect responses to conflict trials, and the less able they will be to distinguish between correct and incorrect responses. The opposite goes for the C parameter: The higher the ability to control the intuitive responses, the better able participants will be to know when they are correct vs. incorrect. In addition, this C parameter might explain the confidence boost that was observed in the previous studies when participants showed greater confidence for correct responses to conflict vs. no-conflict trials. That is, to the extent that conflict problems trigger tricky intuitions but correct responders are able to override them, they should be extra confident in their correct responses, as compared to their confidence in correct responses to no-conflict trials, which do not call for controlling one's intuition.

4.1. Method

4.1.1. Participants

One hundred and forty undergraduates from the University of Lisbon took part in the study (79.1% female, 20.1% male, and one participant who did not indicate their gender; *M* age = 20.86, *SD* = 7.14). One participant gave nonsensical responses (random strings of letters) and was therefore not considered in the analysis.

4.1.2. Procedure

Participants answered 20 problems (the 3 CRT-type problems used in the previous studies, plus 17 new items; see Appendix A). Half of them were tricky conflict problems, which prompted an intuitive but incorrect response, whereas the other half were no-conflict problems, where the intuitive response was correct. There were two versions of the problems, counterbalanced across subjects, such that half of the participants saw a specific problem in its conflict version, whereas the other half saw it without conflict. Conflict problems were transformed into no-conflict versions by changing small details in the premises (see Appendix A). The order of the 20 items was randomized.

As in the previous studies, after each problem/question, participants rated how confident they were that their response was correct, on a scale from 1: "not at all confident" to 9: "very confident".

After participants solved all the problems and made confidence ratings for each of them, they completed additional tasks, related to another project, irrelevant to the present purposes.

4.2. Results

4.2.1. Performance

The number of correct responses was lower for conflict vs. no-conflict problems: $M = 3.19$, $SD = 1.75$ vs. $M = 8.36$, $SD = 1.40$, $t(138) = 27.89$, $p < .001$, $d = 2.37$.

4.2.2. Confidence

Paired-sample *t* tests revealed that (1) correct responses to conflict items were given with greater confidence than incorrect responses to those items, $t(134) = 4.15$, $p < .001$, $d = 0.36$; (2) correct responses to no-conflict items were also associated with greater confidence than incorrect responses to those items, $t(106) = 13.34$, $p < .001$, $d = 1.29$; (3) correct responses to conflict problems were produced with the same confidence as correct responses to no-conflict problems, $t(134) = -1.31$, $p = .191$, $d = 0.12$; (4) incorrect responses to conflict problems were associated with greater confidence than incorrect responses to no-conflict problems, $t(106) = 9.63$, $p < .001$, $d = 0.93$; (5) indeed, confidence for incorrect responses to conflict problems was above the midpoint (5) of the scale (in a one-sample *t* test, $p < .001$; compare that with the confidence shown for incorrect responses to no-conflict problems, which is below the midpoint of the scale, $p < .001$); (6) finally, incorrect responses to conflict problems were given with less confidence than correct responses to no-conflict problems, $t(138) = -8.15$, $p < .001$, $d = 0.70$. Still, and even though this last result suggests some error awareness, the mean confidence for incorrect responses to conflict problems was quite high.

More importantly, and replicating all the previous studies, resolution was lower for conflict vs. no-conflict problems. An ANOVA comparing the confidence ratings of participants who gave all four responses revealed a main effect of conflict, $F(1, 103) = 67.68$, $p < .001$, $\eta_p^2 = .40$, such that confidence was actually higher for conflict problems than for no-conflict problems; a main effect of accuracy, $F(1, 103) = 170.34$, $p < .001$, $\eta_p^2 = .62$, such that confidence in correct responses was higher than confidence in incorrect responses; and the predicted conflict-by-accuracy interaction, $F(1, 103) = 62.88$, $p < .001$, $\eta_p^2 = .38$, such that the difference in confidence for correct vs. incorrect responses was greater for no-conflict than for conflict problems (see Table 4).

Table 4. Mean confidence ratings (and SD) by accuracy and type of problem (Study 4).

	Conflict	No-Conflict
Correct	6.86 (1.67)	7.05 (0.99)
Incorrect	6.27 (1.36)	4.17 (2.26)

4.2.3. PDP

Finally, a PDP analysis was performed by calculating the C parameter as P(correct/non-conflict)-P(incorrect/conflict), and A as P(incorrect/conflict)/(1-C) (Ferreira et al. 2006, 2016; Mata et al. 2013c). After excluding negative C values (which are usually interpreted as resulting from random error and therefore excluded, e.g., Jacoby et al. 1993; Ferreira et al. 2006; Mata et al. 2013c), the C parameter predicts the difference in confidence for correct vs. incorrect responses to conflict problems (i.e., resolution for the critical conflict problems), $r = .22$, $p = .019$; the difference in confidence for correct responses to conflict problems vs. correct responses to no-conflict problems (i.e., the confidence boost when responding correctly to conflict problems), $r = .19$, $p = .046$; the difference in confidence for incorrect responses to conflict problems vs. incorrect responses to no-conflict problems, $r = -.42$, $p < .001$, such that the higher the C scores, the lower this difference (i.e., less overconfidence for errors in conflict vs. no-conflict problems); and the difference in resolution (i.e., the difference in mean confidence for correct vs. incorrect responses) for conflict vs. no-conflict problems, $r = -.43$, $p < .001$. Thus, those responders who were better able to control their intuitive responses were also better at knowing the difference between correct and incorrect responses and proved better able to tell right from wrong for both conflict and no-conflict trials.

The A parameter has opposite effects for the most part (even though C and A are unrelated, $r = .03$, $p = .752$): It correlates negatively with resolution (i.e., the difference in confidence for correct vs. incorrect responses) for conflict trials, $r = -.24$, $p = .011$; it correlates negatively with the difference in confidence for correct responses to conflict vs. no-conflict trials, although this correlation was not significant, $r = -.11$, $p = .251$, suggesting a dissociation, whereby the confidence boost of correct conflict responders is only explained by the C parameter; the A parameter correlates positively with the difference in confidence for incorrect responses to conflict problems vs. incorrect responses to no-conflict problems, $r = .29$, $p = .009$, such that the higher the A scores, the more confidence responders had for errors in conflict vs. no-conflict problems; and it predicts the difference in resolution (i.e., the difference in mean confidence for correct vs. incorrect responses) for conflict vs. no-conflict problems, $r = .25$, $p = .023$, such that the more compelling the intuitive responses, the harder it was to tell right from wrong (i.e., resolution) for conflict trials, as compared to no-conflict trials.

5. General Discussion

Four studies with both online and lab samples from different nationalities (German, US, and Portuguese) assessed overconfidence in reasoning about the CRT and explored its origins. The results of Studies 1–2 were remarkably consistent: Participants showed some ability to discriminate between correct and incorrect answers, but this ability was greater for GK questions, whereas for CRT problems it was poorer. Indeed, the most striking result in these studies is that participants were as confident when they responded incorrectly to CRT problems as when they responded correctly to GK questions. However, even though incorrect responders were hugely (over)confident in their solutions to CRT problems, correct responders were even more confident. In fact, the confidence associated with correct responses was higher for CRT problems than for GK questions. Studies 3–4 provided evidence for the key factor underlying these results: Conflict. Specifically, overconfidence among incorrect responders is greater when problems trigger an intuitive but incorrect response. For those tricky items, people are less able to discriminate between correct and incorrect responses.

5.1. Unskilled and Unaware: On Conflict Detection and Error Monitoring

The overconfidence of incorrect reasoners was patent in several results: Even though confidence for correct responses to conflict problems was higher than for errors in those problems, (1) this discrimination was lower for conflict problems than for no-conflict problems, (2) confidence for incorrect responses to conflict problems was above the midpoint of the scale, whereas for incorrect responses to no-conflict problems it was lower, and (3) it was sometimes as high for incorrect responses to conflict problems as for correct responses to no-conflict problems (in Studies 1–2, though not in Studies 3–4). Altogether, these results suggest that incorrect reasoners do not have a very clear sense of their errors.

This is inconsistent with research on conflict detection (De Neys et al. 2011, 2013), which suggests that incorrect responders are aware of their reasoning shortcomings, as they react differently to different versions of a problem that vary in whether intuition is in conflict or in harmony with deliberation. For instance, reasoners have been shown to be less confident when providing incorrect responses to conflict problems than when they respond correctly to no-conflict problems (e.g., De Neys et al. 2011, 2013; see also Mata 2020).

Despite the evidence for conflict sensitivity when comparing metacognitive reactions across different versions of a problem, one can question whether that qualifies as cogent evidence for true error detection, given how high the confidence ratings usually are for incorrect responses to reasoning problems. Indeed, one question that conflict detection studies evoke is what the criteria for error detection should be, and whether comparing participants' reactions to conflict vs. no-conflict versions is the only type of evidence this research should be concerned with. If participants realize at some level that their responses are incorrect, should they not express low (and not just lower) confidence levels? However, the confidence levels of incorrect reasoners are typically above the midpoint of the scale, as in the present studies, and sometimes as high as 85% or even 91.5% (where 100% means completely confident; De Neys et al. 2013; Johnson et al. 2016).

Thus, the metacognitive judgments that people make for conflict problems should tell us something about whether they are truly aware of their reasoning shortcomings. One thing is for people to detect that there is something different in conflict vs. no-conflict problems. A different thing is to assume that, because of such differences, people realize that they are wrong. Underlying such differences might be true sensitivity to logical principles (for compelling evidence that conflict detection is related to knowledge of logical principles, see Burič and Šrol 2020; Frey et al. 2017; Šrol and De Neys 2021), but sometimes it might be mere sensitivity to peripheral features of a problem, unrelated to logic (Aczel et al. 2016; Ferreira et al. 2022; Ghasemi et al. 2023, 2022; Klauer and Singmann 2013; Mata et al. 2014; Meyer-Grant et al. 2022).

Therefore, the possibility of implicit conflict detection is fascinating, and the amount of evidence supporting it is impressive, but there is a long way from conflict sensitivity to error awareness, and one should be cautious in inferring the latter from the former. Several recent developments are being made in this area of research, which will help to make sense of conflict detection results and what they mean for the possibility of implicit error monitoring (e.g., Šrol and De Neys 2021).

The evidence suggests that the overconfidence of incorrect responders originates from their being generally unaware of alternative responses. For instance, when participants in a study by Mata et al. (2013b) were made aware of the correct alternative response (presented as a possible response; not necessarily the correct response), they dramatically lowered their confidence in the incorrect responses that they had given earlier. Similarly, when participants were asked, after having given their responses, to think of alternative responses that other people might give to the same problem, they grew less confident in their original solutions (Mata 2019a, 2019b). Moreover, studies using measures of decisional conflict, where participants are asked whether they considered alternative responses and how divided they felt between different possible responses before committing to a final response, also show that the overconfidence of incorrect responders is related to being oblivious to alternative responses (Mata et al. 2013b; Mata 2019a, 2020). Thus, past research suggests that confidence is related to reasoners' (un)awareness of alternative responses.

The present Study 4 directly tested this mechanism by manipulating whether the problems presented to participants pose a conflict between intuition or deliberation, i.e., whether the problems prompt an intuitive (but incorrect) response, or not. As predicted, the overconfidence of incorrect responders was greatest when problems brought to mind an intuitive but incorrect response. Moreover, the PDP analysis in Study 4 suggests that the more compelling this intuitive response was (as measured by the A parameter), the less able incorrect responders were to discriminate between correct and incorrect responses, particularly for tricky conflict problems. Thus, the present studies add to previous research by shedding light on the mechanisms underlying the overconfidence of biased reasoners.

5.2. *Skilled and Aware: The Metacognitive Advantage of Deliberative Responders*

These results largely replicate the results of Mata et al. (2013b) in that correct responses to tricky reasoning problems were given with great confidence. Indeed, comparing confidence in correct responses to conflict vs. no-conflict problems shows that the latter were high, but the former was even higher (in Studies 1–3, though not in Study 4). Thus, whereas incorrect responders to conflict problems are overconfident, correct responders can be said to be *superconfident*. I suggest that the reason for this confidence boost is the awareness that the correct responders have of the intuitive but incorrect response. Presumably, they too thought of that response before they came to the correct solution (Mata 2019a, 2020; Šćeta et al. 2022). This awareness that the problem is tricky, but that they were able to override their initial misleading intuition, gives them great confidence in their response.

Several of the studies previously referenced already provide some evidence for this mechanism, either manipulating or measuring whether reasoners consider alternative responses (e.g., Mata 2019a). The present Study 4 adds to those studies in that the PDP analysis provides new evidence for the metacognitive advantage model: To the extent that participants were better able to control and override the intuitive (but incorrect) responses to conflict problems (as measured by the C parameter), they had a confidence boost in their correct responses (as compared to the confidence they showed for their correct responses to no-conflict versions of the same problems).

The present studies were not designed to test the hypotheses with response time analysis (admittedly, a limitation). Indeed, because the conflict and no-conflict problems were often of different lengths (and in one study, they were also of different formats: Open-ended for CRT vs. multiple choice for GK), a response time analysis might not be very rigorous or informative. However, were such an analysis feasible using comparable versions of the problems, one might consider what the prediction would be. On the one hand, previous research shows that correctly solving conflict problems can be more time-consuming than doing so for non-conflict problems (e.g., Bonner and Newell 2010; Ferreira et al. 2016; Stupple et al. 2011), which is compatible with a default-interventionist dual-process perspective whereby correct responses can only be produced after overcoming faulty intuitions. However, other research shows that the relation between response time and accuracy is not as straightforward as such a perspective would predict (Stupple et al. 2017), and some studies suggest that correct responses can be intuitive, with no need for effortful deliberation (e.g., Bago and De Neys 2017). Most likely, the reality is more complicated than simply assuming that correct responding always or never requires deliberation. Future research should continue to pursue the fine-grained exploration of individual differences (and their causes) in the degree to which accuracy is related to effortful deliberation or logical intuitions (Burič and Šrol 2020; Frey et al. 2017; Stupple et al. 2011).

At this point, one might ask the question of when conflict (i.e., having two answers in mind) boosts vs. decreases confidence. The key moderating factor, I believe, is whether there is a verifiable correct response to the problem or not (Trouche et al. 2014). When such a response exists, as is the case for reasoning problems, one can think about it and check that it is correct. Together with being aware of the intuitive but incorrect response, this boosts the confidence of the correct responder (Mata 2019a, 2020; Mata et al. 2013b). However, when there is no objectively correct response, as in the case of moral dilemmas, having competing responses in mind might undermine confidence. There is evidence for this in studies showing that conflict decreases perceptions of consensus, such that those who feel more divided about how to respond to moral dilemmas believe that fewer people share their responses (Mata 2019b). Future studies might test whether the existence (vs. lack) of a response that can be demonstrated to be objectively correct (Trouche et al. 2019) determines whether response conflict and revision boost confidence or rather undermines it.

In conclusion, this metacognitive mechanism whereby confidence judgments are informed by the awareness (or lack thereof) of alternative responses underlies both the curse of the incorrect responders (Dunning 2011) and the advantage of the correct responders (Mata 2019a, 2020; Mata et al. 2013b). Because incorrect responders do not realize that there are other, more valid, possible responses, they do not know the correct response, and they do not know that they do not know. Correct responders, on the other hand, realize that there are alternative solutions that are tempting but incorrect, and so they know, and they know that they know.

Funding: This work received Portuguese national funding from FCT—Fundação para a Ciência e a Tecnologia, I. P., through the Research Center for Psychological Science of the Faculty of Psychology, University of Lisbon (UIDB/04527/2020; UIDP/04527/2020; PTDC/PSI-GER/7592/2020).

Institutional Review Board Statement: The studies were approved by the ethics committee of Faculdade de Psicologia, Universidade de Lisboa.

Informed Consent Statement: Informed consent was obtained from all subjects involved in the study.

Data Availability Statement: Data will be provided upon request.

Conflicts of Interest: The author declares no conflict of interest.

Appendix A

Reasoning problems used in Studies 1–2:

A bat and a ball together cost 110 cents. The bat costs 100 cents more than the ball. How much does the ball cost?

If it takes 5 machines 5 min to make 5 widgets, how long would it take 100 machines to make 100 widgets?

In a lake, there is a patch of lily pads. Every day, the patch doubles in size. If it takes 48 days for the patch to cover the entire lake, how long would it take for the patch to cover half of the lake?

GK questions used in Study 1:

Antibiotics kill what class of pathogen?

What is the largest species of whale?

Who wrote and directed "Kill Bill" Volumes 1 and 2?

What was the full name of the cannibalistic main character in the film "Silence of the Lambs"?

Who painted the Sistine Chapel?

The Italian village of Pompeii was destroyed in 79 AD by what type of natural disaster?

Jim Morrison (lead singer of The Doors), Elvis Presley, and Jimi Hendrix all died from what?

What was the name of the famous 3-day music festival, held in 1969, which featured artists such as Bob Dylan, Janis Joplin, and Jimi Hendrix?

What is the name of the world's largest coral reef, located off the coast of Australia?

What is the highest mountain range in the world?

Table A1. GK questions used in Study 2.

Questions	Response Options		
What is the name for an instant camera?	Canon camera	Polaroid camera	Minolta camera
Where do flounders usually live?	Among the reeds	Amongst coral reefs	On the sea bottom
What is a rollmop made of?	herring	pork	salmon
What country does the Nobel Prize winner in Literature Gabriel García Márquez come from?	Spain	Venezuela	Colombia
What artistic movement does anacreontics belong to?	Rococo	Romanticism	Realism
What is a hot chilli sauce?	Tabasco	Curacao	Macao
How many letters are there in the Russian alphabet?	40	33	26
"Tosca" is an opera by ... ?	G. Puccini	G. Verdi	A. Vivaldi
What is the name of the Greek Goddess of wisdom?	Pallas Athena	Nike	Penelope
What is the most abundant metal on Earth?	iron	Aluminum	Copper
What is a word to describe an unknowing person?	Ignatius	ignorant	ideologue
Who was the first person to fly around the Eiffel Tower in an airship?	Santos-Dumont	Count Zeppelin	Saint-Exupéry
What is the name of Eskimo snow shelter?	wigwam	igloo	tipi
What enterprise belongs to Bill Gates?	Intel	Microsoft	Dell Computers
What is the Islamic month of fasting called?	Sharia	Ramadan	Imam
What language does the term "Fata Morgana" come from?	Arabic	Swahili	Italian
How long does it take for a hen to hatch an egg?	21 days	14 days	28 days
What is ascorbic acid?	apple vinegar	vitamin A	vitamin C

Table A2. Problems used in Study 3.

Conflict Version	No-Conflict Version
A TV and a DVD together cost 88 euros. The TV costs 80 euros more than the DVD. How much does the DVD cost?	A cake and a piece of fruit cost $8.47 in total. The cake costs $6.71. How much does the piece of fruit cost?
If it takes 10 hens 10 days to lay 10 eggs, how long would it take 100 hens to lay 100 eggs?	If it takes 14 min for a turtle to lay 5 eggs, how long would it take for the turtle to lay 100 eggs?
A computer virus is spreading through the system of a computer. Every minute, the number of infected files doubles. If it takes 100 min for the virus to infect all of the system, how long would it take for the virus to infect half of the system?	In a town, there is a flood. If it takes 48 days for the water to cover the entire town, how long would it take for it to cover a town that is 5 times larger?

Table A3. Problems used in Study 4.

Conflict Version (Intuitive and Correct Responses, and Source)	No-Conflict Version
A TV and a DVD together cost 88 euros. The TV costs 80 euros more than the DVD. How much does the DVD cost? (intuitive response = 8; correct response = 4; Mata and Almeida 2014, adapted from Frederick 2005)	A TV and a DVD together cost 88 euros. The TV costs 80 euros. How much does the DVD cost?
If it takes 10 hens 10 days to lay 10 eggs, how long would it take 100 hens to lay 100 eggs? (intuitive response = 100; correct response = 10; Mata and Almeida 2014, adapted from Frederick 2005)	If 10 hens lay 10 eggs in one day, how many eggs would 100 chickens lay in one day?
A computer virus is spreading through the system of a computer. Every minute, the number of infected files doubles. If it takes 100 min for the virus to infect all of the system, how long would it take for the virus to infect half of the system? (intuitive response = 50; correct response = 99; Mata and Almeida 2014, adapted from Frederick 2005)	A computer virus is spreading through the system of a computer. Every minute, one file gets infected. If it takes 100 min for the virus to infect all of the system, how long would it take for the virus to infect half of the system?
A farmer had 15 sheep and all but 8 died. How many are left? (intuitive response = 7; correct response = 8; Thomson and Oppenheimer 2016)	A farmer had 15 sheep and 8 died. How many are left?
How many cubic feet of dirt are there in a hole that is 3′deep x 3′ wide x 3′ long? (intuitive response = 27; correct response = none; Thomson and Oppenheimer 2016)	How many cubic feet of water are there in a fish tank that is 3′deep x 3′ wide x 3′ long?
Steve was standing in a long line. To amuse himself he counted the people waiting, and saw that he stood 15th from the beginning and 15th from the end of the line. How many people stood in the line? (intuitive responses = 31 or 30; correct response = 29; Ackerman 2014)	Steve was standing in a long line. To amuse himself he counted the people waiting, and saw that there were 15 people in front of him and 15 behind him. How many people stood in the line?
Ellen and Kim are running around a track. They run equally fast but Ellen started later. When Ellen has run 5 laps, Kim has run 15 laps. When Ellen had run 30 laps, how many has Kim run? (intuitive response = 90; correct response = 40; Van Dooren et al. 2005)	Ellen and Kim are running around a track. Kim runs 3 times as fast as Ellen. When Ellen has run 5 laps, Kim has run 15 laps. When Ellen had run 30 laps, how many has Kim run?
One month of the year has 28 days. How many of the remaining 11 months have 30 days? (intuitive response = 5; correct response = 11; Weber 2016);	One month of the year has 28 days. How many of the remaining 11 months have 31 days?
You have a book of matches and enter a cold, dark room. You know that in the room there is an oil lamp, a candle, and a heater. What do you light first? (intuitive response = oil lamp, candle, or heater; correct response = matches; Boland 2013)	You enter a cold, dark room. You know that in the room there is an oil lamp, a candle, and a heater. What do you light first?
There are 25 soldiers in a row standing 3 m from each other. How long is the row? (intuitive response = 75 m; correct response = 72 m; Oldrati et al. 2016).	There are 25 vans in a row. Each of them is 3 m long. How long is the row?
Which city is further north: Rome or New York? (intuitive response = New York; correct response = Rome; Juslin et al. 2000)	Which city is further north: Rio de Janeiro or New York?
What is the capital of California: San Francisco, Los Angeles, or Sacramento? (intuitive responses = San Francisco, Los Angeles; correct response = Sacramento; Koriat 1995)	What is the largest city in California: San Francisco, Los Angeles, or Sacramento?

Table A3. *Cont.*

Conflict Version (Intuitive and Correct Responses, and Source)	No-Conflict Version
Would it be ethical for a man to marry the sister of his widow? (intuitive response = yes or no, justified; correct response = impossible, he is dead; Sirota et al. 2021)	Would it be ethical for a man to marry the sister of his wife?
The wind blows west. An electric train runs east. In which cardinal direction does the smoke from the locomotive blow? (intuitive response = west; correct response = there is no smoke in an electric train; Sirota et al. 2021)	The wind blows west. A train runs east. In which cardinal direction does the smoke from the locomotive blow?
What is the capital of Australia: Sydney, Canberra, or Melbourne? (intuitive response = Sidney; correct response = Canberra; Weber 2016)	What is the largest city in Australia: Sydney, Canberra, or Melbourne?
Who was the Mathematician who was the author of the theorem that states that the square of the length of the hypotenuse of a rectangle is equal to the sum of squares of the lengths of other sides? (intuitive response = Pythagoras; correct response = no one; it was a triangle, not a rectangle; Mata et al. 2013a)	Who was the Mathematician who was the author of the theorem that states that the square of the length of the hypotenuse of a triangle is equal to the sum of squares of the lengths of other sides?
When was the first air raid: 1849 or 1937? (intuitive response = 1937; correct response = 1849; Fischhoff et al. 1977; Koriat 2008)	When was the first air raid: 1819 or 1849?
How many animals of each kind did Moses take in the ark? (intuitive response = 2; correct response = zero; Erickson and Mattson 1981)	How many animals of each kind did Noah take in the ark?
What are the names of the Portuguese archipelagos located to the east of continental Portugal? (intuitive response = Madeira and Azores; correct response = the archipelagos are located to the west, not east; Mata et al. 2013a)	What are the names of the Portuguese archipelagos located to the west of continental Portugal?
Potatoes are native to Ireland or Peru? (intuitive response = Ireland; correct response = Peru; Fischhoff et al. 1977)	Potatoes are native to Peru or Japan?

References

Ackerman, Rakefet. 2014. The diminishing criterion model for metacognitive regulation of time investment. *Journal of Experimental Psychology: General* 143: 1349–68. [CrossRef]

Aczel, Balazs, Aba Szollosi, and Bence Bago. 2016. Lax monitoring versus logical intuition: The determinants of confidence in conjunction fallacy. *Thinking & Reasoning* 22: 99–117.

Atir, Stav, Emily Rosenzweig, and David Dunning. 2015. When knowledge knows no bounds: Self-perceived expertise predicts claims of impossible knowledge. *Psychological Science* 26: 1295–303. [CrossRef]

Bago, Bence, and Wim De Neys. 2017. Fast logic?: Examining the time course assumption of dual process theory. *Cognition* 158: 90–109. [CrossRef] [PubMed]

Bialek, Michal, and Gordon Pennycook. 2018. The cognitive reflection test is robust to multiple exposures. *Behavior Research Methods* 50: 1953–59. [CrossRef] [PubMed]

Boland, Sam. 2013. Reflection expanded: Expanding the Cognitive Reflection Task. Available online: https://pt.scribd.com/document/138024235/Reflection-Expanded-Expanding-the-Cognitive-Reflection-Task (accessed on 21 November 2022).

Bonner, Carissa, and Ben R. Newell. 2010. In conflict with ourselves? An investigation of heuristic and analytic processes in decision making. *Memory & Cognition* 38: 186–96.

Branas-Garza, Pablo, Praveen Kujal, and Balint Lenkei. 2019. Cognitive reflection test: Whom, how, when. *Journal of Behavioral and Experimental Economics* 82: 101455. [CrossRef]

Burič, Roman, and Jakub Šrol. 2020. Individual differences in logical intuitions on reasoning problems presented under two-response paradigm. *Journal of Cognitive Psychology* 32: 460–77. [CrossRef]

De Neys, Wim. 2012. Bias and conflict: A case for logical intuitions. *Perspectives on Psychological Science* 7: 28–38. [CrossRef]

De Neys, Wim, Sandrine Rossi, and Olivier Houdé. 2013. Bats, balls, and substitution sensitivity: Cognitive misers are no happy fools. *Psychonomic Bulletin & Review* 20: 269–73.

De Neys, Wim, Sofie Cromheeke, and Magda Osman. 2011. Biased but in doubt: Conflict and decision confidence. *PLoS ONE* 6: e15954. [CrossRef] [PubMed]

Dunning, David. 2011. The Dunning–Kruger effect: On being ignorant of one's own ignorance. In *Advances in Experimental Social Psychology*. New York: Academic Press, vol. 44, pp. 247–96.

Erickson, Thomas D., and Mark E. Mattson. 1981. From words to meaning: A semantic illusion. *Journal of Memory and Language* 20: 540–51. [CrossRef]

Ferreira, Mário B., André Mata, Christopher Donkin, Steven J. Sherman, and Max Ihmels. 2016. Analytic and heuristic processes in the detection and resolution of conflict. *Memory & Cognition* 44: 1050–63.

Ferreira, Mário. B., Jerônimo C. Soro, Joana Reis, André Mata, and Valerie A. Thompson. 2022. When Type 2 Processing Misfires: The Indiscriminate Use of Statistical Thinking about Reasoning Problems. *Journal of Intelligence* 10: 109. [CrossRef]

Ferreira, Mário B., Leonel Garcia-Marques, Steven J. Sherman, and Jeffrey W. Sherman. 2006. Automatic and controlled components of judgment and decision making. *Journal of Personality and Social Psychology* 91: 797–813. [CrossRef]

Fischhoff, Baruch, Paul Slovic, and Sarah Lichtenstein. 1977. Knowing with certainty: The appropriateness of extreme confidence. *Journal of Experimental Psychology: Human Perception and Performance* 3: 552–64.

Frederick, Shane. 2005. Cognitive reflection and decision making. *Journal of Economic Perspectives* 19: 25–42. [CrossRef]

Frey, Darren, Eric D. Johnson, and Wim De Neys. 2017. Individual differences in conflict detection during reasoning. *The Quarterly Journal of Experimental Psychology* 71: 1188–1208. [CrossRef]

Ghasemi, Omid, Simon Handley, Stephanie Howarth, Ian R. Newman, and Valerie A. Thompson. 2022. Logical intuition is not really about logic. *Journal of Experimental Psychology: General* 151: 2009–28. [CrossRef]

Ghasemi, Omid, Simon. J. Handley, and Stephanie Howarth. 2023. Illusory intuitive inferences: Matching heuristics explain logical intuitions. *Cognition* 235: 105417. [CrossRef]

Jacoby, Larry L. 1991. A process dissociation framework: Separating automatic from intentional uses of memory. *Journal of Memory and Language* 30: 513–41. [CrossRef]

Jacoby, Larry L., Jeffrey P. Toth, and Andrew P. Yonelinas. 1993. Separating conscious and unconscious influences of memory: Measuring recollection. *Journal of Experimental Psychology: General* 122: 139–54. [CrossRef]

Johnson, Eric D., Elisabet Tubau, and Wim De Neys. 2016. The Doubting System 1: Evidence for automatic substitution sensitivity. *Acta Psychologica* 164: 56–64. [CrossRef]

Juslin, Peter, Anders Winman, and Henrik Olsson. 2000. Naive empiricism and dogmatism in confidence research: A critical examination of the hard–easy effect. *Psychological Review* 107: 384–96. [CrossRef]

Klauer, Karl Christoph, and Henrik Singmann. 2013. Does logic feel good? Testing for intuitive detection of logicality in syllogistic reasoning. *Journal of Experimental Psychology: Learning, Memory, and Cognition* 39: 1265–73. [CrossRef] [PubMed]

Koriat, Asher. 1995. Dissociating knowing and the feeling of knowing: Further evidence for the accessibility model. *Journal of Experimental Psychology: General* 124: 311–33. [CrossRef]

Koriat, Asher. 2008. Subjective confidence in one's answers: The consensuality principle. *Journal of Experimental Psychology: Learning, Memory, and Cognition* 34: 945–59. [CrossRef] [PubMed]

Kruger, Justin. 1999. Lake Wobegon be gone! The "below-average effect" and the egocentric nature of comparative ability judgments. *Journal of Personality and Social Psychology* 77: 221–32. [CrossRef]

Kruger, Justin, and David Dunning. 1999. Unskilled and unaware of it: How difficulties in recognizing one's own incompetence lead to inflated self-assessments. *Journal of Personality and Social Psychology* 77: 1121–34. [CrossRef] [PubMed]

Liberman, Varda. 2004. Local and global judgments of confidence. *Journal of Experimental Psychology: Learning, Memory, and Cognition* 30: 729–34. [CrossRef]

Mata, André. 2019a. Further tests of the metacognitive advantage model: Counterfactuals, confidence and affect. *Psychological Topics (Special Issue on Meta-Reasoning)* 28: 155–24. [CrossRef]

Mata, André. 2019b. Social metacognition in moral judgment: Decisional conflict promotes perspective taking. *Journal of Personality and Social Psychology* 117: 1061–82. [CrossRef] [PubMed]

Mata, André. 2020. Metacognition and social perception: Bringing meta-reasoning and social cognition together. *Thinking and Reasoning* 26: 140–49. [CrossRef]

Mata, André, and Tiago Almeida. 2014. Using metacognitive cues to infer others' thinking. *Judgment and Decision Making* 9: 349–59. [CrossRef]

Mata, André, Anna-Lena Schubert, and Mário B. Ferreira. 2014. The role of language comprehension in reasoning: How "good enough" representations induce biases. *Cognition* 133: 457–63. [CrossRef]

Mata, André, Klaus Fiedler, Mário B. Ferreira, and Tiago Almeida. 2013a. Reasoning about others' reasoning. *Journal of Experimental Social Psychology* 49: 486–91. [CrossRef]

Mata, André, Mário B. Ferreira, and Joana Reis. 2013b. A process-dissociation analysis of semantic illusions. *Acta Psychologica* 144: 433–43. [CrossRef] [PubMed]

Mata, André, Mário B. Ferreira, and Steven J. Sherman. 2013c. The metacognitive advantage of deliberative thinkers: A dual-process perspective on overconfidence. *Journal of Personality and Social Psychology* 105: 353–73. [CrossRef]

Meyer-Grant, Constatin G., Nicole Cruz, Henrik Singmann, Samuel Winiger, Spriha Goswami, Brett K. Hayes, and Karl Christoph Klauer. 2022. Are logical intuitions only make-believe? Reexamining the logic-liking effect. *Journal of Experimental Psychology: Learning, Memory, and Cognition*. [CrossRef]

Michailova, Julija, and Joana K. Katter. 2014. Quantifying overconfidence in experimental finance. *International Journal of Behavioural Accounting and Finance* 4: 351–67. [CrossRef]

Moore, Don, and Paul J. Healy. 2008. The trouble with overconfidence. *Psychological Review* 115: 502–17. [CrossRef] [PubMed]

Oldrati, Viola, Jessica Patricelli, Barbara Colombo, and Alessandro Antonietti. 2016. The role of dorsolateral prefrontal cortex in inhibition mechanism: A study on cognitive reflection test and similar tasks through neuromodulation. *Neuropsychologia* 91: 499–508. [CrossRef] [PubMed]

Pennycook, Gordon, Robert M. Ross, Derek J. Koehler, and Jonathan A. Fugelsang. 2017. Dunning–Kruger effects in reasoning: Theoretical implications of the failure to recognize incompetence. *Psychonomic Bulletin & Review* 24: 1774–84.

Pulford, Briony D., and Andrew M. Colman. 1997. Overconfidence: Feedback and item difficulty effects. *Personality and Individual Differences* 23: 125–33. [CrossRef]

Šćeta, Lamija, Josip Sliško, and Nikola Erceg. 2022. Predicting the most common incorrect response: Metacognitive advantage of deliberative over intuitive responders on Cognitive Reflection Test. *Studia Psychologica* 64: 256–67. [CrossRef]

Schraw, Gregory, and Teresa Debacker Roedel. 1994. Test difficulty and judgment bias. *Memory & Cognition* 22: 63–69.

Sirota, Miroslav, Chris Drewberry, Marie Juanchich, Lenka Valus, and Amanda C. Marshall. 2021. Measuring Cognitive Reflection without Maths: Developing and Validation of the Verbal Cognitive Reflection Test. *Journal of Behavioral Decision Making* 34: 322–343. [CrossRef]

Šrol, Jakub, and Wim De Neys. 2021. Predicting individual differences in conflict detection and bias susceptibility during reasoning. *Thinking and Reasoning* 27: 38–68. [CrossRef]

Stupple, Edward J., Linden J. Ball, Jonathan S. B. Evans, and Emily Kamal-Smith. 2011. When logic and belief collide: Individual differences in reasoning times support a selective processing model. *Journal of Cognitive Psychology* 23: 931–41. [CrossRef]

Stupple, Edward J., Melanie Pitchford, Linden J. Ball, Thomas E. Hunt, and Richard Steel. 2017. Slower is not always better: Response-time evidence clarifies the limited role of miserly information processing in the Cognitive Reflection Test. *PLoS ONE* 12: e0186404. [CrossRef] [PubMed]

Thomson, Kella S., and Daniel M. Oppenheimer. 2016. Investigating an alternate form of the cognitive reflection test. *Judgment and Decision Making* 11: 99–113. [CrossRef]

Trouche, Emmanuel, Emmanuel Sander, and Hugo Mercier. 2014. Arguments, more than confidence, explain the good performance of reasoning groups. *Journal of Experimental Psychology: General* 143: 1958–71. [CrossRef] [PubMed]

Trouche, Emmanuel, Jing Shao, and Hugo Mercier. 2019. Objective evaluation of demonstrative arguments. *Argumentation* 33: 23–43. [CrossRef]

Van Dooren, Wim, Dirk De Bock, An Hessels, Dirk Janssens, and Lieven Verschaffel. 2005. Not everything is proportional: Effects of age and problem type on propensities for overgeneralization. *Cognition and Instruction* 23: 57–86. [CrossRef]

Weber, Mel. 2016. The Role of Metacognition and Anxiety in College Students' Performance on a General-Knowledge Test (Honors Thesis). Available online: https://pdfs.semanticscholar.org/7ae5/ee960a080a62fe15cd9e17fe01265ec1e522.pdf?_ga=2.29673148.2064039678.1582733636-1459196501.1522233394 (accessed on 6 December 2022).

Journal of
Intelligence

Article

When Type 2 Processing Misfires: The Indiscriminate Use of Statistical Thinking about Reasoning Problems

Mário B. Ferreira [1,*,†], Jerônimo C. Soro [2,*,†], Joana Reis [1], André Mata [1] and Valerie A. Thompson [3]

1 CICPSI, Faculdade de Psicologia, Universidade de Lisboa, 1649-013 Lisboa, Portugal
2 Digital Human-Environment Interaction Labs, Lusófona University, HEI-Lab, 1749-024 Lisboa, Portugal
3 Department of Psychology, University of Saskatchewan, 9 Campus Drive, Saskatoon, SK S7N 5A5, Canada
* Correspondence: mferreira@fp.ul.pt (M.B.F.); jeronimo.soro@ulusofona.pt (J.C.S.)
† These authors contributed equally to this work.

Abstract: Research on dual-process theories of judgment makes abundant use of reasoning problems that present a conflict between Type 1 intuitive responses and Type 2 rule-based responses. However, in many of these reasoning tasks, there is no way to discriminate between the adequate and inadequate use of rules based on logical or probabilistic principles. To experimentally discriminate between the two, we developed a new set of problems: rule-inadequate versions of standard base-rate problems (where base rates are made irrelevant). Across four experiments, we observed conflict sensitivity (measured in terms of response latencies and response confidence) in responses to standard base-rate problems but also in responses to rule-inadequate versions of these problems. This failure to discriminate between real and merely apparent (or spurious) conflict suggests that participants often misuse statistical information and draw conclusions based on irrelevant base rates. We conclude that inferring the sound use of statistical rules from normatively correct responses to standard conflict problems may be unwarranted when this kind of reasoning bias is not controlled for.

Keywords: dual-process theory; reasoning; judgment; bias; metacognition

1. Introduction

One of the pioneering contributions of the initial work of the research program on heuristics and biases (Tversky and Kahneman 1974, 1983; Kahneman et al. 1982) was the careful development of several reasoning tasks that were used to demonstrate that, more often than not, people's judgments seem to violate basic principles of probability and logic. One of these tasks was the so-called "lawyer-engineer problem" (Kahneman and Tversky 1972):

> *In a study, several psychologists interviewed a group of people. The group included 5 engineers and 995[1] lawyers. The psychologists prepared a brief summary of their impression of each interviewee. The following description was drawn randomly from the set of descriptions: Dan is 45. He is conservative, careful, and ambitious. He shows no interest in political issues and spends most of his free time on his many hobbies, which include carpentry, sailing, and mathematical puzzles. Which of the following is more likely?*
>
> *(a) Dan is an engineer*
>
> *(b) Dan is a lawyer*

When facing the problem above, many people might rely on the similarity between Dan's description and the stereotype of engineer to infer that Dan is an engineer (without taking into account the prior odds of being an engineer or a lawyer; 5/995). In other words, people's judgments often rely on intuitive processes (e.g., judgment by representativeness; Kahneman and Tversky 1972) rather than analytic ones (reasoning considering the initial base rates).

In accordance with dual-process theories of reasoning, this kind of fast and intuitive process (Type 1 or T1) underpins a default response that can only be remediated by the effortful and more demanding Type 2 (T2) process (Evans and Stanovich 2013). Indeed, there is a well-known research tradition in which T2 processes are viewed as being designed to correct or counteract T1 outputs, enabling the reasoner to apply relevant rules of logic and probability that will lead to less harmful, undesirable, irrational, or impulsive courses of actions (Kahneman 2003). However, this research tradition has been focusing on T1 processes as the source of the decision-making error, mostly ignoring the potential contribution of decision-making errors associated with T2.

In this paper, we aimed to uncover judgment errors associated with T2. In particular, we are interested in an overgeneralization bias, whereby the agent continues to deliberatively apply rules of probability in circumstances in which it may no longer be appropriate to do so. We suggest that the misuse of probability rules is more likely to occur when T2 processing is based on a superficial understanding of normative principles. For instance, when one inhibits T1 intuitive responses to make deliberate use of biased statistics.

Revisiting our opening example, if the problem included an additional premise stating that in the first day of the study the psychologists had randomly chosen, say, two lawyers and two engineers to interview (from the total sample of lawyers and engineers), then the same number of lawyers and engineers would have been interviewed, and therefore the base rates corresponding to the initial group composition (995 lawyers and 5 engineers) are not relevant. Hence, considering those base rates would be a mistake.

In what follows, we begin by briefly reviewing relevant research on dual-process models. Next, we focus on how the overuse of rules (i.e., relying on rules of probability in circumstances where it is inappropriate to do so) may lead to errors and biases (Elqayam and Evans 2011; Fong et al. 1986). We will then describe and test a way to experimentally discriminate between the suitable and unsuitable use of rules, using the conflict-detection paradigm (e.g., De Neys and Pennycook 2019) with a variation of base-rate problems better capable of capturing unwarranted, rule-based thinking.

A major source of experimental evidence to test and contrast different theories of reasoning comes from research using reasoning tasks, where a stereotypical or belief-based heuristic response conflicts with a normatively correct response that requires the use of probabilistic or logical principles. One classic example of these tasks is the base-rate neglect task, where the description of an individual target is inconsistent with the majority of the sample from which the description was drawn (e.g., Kahneman and Tversky 1973; De Neys and Glumicic 2008).

In responding to these tasks, T2 thinking is often seen as being responsible for the application of relevant rules of logic and probability and is thus associated with normatively correct responses, whereas belief biases are attributed to T1 processes and to failures in engaging in T2 thinking (necessary to override the biases; Kahneman 2003; Kahneman and Frederick 2002; Stanovich 1999; Stanovich and West 1998, 2008).

Other dual-process approaches present an alternative view. They argue that conflict is usually implicitly detected between different T1 outputs. That is, in addition to the heuristic intuitions that lead to erroneous responses in the "cognitive illusions" tradition (e.g., Gilovich et al. 2002), reasoners also have logical intuitions (i.e., implicit knowledge of the logical and probabilistic principles) that are automatically activated when they respond to reasoning tasks. Response bias may still prevail because reasoners might fail to override the stronger intuitive appeal of the heuristic outputs, eventually leading them to behave against their logical intuitions (De Neys 2012; De Neys and Glumicic 2008; Morsanyi and Handley 2012).

Several authors (Ball et al. 2017; De Neys 2012, 2018; Handley et al. 2011; Pennycook et al. 2015; Thompson et al. 2018) have begun to successfully integrate the conceptual distinction brought by these different dual-process views in a process model of analytic engagement. According to this model, classic reasoning problems are likely to generate one or more competing T1 outputs (De Neys 2012, 2014). If no conflict is detected (either because T1 processes generate just one output or because the difference in activation strength between the competing intuitions is large enough to avoid conflict detection), heuristic responses will be accepted with cursory T2 processing. If more than one T1 outputs are generated and conflict between them is detected, more substantive T2 reasoning will be engaged, which may lead to confirmation of one of the T1 outputs or the replacement of the initial T1 outputs by a T2 normatively correct response.

Despite the substantial progress accomplished so far, most previous research and theorizing tends to assume that correct responses to reasoning tasks are logical or probabilistic sound responses either cued by logical intuitions or generated by T2 thinking.

However, as noted by Elqayam and Evans (2011) (see also, Evans and Stanovich 2013; Stanovich 2009; Thompson et al. 2011), the temptation to treat correct responses as being diagnostic of T2 reasoning is a dangerous "ought-is" fallacy (i.e., inferring from normative responses that T2 processing is involved), which has been described as "perhaps the most persistent fallacy in the perception of dual-process theories" (Evans and Stanovich 2013, p. 229).

The ought-is fallacy is particularly problematic in classic reasoning tasks where there are just two alternative answers: one considered biased, and one considered correct. Choosing the correct answer in such forced-choice paradigms is often interpreted as the T2 suitable use of normative principles. Furthermore, longer response times and lower response confidence are taken as indicators of conflict detection between the two alternative answers, often followed by an increased level of T2 engagement (e.g., De Neys and Pennycook 2019; Pennycook et al. 2015; Thompson et al. 2011).

However, engagement in T2 processing is not a sufficient condition for the suitable use of normative principles because T2 processes are not immune to error (Evans 2018; Pennycook et al. 2015; Stanovich 2009; Thompson et al. 2011). T2 thinking may simply be normatively incorrect as it happens, for example, when people draw upon the "law of averages" to justify their belief in the gambler's fallacy (Elqayam and Evans 2011). Furthermore, the overuse of rules may lead to false alarms. That is, relying on these rules in situations where it is inappropriate to do so (Elqayam and Evans 2011; Fong et al. 1986).

Given the aforementioned one-to-one correspondence in classic reasoning tasks between T1 heuristic-based processes and incorrect responses, and sound logic (either cued by logic intuitions or generated by T2 rule-based processes) and correct responses, it is often difficult to evaluate whether participants who give the rule-based response are actually making an appropriate use of the underlying normative principles. This is because there is no way to discriminate between the suitable use and the misuse of such principles.

The main goal of the research here reported is to evaluate errors associated with T2 thinking. In particular, we are interested in an overgeneralization bias, whereby reasoners apply rules of probability in circumstances where is not appropriate to do so. This may occur if superficial features of a judgment problem cue the use of rules even when the underlying normative principle is not applicable (e.g., references to base rates in base-rate problems may cue its use even when these are non-applicable).

To experimentally discriminate between the suitable and unsuitable use of rules, we developed a new set of base-rate problems that may give rise to unwarranted, rule-based thinking, henceforth referred to as rule-inadequate versions of standard problems. These are described in the following section.

Our approach was inspired by previous work that also made use of new (rule-inadequate) versions of standard reasoning problems (Fong et al. 1986; Klauer and Singmann 2013). These versions are typically made by introducing changes to the deep structure of standard problems (e.g., shifting the logical validity of syllogisms or rendering statistical information of a probabilistic problem invalid) while leaving the problems' surface features largely unchanged.

Specifically, to test the effects of statistical training in the use of normative principles in reasoning, such as the law of large numbers, Fong et al. (1986) developed standard and so-called false-alarm versions of reasoning problems opposing statistical information to stereotype-based information. In this case, false-alarm answers involved (inadequate) conclusions drawn from large but biased samples. This was crucial to show that even brief sessions of statistical training increased the use of the law of large numbers in standard problems without increasing its use in situations where it would be inappropriate to do so (i.e., rule-inadequate problems).

In a similar vein, Klauer and Singmann (2013) developed "pseudo-syllogisms", which may be described as rule-inadequate versions of standard syllogism problems. Indeed, pseudo-syllogisms modify the logical status of the original syllogisms while keeping the superficial structure of the problems (formal features and contents) unchanged. The use of "pseudo-syllogisms" allowed Klauer and Singmann to reinterpret previous findings that people can intuitively detect the logicality even of difficult syllogisms (Morsanyi and Handley 2012). We will return to this point in the General Discussion.

How to Discriminate between the Suitable and Unsuitable Use of Normative Rules in the Base-Rate Neglect Task

To experimentally distinguish between reasoning that makes an appropriate use of rules and reasoning that uses rules in an indiscriminate fashion, we used the conflict-detection paradigm and invited participants to respond not only to standard (conflict and no-conflict) versions of base-rate problems but also to rule-inadequate versions of these problems. The latter present a conflict between stereotypical information and probabilistic information in the form of invalid or irrelevant base rates that do not provide a valid basis for decisions. That is to say, the conflict in rule-inadequate problems is only apparent or spurious.

To illustrate, take the well-known lawyer–engineer problem (Kahneman and Tversky 1972) presented in the beginning. A rule-inadequate version of the same problem would be as follows:

> *In a study, several psychologists interviewed a group of people. The group included 5 engineers and 995 lawyers.* ***In the first day of the study, an equal number of engineers and lawyers were interviewed****, and the psychologists prepared a brief summary of their impression of each interviewee. The following description was drawn randomly from this set of descriptions: Dan is 45. He is conservative, careful, and ambitious. He shows no interest in political issues and spends most of his free time on his many hobbies, which include carpentry, sailing, and mathematical puzzles. Which of the following is more likely?*
>
> *(a) Dan is an engineer*
>
> *(b) Dan is a lawyer*

In this version, there is additional information indicating that the base rates (stemming from the group composition: 5 engineers and 995 lawyers) may not be reliable in the current context. For those such as Dan, who were tested on the first day, the prior probability of being an engineer or a lawyer is the same (i.e., 50/50). If participants go ahead and still use the initial base rates to respond, this would be an inappropriate use of the base-rate information.

Sound statistical thinking should discriminate between standard and rule-inadequate versions of base-rate problems, taking into consideration the 5/995 group composition in the former but discarding them in the latter (leaving Dan's description as the only source of diagnostic information).

In contrast, an indiscriminate reliance on the base rates of the group composition (5/995) is expected to lead to responding according to them both in the standard and rule-inadequate versions of the problem.

Importantly, the conflict-detection paradigm further allowed us to look into the participants' metacognition. More specifically, whether participants show some sensitivity to the opposing information in standard conflict problems (e.g., De Neys and Glumicic 2008).

If participants do not take logical principles into account (e.g., neglecting base-rate information), then conflict should be irrelevant and have no impact on reasoning. However, previous findings indicate that biased reasoners often do show conflict sensitivity by displaying increased response doubt. This is reflected in longer response latencies and lower confidence in their incorrect answers to conflict problems compared to correct answers to no-conflict problems (De Neys 2012; De Neys and Glumicic 2008; Pennycook et al. 2015).

We expect to observe the same kind of conflict effects with rule-inadequate versions of base-rate problems as have been observed with standard base-rate problems. That is, we predict that the irrelevant base rates in rule-inadequate problems are going to trigger (inaccurate) logical intuitions that oppose other T1 intuitions (based on the similarity between the target's description and a professional/group stereotype). This may lead to the detection of a spurious conflict between invalid base rates and stereotype-based information that will be reflected by longer responses latencies (Experiments 1 to 4) and lower response confidence (Experiment 4).

Finally, individual differences in reflective thinking ability have been shown to be associated with judgment biases and errors (e.g., Toplak et al. 2011). We thus used the cognitive reflection test (CRT; Frederick 2005) to assess participants' susceptibility to the overgeneralization bias. We expect higher reflective ability to be associated with less overgeneralization bias.

2. Experiment 1

Experiment 1 is a first attempt at examining the extent to which base-rate responses to standard problems stem from adequate statistical thinking or from an indiscriminate use of base rates (i.e., using the base rates both when they are valid and invalid). In order to obtain measures of processing time, participants responded to a "moving window" paradigm (De Neys and Glumicic 2008; Just et al. 1982), in which there were three kinds of problems: standard base-rate problems—conflict and no-conflict versions—and conflict versions of rule-inadequate problems, which present a spurious conflict between base rates and stereotype-based information. In this paradigm, the base-rate information and the stereotypical description are presented separately. Participants were first presented the base-rate information on a computer screen, which was then substituted by the stereotypical description and the question. Before giving their answer, participants had the option to visualize the base-rate information again by pressing and holding down a specific computer key.

De Neys and Glumicic (2008) showed that not only normatively correct responses but also incorrect responses to problems that presented a conflict between the description and the previously presented base rates took longer (and were associated with more revisits of the initial base-rate information) than responses to no-conflict problems (where base rates were congruent with the stereotype description), thus providing evidence consistent with conflict detection even for participants who gave the heuristic response.

Our main predictions are as follows. An indication of the indiscriminate use of base-rate information might be found in the form of a positive correlation between base-rate responses for standard and rule-inadequate problem versions. This would show that the more participants give base-rate responses to standard problems, the more they overgeneralize the use of base rates to cases where the base rates are invalid (i.e., rule-inadequate problems)[2]. Furthermore, detection of (a spurious) conflict between the stereotype-based information and irrelevant base rates (followed by unwarranted use of these base rates) should be apparent if base-rate responses to rule-inadequate problems are also coupled with longer response times and a higher frequency of base rate reviews (when compared to no-conflict problems). In addition, participants' responses to standard versions of base-rate problems are expected to replicate De Neys and Glumicic's (2008) results.

2.1. Method

2.1.1. Participants

Eighty-six psychology undergraduate students (70 female; M_{age} = 21.2, SD = 6.02) participated in the study in exchange for course credits[3]. Sensitivity analysis with this sample size (at α = .05, and power = .80) showed that the experimental design could reliably detect medium or larger effect sizes ($f \geq .34$).

2.1.2. Material

The material included 18 problems (translated and adapted to Portuguese from De Neys and Glumicic 2008). Three were presented during practice trials, while the remaining fifteen were used in the experimental task. The problems followed the common design of base-rate problems, with a description of the base rates for two groups of people and a description of characteristics of a subject selected randomly from those groups. The description was either stereotypical of the larger group (no-conflict problems) or of the smaller group (conflict problems). The original problems were changed to include information that rendered the described base rates irrelevant (rule-inadequate problems) or kept these base rates relevant (standard problems). The following is an example of one of the problems, with the text that renders the base rates irrelevant in italics:

> In a study 1000 people were tested. Among the participants there were 996 women and 4 men. *In the first week of the study, as many men as women were interviewed. Jo was one of the people interviewed in the first week.*
>
> Jo is 23 years old and is finishing a degree in engineering. On Friday nights, Jo likes to go out cruising with friends while listening to loud music and drinking beer.
>
> What is most likely?
>
> a. Jo is a man
>
> b. Jo is a woman

In this case, the tested subsample comprises the same number of men and women, so the *relevant* base rates indicate a 50/50 chance (rather than the initial 996/4) for the described person to be a woman or a man.

For standard problems, the phrase in italics reads, "*In the first week of the study, all men and women were interviewed.*" As such, the relevant base rate is still 996/4, making "b" ("Jo is a woman") the most likely correct response (examples for all variations of problems in Experiment 1 are displayed in Appendix B, Table A3).

The fifteen problems were separated into three sets of five problems each. In each set, the problems were all of one type: standard conflict problems, standard no-conflict problems, or rule-inadequate problems.

Problem content was counterbalanced across problem types and not repeated for an individual participant. Conflict and no-conflict versions of standard problems were exactly the same except for the base rates, which were inverted. Rule-inadequate versions were the same as conflict versions, except that the original base rates were rendered irrelevant with the addition of the subsample information (each participant responded to only one of the three versions of each specific problem). All problems presented extreme base rates with three variations between them (997/3, 996/4, or 995/5), which were assigned evenly between the fifteen problems.

2.1.3. Procedure

Participants performed the experiment on a computer, where they received instructions to read the texts and to solve the problems at their own pace. Participants were presented 3 practice trials.

Participants were informed that, in each trial, they would see the first part of the problem (containing the base-rate information). After reading the text and whenever they felt ready to move on, they should press the space key on the keyboard to replace the base-rate information with the second part of the problem (containing the stereotypical description of the selected individual and the response options). During the display of this second part, participants could again have access to and review the first part with the crucial base-rate information by pressing the space key. As long as they held down the space key, the first part remained visible. Once the space key was released, the information disappeared again. The second part of the text with the description always remained visible after the initial presentation.

The main dependent variables included the proportion of base-rate responses for each type of problem: the base-rate reading time (when this information was first presented without the stereotype information) and the decision-making time (i.e., the time elapsed between the display of stereotype information and response). The mean number of problems where base-rate information was reviewed was also analyzed for each type of problem.

2.2. Results

In this and the remaining experiments, we begin by describing and analyzing the proportion of base-rate responses across the experimental conditions (see Appendix B, Table A1), as well as the correlation of base-rate responses between standard and rule-inadequate versions of the conflict problems (see Table 1). Evidence for the overgeneralization bias is provided by positive and significant correlations between the two.

Table 1. Correlation of base-rate responses between standard and rule-inadequate problems (Experiments 1 to 4).

	r (p)	**High Reflective** *r (p)*	**Low Reflective** *r (p)*
Experiment 1	.46 (<.001)		
Experiment 2	.50 (<.001)	.05 (.773)	.70 (<.001)
Experiment 3	.53 (<.001)	.32 (.138)	.65 (<.001)
Experiment 4	.71 (<.001)		

Furthermore, to test for conflict detection, we then analyze response times (Experiment 1 to 4; see Appendix B, Table A2), mean base-rate reviewing (only Experiment 1), and response confidence (only Experiment 4) separately for standard and rule-inadequate problems. If increased response doubt occurs, not only for standard, but also for rule-inadequate problems, this will be an indication of sensitivity to a spurious conflict between response outputs based on stereotypical descriptions and opposing (but irrelevant) base rates (see Table 2 for a summary of the conflict detection results).

Table 2. Evidence of conflict detection for standard and rule-inadequate problems (Experiments 1 to 4).

	Response Time Higher than Response Time for Problems without Conflict		Confidence in Responses Lower than for Responses to Problems without Conflict (only Experiment 4)		Base-Rate Reviewing Time (only Experiment 1)	
	Responses Based on Base Rate	Responses Based on Stereotypes	Responses Based on Base Rate	Responses Based on Stereotypes	Responses Based on Base Rate	Responses Based on Stereotypes
Experiment 1						
Standard	X	X				
Rule-inadequate					X	
Experiment 2						
Standard	X	X				
Rule-inadequate	X	X				
Experiment 3						
Standard	X	X				
Rule-inadequate	X	X				
Experiment 4						
Standard	X	X	X	X		
Rule-inadequate	X	*	X	X		

* Stereotype-based responses were slower than responses to no-conflict trials only when information regarding base rates was presented last.

2.2.1. Base-Rate Responses

Figure 1 presents the mean proportion of responses according to the base rates in no-conflict problems, and standard and rule-inadequate versions of conflict problems.

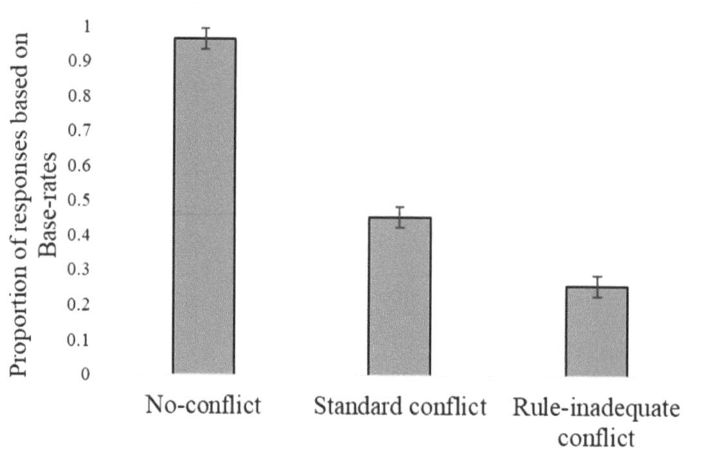

Figure 1. Proportion of correct responses to no-conflict problems[4] and base-rate responses to the standard and rule-inadequate versions of conflict problems (Experiment 1).

Differences in responses across problems were significant ($F(2, 170) = 169.36, p < .001, \eta_p^2 = .67$). Almost all responses were according to the base rates (and stereotype-based information) in no-conflict problems ($M = .97$; $SE = .03$). About half of the responses to conflict problems ($M = .46$; $SE = .03$) and one-fourth of the responses to rule-inadequate problems ($M = .26$; $SE = .03$) were according to base rates. This greater reliance on valid (compared to invalid) base rates may indicate some sensitivity to the quality of the base-rate information at the aggregate level of analysis.

However, and as predicted, the correlation between standard and rule-inadequate problems was positive ($r = .46, p < .001$), such that the more participants decided according to base rates in standard problems, the more they did so in rule-inadequate problems. This suggests an overgeneralization bias in the use of base-rate information.

2.2.2. Base-Rate Reading Time

The initial base-rate reading time (i.e., the time people initially spent reading the first part of the problem before the description was presented) did not vary for the three types of decisions ($F < 1$), which indicates that the initial presentation of the base-rate information was not processed differently across problem versions.

2.2.3. Response Time for Standard Problems

A repeated-measures ANOVA on response times, for the correct responses to no-conflict problems, and stereotype-based responses and base-rate responses to conflict problems as a within-participants factor, revealed a significant effect ($F(2, 86) = 6.23$, $p = .002$, $\eta_p^2 = .12$). As expected, base-rate responses for conflict problems ($M = 12{,}386.22$, $SE = 839.41$) were slower than base-rate responses to no-conflict problems ($M = 9615.04$, $SE = 839.41$; $F(1, 43) = 10.96$, $p = .001$, $\eta_p^2 = .20$).

Stereotype responses for conflict problems ($M = 12{,}052.92$, $SE = 839.41$) were also slower than stereotype responses to no-conflict problems ($F(1, 43) = 10.46$, $p = .002$, $\eta_p^2 = .19$), providing evidence consistent with conflict detection for participants who gave the heuristic response (see Appendix B, Table A2).

2.2.4. Response Time for Rule-Inadequate Problems

A repeated measure ANOVA on response times, with correct responses to no-conflict problems, and stereotype responses and base-rate responses to rule-inadequate problems as a within-participants factor, did not yield a significant effect ($F(2, 66) = 1.91$, $p = .157$, $\eta_p^2 = .05$). Likewise, the tendency for longer times in base-rate responses to conflict problems ($M = 11{,}995.65$, $SE = 857.88$) than to correct responses to no-conflict problems ($M = 10{,}117.63$, $SE = 857.88$) did not reach conventional levels of statistical significance ($F(1, 33) = 3.89$, $p = .057$, $\eta_p^2 = .10$). Finally, the difference between stereotype-based responses to conflict problems ($M = 10{,}987.56$, $SE = 857.88$) and no-conflict problems was not significant ($F(1, 33) = 1.84$, $p = .184$, $\eta_p^2 = .05$) (see Appendix B, Table A2).

2.2.5. Base-Rate Reviewing for Standard Problems

A repeated-measures ANOVA, with an average number of problems reviewed for correct responses to no-conflict problems and base-rate and stereotype-based responses to standard conflict problems, did not reach significance ($F(2, 90) = 2.59$, $p = .080$, $\eta_p^2 = .05$). The reviewing tendency of base-rate information was higher for base-rate responses to conflict problems ($M = .43$, $SE = .06$) than for correct responses to no-conflict problems ($M = .33$, $SE = .49$) but was also non-significant ($F(1, 45) = 3.35$, $p = .074$, $\eta_p^2 = .06$). There was no difference in reviewing between stereotype-based responses to conflict problems ($M = .31$, $SE = .05$) compared to correct responses to no-conflict problems ($F < 1$).

2.2.6. Base-Rate Reviewing Rule-Inadequate Problems

An equivalent ANOVA, performed with rule-inadequate problems, produced a significant effect ($F(2, 68) = 5.41$, $p = .006$, $\eta_p^2 = .13$). Base-rate responses to conflict problems ($M = .45$, $SE = .08$) had more reviews than correct responses to no-conflict problems ($M = .36$, $SE = .06$; $F(1, 34) = 4.38$, $p = .043$, $\eta_p^2 = .11$).

Reviews of rule-inadequate problems with stereotype responses ($M = .26$, $SE = .04$) did not significantly differ from reviews of no-conflict problems ($F(1, 34) = 2.32$, $p = .136$, $\eta_p^2 = .06$).

2.3. Discussion

A larger number of participants used base rates when they were a valid source of information (standard problems) than when they were not (rule-inadequate problems), suggesting some discrimination between relevant and irrelevant base-rate information. Nevertheless, the mean frequency of responses in the latter category (26%) is non-negligible. Importantly, the positive correlation between base-rate responses to standard and rule-inadequate problems is congruent with an overgeneralization bias in the use of base-rate information. Indeed, those participants who relied more on base rates to respond to standard problems also relied more on base rates to respond to rule-inadequate problems.

In addition, response latencies (but not review frequency) provided some indication of conflict detection and possible engagement of T2 reasoning for responses to standard problems, whereas review frequency (but not response latencies) provided some indication of conflict detection and possible engagement of T2 for responses to rule-inadequate problems (see Table 2).

Taken together, these initial findings provide preliminary indications of an overgeneralization bias in the use of base rates and may be seen as questioning the extent to which base-rate responses to the base-rate problems rely on the proper use of the statistical information (as is often assumed).

After having established the basic phenomenon, we sought to increase the number of observations that we could make per person, as well as to potentially reduce the large variability in response time that arose because of the lengthy text passages that we used. Indeed, response times in classic base-rate problems tend to be quite noisy, with the mean RT range varying between 3.4 s and 21.80 s. This variability may have made it difficult to detect the subtle conflict effects.

Furthermore, standard no-conflict problems were used as a baseline for standard conflict and rule-inadequate problems. A better baseline for the latter would be no-conflict versions of rule-inadequate problems, which were included in the following experiments.

3. Experiment 2

Experiments 2 used a rapid-response base-rate paradigm (Pennycook et al. 2015). In this paradigm, base rates and stereotype-based information are presented in a simplified manner in a fast-paced sequence of computer screens.

On the first screen, the two groups involved in the problem were presented (e.g., "politicians and nannies"); the second screen presented an attribute (e.g., "kind") pertaining to the target subject who was randomly drawn from the total sample of the two groups (e.g., "politicians and nannies"); the third screen presented the groups' composition (e.g., "995 politicians and 5 nannies"); and the fourth screen presented the question and response options (e.g., "is the target person more likely to be a politician or a nanny?").

This paradigm has been shown to be sensitive to conflict detection and subtle increases in T2 thinking (Pennycook et al. 2015) and allows us to substantially increase the number of trials (base-rate problems) per participant.

Furthermore, whereas Experiment 1 used three types of problems (i.e., conflict and no-conflict versions of standard problems and conflict versions of rule-inadequate problems), this experiment and the following (i.e., Experiments 3 and 4) also included no-conflict versions of rule-inadequate problems. No-conflict rule-inadequate problems have the same structure as the conflict rule-inadequate problems (i.e., inclusion of additional information that makes the initial base rates irrelevant). However, the initial base rates are aligned with the stereotype-based information. For an outline of these different kinds of problems (standard and rule-inadequate versions of conflict and no-conflict problems), see Table 3. For examples of the problems, see Appendix B, Figure A1.

Table 3. Information displayed in conflict/no-conflict and standard/rule-inadequate problems in the base-rate neglect tasks of the present studies.

	For a Person Randomly Selected from the Sample with the Trait "Kind":	
	Base Rate of the Groups in the Study	**Base Rate from Which a Person Was Randomly Selected**
Standard conflict	995 politicians and 5 nannies	All politicians and nannies
Rule-inadequate conflict	995 politicians and 5 nannies	An equal sample of politicians and nannies/5 politicians and 5 nannies *
Standard no-conflict	5 politicians and 995 nannies	All politicians and nannies
Rule-inadequate no-conflict	5 politicians and 995 nannies	An equal sample of politicians and nannies/5 politicians and 5 nannies *

* Format used in Experiment 4.

As in Experiment 1, we predict a positive correlation between base-rate responses for standard and rule-inadequate problem versions (i.e., the more participants give base-rate responses to standard problems, the more they overgeneralize the use of base rates to cases where the base rates are invalid). An unwarranted reliance on irrelevant base rates should also be apparent if base-rate responses to rule-inadequate problems are coupled with longer response times (when compared to no-conflict problems), suggesting the detection of a spurious conflict.

Finally, in Experiments 2 to 4, the CRT was included as a measure of participants' analytical skills. Performance in the CRT should predict a more appropriate use of base-rate information.

3.1. Method

3.1.1. Participants

One hundred-sixteen students from the University of Heidelberg (89 female) performed the experiment for course credits. Sensitivity analysis with this sample size (at $\alpha = .05$, and power = .80) showed that the experimental design could reliably detect medium or larger effect sizes ($f \geq .31$).

3.1.2. Material

Sixty-six pairs of social groups, with opposite stereotype personality traits associated with each group pair, were obtained through pre-testing. In the pre-test, an independent sample of 40 students was presented with a sample of groups (including professions and sociodemographic categories) and a sample of personality traits[5]. Participants were asked to indicate the two traits most stereotypical of each group. The groups were subsequently paired, and two personality traits were selected so that one trait was frequently associated with one group and rarely (if ever) associated with the other, and vice versa for the other selected trait. Twenty-two pairs were presented in the rule-inadequate version and forty-four pairs were presented in the standard version[6]. In each version, half were conflict trials, and the other half were no-conflict trials. Content was counterbalanced across problem type by creating two lists of trials that varied only in the way the base rates were presented, so that pairs of a conflict type of problem in one list were of a no-conflict type of problem in the other, and vice versa. Participants were randomly presented with one of the two lists of problems.

3.1.3. Procedure

The procedure was based on Pennycook et al.'s (2015) rapid-response base-rate task, with minor adjustments to accommodate the inclusion of rule-inadequate versions of both no-conflict and conflict problems. Specifically, the base rates were qualified by further information referring to how many people from each of the two groups (i.e., initial base rates) were actually tested: the whole sample (in the case of standard trials) or a subsample composed of an equal number of individuals from each of the two groups (in the case of rule-inadequate trials). Participants were also informed that the individual described in each trial was randomly selected from the tested sample.

Participants began by reading the following instructions taken from Pennycook et al. (2015; text in italics corresponds to additional sentences introduced to accommodate the use of rule-inadequate problems):

> In a big research project, a large number of studies were carried out where short personality descriptions of the participants were made. In every study, there were participants from two population groups (e.g., carpenters and policemen). In each study, one participant was drawn at random from *a sample of tested participants*. In each trial, you will be presented with a fixation dot where you should look at. After a few seconds, this dot will be replaced by a personality trait for the randomly chosen participant and finally by some information about the composition of the population groups *and how many participants from each group were tested in the study in question*. After that, you will be asked to indicate to which population group the participant most likely belongs. Please answer the problems as quickly and accurately as possible. Once you have made up your mind, you must enter your answer ('a' or 'b') immediately and then the next problem will be presented.

Following Pennycook et al. (2015), each trial began with the presentation of a fixation point (500 ms), followed by (a) the information concerning the groups involved in the trial (2000 ms), (b) the number of individuals actually tested (2000 ms), (c) the trait describing one of the individuals tested (2000 ms), and (d) the groups' base rates (2000 ms). Participants were then asked to which group the individual most likely belonged (see Figure 2).

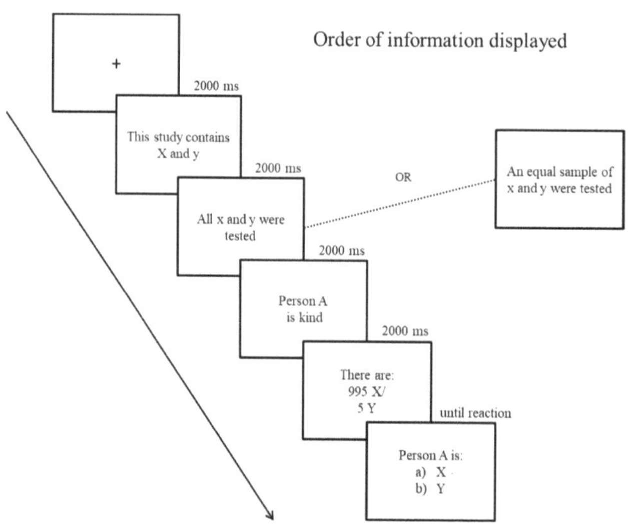

Figure 2. Procedure for rapid-response base-rate task (Experiment 2).

Participants began by responding to three practice trials, after which they responded to the block of sixty-six trials (each trial corresponding to a different group pair). After the experimental task, participants responded to the CRT (see Appendix A).

Dependent measures included base-rate responses for each type of problem, as well as response time (RT) of correct responses to no-conflict problems and of stereotype-based and base-rate responses to conflict problems.

3.2. Results

3.2.1. Base-Rate Responses

Three participants with accuracies below .80 in the no-conflict standard trials were removed from the analysis (see Pennycook et al. 2015). The mean proportion of base-rate responses for conflict and no-conflict problems for standard and rule-inadequate problems is presented in Figure 3.

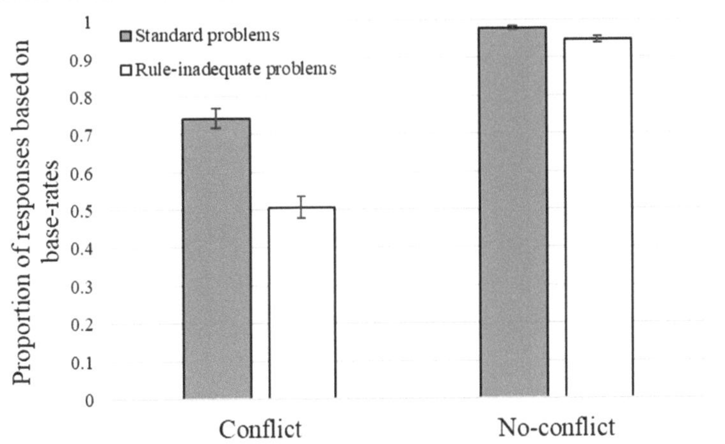

Figure 3. Proportion of correct responses to no-conflict problems and base-rate responses to the standard and rule-inadequate versions of conflict problems in Experiment 2.

A 2X2 ANOVA, with trial version (standard and rule-inadequate) and trial type (conflict and no-conflict) as within-participant factors, and responses based on base rates as the dependent variable, showed a significant main effect of version ($F(1, 112) = 85.65$, $p < .001$, $\eta_p^2 = .43$), with a higher frequency of base-rate responses for standard problems ($M = .86$, $SE = .02$) than for rule-inadequate ones ($M = .73$, $SE = .02$). A significant main effect of trial type was also observed ($F(1, 112) = 211.78$, $p < .001$, $\eta_p^2 = .65$), with a higher proportion of base-rate responses for no-conflict problems ($M = .96$, $SE = .02$) than conflict ones ($M = .62$, $SE = .02$). Additionally, there was a significant interaction between the factors ($F(1, 112) = 50.75$, $p < .001$, $\eta_p^2 = .31$), indicating that the difference between standard and rule-inadequate problems was larger for conflict problems ($M = .74$, $SE = .03$ for standard problems, and $M = .51$, $SE = .03$ for rule-inadequate problems) than for no-conflict problems ($M = .98$, $SE < .01$ for standard problems, and $M = .95$, $SE = .01$ for rule-inadequate problems (see Appendix B, Table A1).

Overall, the greater reliance on valid (compared to invalid) base rates indicates some sensitivity to the quality of the base-rate information at the aggregate level of analysis.

However, there was a large positive correlation between standard and rule-inadequate trial versions ($r = 50$; see Table 1), indicating that the more participants decided according to base rates in standard trials, the more they did so in rule-inadequate trials, suggesting an overgeneralization bias in the use of base-rate information.

Furthermore, in order to verify whether more reflective participants were better able to discriminate between trials where base rates were a relevant source of information (standard trials) and trials where they were not (rule-inadequate trials), participants were divided in two subgroups based on their CRT performance.

The low-reflective reasoners included participants who gave one or more incorrect responses to the CRT ($N = 70$) and the high-reflective reasoners included participants who responded correctly to all three CRT problems ($N = 43$).

The rationale for this partition was that those participants who respond to all problems correctly were inhibiting and replacing the highly appealing intuitive responses with the analytical correct responses in a consistent way (whereas the remaining participants did it inconsistently or not at all across the three CRT problems). As such, these are the participants with better chances to discriminate between standard and rule-inadequate base-rate problems (i.e., whether or not to inhibit the stereotype-based response).

For low-reflective reasoners, we found a strong positive correlation between base-rate responses to conflict standard and rule-inadequate trials ($r = .70$), whereas for high-reflective reasoners there was no correlation ($r = .05$). High-reflective participants showed a higher frequency of base-rate responses for standard conflict problems compared to rule-inadequate conflict problems (difference of .30) than low-reflective participants (.19). See Table 4 for details.

Table 4. Means and standard deviations of base-rate responses for different types and versions of problems for low- and high-reflective reasoners in Experiments 2 and 3.

		Low-Reflective Reasoners (N = 70)		**High-Reflective Reasoners (N = 43)**	
Type of Problem	**Version of Problem**	*M*	*SD*	*M*	*SD*
Experiment 2					
No-conflict	Standard	.98	.04	.98	.03
	Rule-inadequate	.94	.10	.97	.07
Conflict	Standard	.68	.29	.84	.20
	Rule-inadequate	.49	.32	.54	.31
Difference between conflict problems		.19		.30	
Experiment 3					
No-conflict	Standard	.97	.06	.98	.07
	Rule-inadequate	.95	.08	.94	.10
Conflict	Standard	.60	.33	.74	.27
	Rule-inadequate	.34	.32	.23	.25
Difference between conflict problems		.26		.51	

3.2.2. Response Times

Response times (converted to $\log_{10}$) were analyzed separately for standard and rule-inadequate trials (see Appendix B, Table A2 for means in ms).

Standard trials. A repeated-measures one-way ANOVA, with base-rate responses to no-conflict trials, and base-rate and stereotype-based responses to conflict trials, yielded a significant main effect ($F(2, 166) = 41.94$, $p < .001$, $\eta_p^2 = .34$). Planned comparisons showed that the RT for no-conflict trials ($M = 2.96$, $SE = .02$) was faster than base-rate responses to standard conflict trials ($M = 3.08$, $SE = .03$; $F(1, 83) = 40.18$, $p < .001$, $\eta_p^2 = .33$) and faster than stereotype-based responses to standard conflict trials ($M = 3.17$, $SE = .03$; $F(1, 83) = 86.40$, $p < .001$, $\eta_p^2 = .51$).

This pattern of results replicates previous findings by Pennycook et al. (2015), suggesting that responding according to base rates involved the engagement in time-consuming, deliberate reasoning. Furthermore, stereotype-based responses to conflict trials were also slower than to no-conflict trials, providing evidence consistent with conflict detection (between stereotype and base-rate information) for participants who ended up giving the heuristic response.

Rule-inadequate trials. A repeated-measures ANOVA, with base-rate responses for no-conflict trials, and base-rate and stereotype-based responses for conflict trials, yielded a significant main effect ($F(2, 186) = 13.54$, $p < .001$, $\eta_p^2 = .13$). Planned comparisons showed that responses to no-conflict trials ($M = 3.02$, $SE = .03$) were faster than base-rate responses to rule-inadequate conflict trials ($M = 3.16$, $SE = .04$; $F(1, 93) = 24.78$, $p < .001$, $\eta_p^2 = .21$) and faster than stereotype-based responses to rule-inadequate conflict trials ($M = 3.14$, $SE = .03$; $F(1, 93) = 20.01$, $p < .001$, $\eta_p^2 = .18$).

In other words, the response-time pattern that was obtained for the standard trials was also obtained for rule-inadequate trials. This suggests that responding according to invalid base rates also involved conflict detection and possible engagement in T2 thinking. Furthermore, there is also indication that participants detected a conflict between traits and opposite (but invalid) base-rate information, even when they opted for the heuristic response.

3.3. Discussion

The number of participants who responded according to the base rates when they were a valid source of information (standard problems) was larger than when the base rates were made irrelevant (rule-inadequate problems). This suggests some discrimination between relevant and irrelevant base-rate information. However, the mean frequency of responses in the latter category (51%) is considerable. In addition, the positive correlation between base-rate responses to standard and rule-inadequate problems is congruent with an overgeneralization bias in the use of base-rate information.

The positive correlation in the tendency to respond according to base rates observed between standard and rule-inadequate conflict problems was, however, weaker for highly reflective participants compared to less reflective participants. This could suggest individual differences in the expected direction concerning the relation between rationality (as measured by the CRT) and the discriminate use of statistical information.

Response-time analysis of standard problems provided evidence of conflict detection, both when participants gave the stereotype-based response and when they responded according to the base rates. The same results pattern emerged for rule-inadequate problems, where there is no real conflict between statistical information and stereotype-based information. Taken together, these results suggest that participants failed to discriminate between real and merely apparent (or spurious) conflict and misused statistical information to draw conclusions based on irrelevant base rates.

4. Experiment 3

Experiment 2 presented the group compositions last, as in the original studies with this rapid-response paradigm (Pennycook et al. 2015, Experiments 1 to 3).

However, previous research has shown that presenting a piece of information last, just before judgment, increases the likelihood of its use (Krosnick et al. 1990; Pennycook et al. 2015, Experiment 4). It is thus possible that including the subsample (vs. the group composition) information as the last piece of information prior to judgment would increase the adequate use of base rates. In Experiments 3 and 4, the information about the subsample actually tested and the group compositions was presented in the last two screens (before the screen with the question) and its order of presentation was manipulated.

4.1. Method

4.1.1. Participants

Eighty-six students from the University of Heidelberg (71 female) completed the experiment in exchange for course credit. Sensitivity analysis with this sample size (at $\alpha = .05$, and power = .80) showed that the experimental design could reliably detect medium or larger effect sizes ($f \geq .36$).

4.1.2. Material

Forty-four pairs of social groups were selected from the material used in Experiment 2, with opposite stereotypical traits associated with each group pair. Half of the pairs were assigned to standard problems and the other half to rule-inadequate problems. Both kinds of problems had conflict and no-conflict versions (obtained by changing the personality trait associated with the trial).

4.1.3. Procedure

The procedure was the same of Experiment 2. However, the order of presentation of the base rates for each trial and the information referring to how many people from each of the two groups were actually tested—the whole sample (in the case of standard base-rate problems) or a subsample composed of an equal number of individuals from each of the two groups (in the case of rule-inadequate problems)—was manipulated. Participants in the present study either saw the base rates followed by the subsample of individuals who were actually tested or the other way around (subsample information followed by the base rates from where the subsample was taken) (see Figure 4).

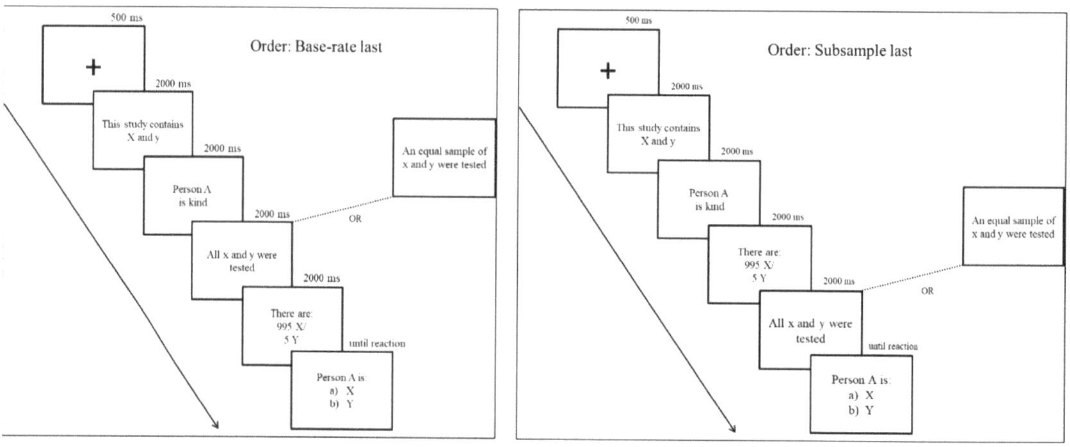

Figure 4. Procedure for rapid-response base-rate task (Experiment 3).

Participants began by responding to three practice trials, after which they responded to 2 blocks of 44 trials (each trial corresponding to a different group pair). Between blocks, the pairs were the same, but the trait presented for each pair varied so that a conflict problem in one block would be a no-conflict problem in the other block and vice versa. After the experimental task, participants responded to the CRT.

4.2. Results

4.2.1. Base-Rate Responses

One participant with an accuracy below .80 in the no-conflict standard problems was removed from the analysis. The mean proportion of base-rate responses for conflict and no-conflict problems for standard and rule-inadequate problems is presented in Figure 5.

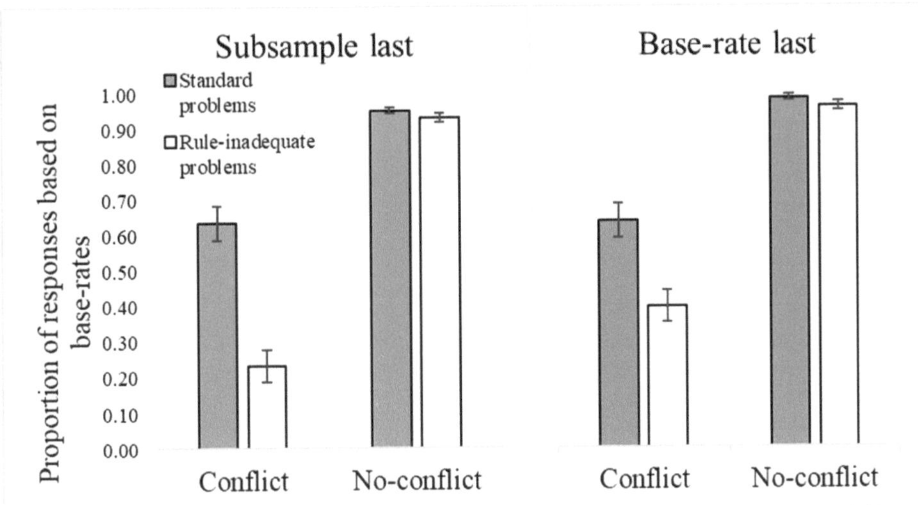

Figure 5. Proportion of correct responses to no-conflict problems and base-rate responses to the standard and rule-inadequate versions of conflict problems in both orders of information presentation (Experiment 3).

A 2X2X2 ANOVA, with problem version (standard and rule-inadequate problems) and trial type (conflict and no-conflict) as within-participant factors, with order (subsample last; base rate last) as the between-participant factor, and with the mean proportion of base-rate responses as the dependent variable, showed a main effect of trial type, with a higher proportion of base-rate responses for no-conflict problems (M = .96, SE = .01) than conflict ones (M = .48, SE = .03; $F(1, 83) = 261.09$, $p < .001$, $\eta_p^2 = .76$) and a main effect of version ($F(1, 83) = 112.22$, $p < .001$, $\eta_p^2 = .58$), with a higher proportion of base-rate responses to standard problems (M = .80, SE = .02) than to rule-inadequate problems (M = .63, SE = .02). There was a version x order interaction ($F(1, 83) = 5.86$, $p = .018$, $\eta_p^2 = .07$). While in standard problems, the frequency of base-rate responses did not differ significantly between different orders of information ($F < 1$) for rule-inadequate problems, the frequency of base-rate responses was lower when the subsample information was presented last (M = .44, SE = .04) compared to when base rates were presented last (M = .52, SE = .04; $F(1, 83) = 8.68$, $p = .004$, $\eta_p^2 = .09$). There was also a version x type interaction ($F(1, 83) = 84.82$, $p < .001$, $\eta_p^2 = .51$). The difference in base-rate responses between standard (M = .64, SE = .03) and rule-inadequate problems (M = .32, SE = .03) was higher for conflict problems ($F(1, 83) = 104.84$, $p < .001$, $\eta_p^2 = .56$) than for no-conflict problems (M = .97, SE = .01 for standard problems, and M = .95, SE = .01 for rule-inadequate problems; $F(1, 83) = 7.70$, $p = .007$, $\eta_p^2 = .08$). Finally, there was a 3-way interaction between version, type, and order ($F(1, 83) = 6.47$, $p = .013$, $\eta_p^2 = .07$). Planned comparisons showed that the proportion of base-rate responses in conflict problems was lower when the subsample information was presented last for rule-inadequate problems ($F(1, 83) = 6.62$, $p = .012$, $\eta_p^2 = .07$), but not for standard problems ($F < 1$). Regarding no-conflict problems, the difference in base-rate responses was also lower when the subsample information was presented last for standard problems ($F(1, 83) = 6.13$, $p = .015$, $\eta_p^2 = .07$), but not for rule-inadequate problems ($F(1, 83) = 2.55$, $p = .114$, $\eta_p^2 = .03$) (see Appendix B, Table A1).

In sum, the analysis of base-rate responses replicated that of Experiment 2. Furthermore, when the subsample of participants actually tested was the last information presented, it reduced participants' reliance on irrelevant base rates (in rule-inadequate problems).

Regardless, the proportion of base-rate responses to standard and rule-inadequate conflict problems were positively and strongly correlated ($r = .53$, $p < .001$), indicating an overgeneralization bias (see Table 1). The same positive correlation was observed for participants in the subsample-last condition ($r = .40$, $p = .008$) and in the base-rate last condition ($r = .65$, $p < .001$).

In order to verify whether more reflective participants were better able to discriminate between trials where base rates were a relevant source of information (standard problems) and trials where they were not (rule-inadequate problems), participants were divided in two subgroups based on their CRT performance. The low-reflective thinking group included participants who gave one or more incorrect responses to the CRT ($N = 63$) and the high-reflective thinking group included participants who responded correctly to all three CRT problems ($N = 23$).

For low-reflective participants, there was a positive and large correlation ($r = .65$, $p < .001$) in the proportion of base-rate responses for standard and rule-inadequate conflict problems, whereas for high-reflective participants, this correlation was smaller and not statistically significant ($r = .32$, $p = .138$) (see Table 1). Hence, high-reflective thinking (as measured by the CRT) seems to discriminate slightly better between trials where base rates should (standard problems) and should not be used (rule-inadequate problems). Furthermore, when comparing base-rate responses between standard and rule-inadequate problems, high-reflective participants showed a higher frequency of base-rate responses for standard conflict problems compared to rule-inadequate conflict problems (difference of .51) than low-reflective participants (.26). See Table 4 for details.

4.2.2. Response Times

Response times (converted to $\log_{10}$) were analyzed separately for each kind of problem (see Appendix B, Table A2 for means in ms).

Standard Problems. A 3X2 ANOVA, with response (correct responses to no-conflict trials, base-rate or stereotype-based responses to conflict trials) as within-participants factor and order (subsamples last; base rates last) as between-participants factor, yielded just a main effect of response ($F(2, 136) = 24.54$, $p < .001$, $\eta_p^2 = .26$). Planned comparisons showed that responses to no-conflict trials ($M = 2.974$, $SE = .03$) were faster than base-rate responses to conflict trials ($M = 3.06$, $SE = .03$; $F(1, 68) = 12.62$, $p < .001$, $\eta_p^2 = .16$) and faster than stereotype-based responses to conflict trials ($M = 3.19$, $SE = .04$; $F(1, 68) = 38.70$, $p < .001$, $\eta_p^2 = .36$).

Rule-inadequate Problems. A 3X2 ANOVA, with response (correct responses to no-conflict trials, base-rate or stereotype-based responses to conflict trials) as within-participants factor and order (subsamples last; base rates last) as between-participants factor, yielded just a main effect of response ($F(2, 140) = 13.19$, $p < .001$, $\eta_p^2 = .16$). Planned comparisons showed that responses to no-conflict trials ($M = 2.99$, $SE = .03$) had shorter RTs than base-rate responses to conflict trials ($M = 3.14$, $SE = .04$; $F(1, 70) = 25.74$, $p < .001$, $\eta_p^2 = .27$) and stereotype-based responses to conflict trials ($M = 3.07$, $SE = .03$; $F(1, 70) = 12.30$, $p < .001$, $\eta_p^2 = .15$).

4.3. Discussion

In Experiment 3, we found differences in the proportion of base-rate responses as a function of the presentation order of the base rates. Specifically, presenting the crucial information about the number of individuals actually tested after the base-rate information and just before participants were prompted to give their response increased the likelihood of its use and thus helped reduced the reliance on irrelevant base rates. In contrast, there were no changes in the proportion of base-rate responses as a function of order of presentation for standard problems. This is also expected since, in these problems, the last two pieces of information converge in establishing the initial group composition as the problems' relevant base rates (it makes no difference to say "all were tested/from a group of X and Y" vs. "from a group of X and Y/all were tested).

Still, the positive correlation observed in the previous studies between the proportion of base-rate responses to standard and rule-inadequate trials was replicated for participants in the different information orders, which is congruent with an overgeneralization bias. In addition, highly reflective participants showed this tendency to a lesser degree, which suggests individual differences in the expected direction concerning the relation between rationality (as measured by the CRT) and the discriminate use of statistical information.

As in Experiment 2, response-time analysis of standard trials provided evidence of conflict detection, both when participants gave the stereotype-based responses and when they responded according to base rates. Interestingly, the same pattern of results emerged for rule-inadequate problems. Participants detected a conflict between stereotype-based and base-rate information, even when base rates were made irrelevant to respond to the problem. In other words, conflict detection does not seem to discriminate between valid and irrelevant statistical information.

5. Experiment 4

Experiment 4 was designed to replicate and extend the results of Experiments 2 and 3, with the following modifications. First, to make sure that all participants understood the group compositions of the people actually tested in the rule-inadequate trials, the numbers of Xs and Ys in the subsample were explicitly stated (e.g., "in the first day of the study only 5 politicians and 5 nannies were interviewed"), thus overcoming any potential ambiguity of the formulations used in the previous experiments.

Furthermore, a measure of confidence was added as an additional indicator of conflict detection. After responding to each trial, participants were asked to express the degree to which they were confident in their response. If the opposition between stereotype-based information (traits) and base rates was detected, then the confidence in responses to conflict trials should be lower than the confidence in responses to no-conflict trials (e.g., Bago and De Neys 2017).

5.1. Method

5.1.1. Participants

Sixty-four participants from the University of Lisbon (54 females) performed the experiment in exchange for course credit. Sensitivity analysis with this sample size (at $\alpha = .05$, and power = .80) showed that the experimental design could reliably detect medium or larger effect sizes ($f \geq .42$).

5.1.2. Material

The same 44 pairs of social groups from the previous experiment were used (translated to Portuguese).

5.1.3. Procedure

The procedure and design were the same as in Experiment 3, except that (a) the distinction between standard and rule-inadequate problems was created by the inclusion of one screen that established whether the target person in the problem was randomly selected from the total sample described in the group composition information screen ("in the first day of the study, all politicians and nannies were interviewed"), or rather from a subset of the sample where base rates are equal ("in the first day of the study only 5 politicians and 5 nannies were interviewed"); (b) after each trial, participants had to indicate how confident they were in their responses on a 9-point rating scale (from 1—not at all confident; to 9—totally confident); and (c) the final CRT task had one extra problem (see Appendix A).

5.2. Results

Six participants who responded to standard no-conflict problems with accuracies below .80 were removed from the analysis. The mean proportion of base-rate responses for conflict and no-conflict problems for standard and rule-inadequate problems is presented in Figure 6.

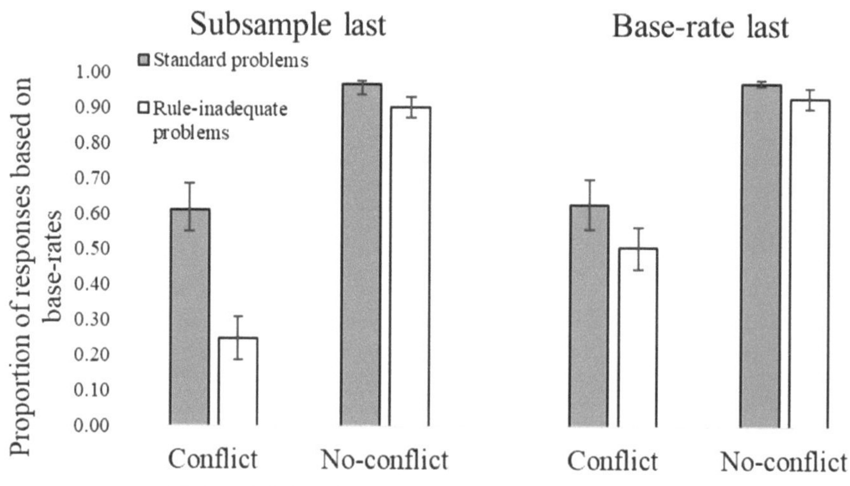

Figure 6. Proportion of correct responses to no-conflict problems and base-rate responses to the standard and rule-inadequate versions of conflict problems in both orders of information presentation (Experiment 4).

5.2.1. Base-Rate Responses

A 2X2X2 ANOVA, with problem Version (standard and rule-inadequate problems) and problem type (conflict and no-conflict) as within-participant factors, order (subsample last; base rate last) as between-participant factor, and mean proportion of base-rate responses as the dependent variable, showed a main effect of version ($F(1, 56) = 46.91$, $p < .001$, $\eta_p^2 = .46$), such that standard problems led to a higher proportion of base-rate responses ($M = .79$, $SE = .03$) than rule-inadequate problems ($M = .65$, $SE = .02$). There was also a main effect of type, with more responses according to base rates to no-conflict trials ($M = .94$, $SE = .01$) than conflict trials ($M = .50$, $SE = .04$; $F(1, 58) = 87.98$, $p < .001$, $\eta_p^2 = .61$). Additionally, there was an interaction between version and order ($F(1, 56) = 9.17$, $p = .004$, $\eta_p^2 = .14$). While the proportion of base-rate responses was lower for rule-inadequate problems ($M = .58$, $SE = .03$) than for standard problems ($M = .79$, $SE = .04$) when the subsample information was presented last ($F(1, 56) = 9.08$, $p = .004$, $\eta_p^2 = .14$), there were no differences in base-rate responses for standard problems as a function of order ($F < 1$). Furthermore, there was also a version x type interaction ($F(1, 56) = 34.01$, $p < .001$, $\eta_p^2 = .38$). The difference in proportion of base-rate responses between rule-inadequate and standard problems was significant for both types of problems, but was higher in conflict problems ($M = .62$, $SE = .05$ for standard problems, and $M = .38$, $SE = .04$ for rule-inadequate problems; $F(1, 56) = 52.22$, $p < .001$, $\eta_p^2 = .48$) than in no-conflict problems ($M = .97$, $SE = .01$ for standard problems and $M = .92$, $SE = .02$ for rule-inadequate problems; $F(1, 56) = 8.37$. $p = .005$, $\eta_p^2 = .13$). Finally, an interaction between Version, Type, and Order ($F(1, 56) = 11.37$, $p = .001$, $\eta_p^2 = .17$) showed that the proportion of base-rate responses to conflict rule-inadequate trials was lower when the subsample information was presented last ($F(1, 56) = 8.81$, $p = .004$, $\eta_p^2 = .14$), but there were no differences in base-rate responses for standard conflict trials ($F < 1$) (see Appendix B, Table A1).

Summing up, the analysis of base-rate responses replicated that of the previous experiments. As in Experiment 3, when the subsample of participants actually tested was the last information presented, participants relied less on irrelevant base rates (in rule-inadequate problems).

Notably, however, the proportion of base-rate responses to standard and rule-inadequate trials showed a positive and large correlation ($r = .71$, $p < .001$), suggesting an overgeneralization bias (see Table 1). The same positive correlation was observed for participants in the subsample-last condition ($r = .58$, $p = .001$) and in the base-rates last condition ($r = .86$, $p < .001$).

Unfortunately, it was not possible to compare subgroups of high and low cognitive reflection because thirty-three participants (56.90%) erred all CRT problems and only two answered accurately to all problems.

5.2.2. Response Times

RTs (converted to log^{10}) were analyzed separately for standard and rule-inadequate problems (see Appendix B, Table A2 for means in ms).

Standard Trials. A 3X2 ANOVA on the mean RT of trials, with response (base-rate response in no-conflict problems, base-rate or stereotype-based in conflict problems) as within-participants factor and order (base rates last, subsamples last) as between-participants factor, yielded a main effect of response ($F(2, 90) = 7.29$, $p = .001$, $\eta_p^2 = .14$). Responses to no-conflict problems ($M = 2.99$, $SE = .02$) were faster than stereotype-based responses to conflict problems ($M = 3.14$, $SE = .04$; $F(1, 45) = 21.57$, $p < .001$ $\eta_p^2 = .32$) and were faster than base-rate responses to conflict problems ($M = 3.05$, $SE = .33$), although the difference did not reach statistical significance ($F(1, 45) = 3.47$, $p = .069$, $\eta_p^2 = .07$).

Rule-inadequate Trials. A 3X2 ANOVA on the mean RT of trials, with response (base-rate responses to no-conflict trials, base-rate or stereotype-based responses to conflict trials) as within-participants factor and order (base rates last, subsamples last) as between-participants factor, showed a main effect of response ($F(2, 92) = 6.17$, $p = .003$, $\eta_p^2 = .12$). Planned comparisons showed that response to no-conflict trials ($M = 2.97$, $SE = .02$) had shorter RTs when compared to base-rate responses to conflict trials ($M = 3.07$, $SE = .03$; $F(1, 46) = 14.16$, $p < .001$, $\eta_p^2 = .23$), but not when compared to stereotype-based responses to conflict trials ($M = 3.02$, $SE = .33$; $F(1, 46) = 3.12$, $p = .083$, $\eta_p^2 = .06$). There was also a significant interaction between response and order ($F(2, 92) = 5.70$, $p = .004$, $\eta_p^2 = .11$). Base-rate responses in conflict trials were significantly slower than responses in no-conflict trials both when subsamples were presented last ($F(1, 46) = 7.28$, $p = .009$, $\eta_p^2 = .14$), and when base rates were presented last ($F(1, 46) = 6.89$, $p = .011$, $\eta_p^2 = .13$). Stereotype-based responses in conflict trials, however, were slower than normative responses in no-conflict trials when base rates were presented last ($F(1, 46) = 11.93$, $p = .001$, $\eta_p^2 = .21$), but not when subsamples were presented last ($F(1, 46) = 1.10$, $p = .300$, $\eta_p^2 < .21$).

5.2.3. Confidence

Confidence ratings were analyzed separately for standard and rule-inadequate problems.

Standard Trials. A 3X2 ANOVA on confidence ratings of standard trials, with response (base-rate responses to no-conflict problems, base-rate or stereotype-based responses to conflict problems) as within-participant factor and order (base rate last, subsample last) as between-participants factor, showed only a main effect of response ($F(2, 90) = 13.16$, $p < .001$, $\eta_p^2 = .23$), such that responses to no-conflict trials ($M = 6.88$, $SE = .17$) were given with greater confidence than both base-rate responses ($M = 6.10$, $SE = .20$; $F(1, 45) = 27.37$, $p < .001$, $\eta_p^2 = .38$) and stereotype-based responses ($M = 5.78$, $SE = .30$; $F(1, 45) = 22.88$, $p < .001$, $\eta_p^2 = .34$) to conflict trials.

Rule-inadequate Trials. A 3X2 ANOVA on confidence ratings, with response (base-rate responses to no-conflict problems, base-rate or stereotype-based responses to conflict problems) as within-participant factor and order (base rate last, subsample last) as between-participants factor, showed a main effect of response ($F(2, 92) = 11.36$, $p < .001$, $\eta_p^2 = .20$). Responses to no-conflict trials ($M = 6.74$, $SE = .22$) were given with more confidence than both base-rate responses to conflict trials ($M = 5.82$, $SE = .22$; $F(1, 46) = 25.60$, $p < .001$, $\eta_p^2 = .36$) and stereotype-based responses to conflict trials ($M = 6.23$, $SE = .26$; $F(1, 46) = 7.92$, $p = .007$, $\eta_p^2 = .15$). There was also a response x order interaction ($F(2, 92) = 5.50$, $p = .006$, $\eta_p^2 = .11$). When base-rate information was presented last, confidence in correct responses to no-conflict trials was higher than both base-rate responses to conflict trials ($F(1, 46) = 19.85$, $p < .001$, $\eta_p^2 = .30$) and stereotype-based responses to conflict trials ($F(1, 46) = 19.46$, $p < .001$, $\eta_p^2 = .30$). However, when subsample information was presented last, confidence in correct responses to no-conflict trials was higher than confidence in base-rate responses to conflict trials ($F(1, 46) = 7.10$, $p = .011$, $\eta_p^2 = .13$) but did not differ from confidence in stereotype-based responses to conflict problems ($F < 1$).

5.3. Discussion

The results pattern replicated that of Experiment 3. Presenting the number of individuals tested after the group composition information and just before participants were prompted to give their response seems to have made this crucial information more salient to the participants, improving their performance in rule-inadequate trials. Importantly, the correlation between the proportion of base-rate responses to standard and rule-inadequate trials observed in Experiments 1 to 3 was even more pronounced in Experiment 4 (in terms of Cohen's effect sizes, three of the reported correlations were large, and one—Experiment 1—was medium-to-large).

Response-time and confidence-rating analyses of standard trials provided partial evidence of conflict detection both when participants opted for the stereotype-based response and when they responded according to base rates. A similar pattern of results emerged for rule-inadequate trials.

In sum, these results converge with those of Experiments 1 to 3 and show that participants have difficulties in discriminating between relevant and irrelevant base-rate information. This may lead participants to misuse statistical information and draw conclusions based on irrelevant base rates.

6. General Discussion

In four experiments using two different experimental procedures (moving windows and rapid-response paradigm), we partially replicated the key results of previous research (De Neys and Glumicic 2008; Pennycook et al. 2015). Specifically, we found increased response doubt reflected in terms of longer response latencies, a stronger tendency to revisit the initial base-rate information, and lower confidence ratings (lower feeling of confidence) for responses to conflict versions of standard base-rate problems (when compared to no-conflict problems). This occurred both for base-rate responses and for stereotype-based responses.

These results support the notion that the opposition, in conflict problems, between stereotype-based and statistical information (i.e., base rates) is detected and possibly triggers more deliberate processing.

Interestingly, the results pattern for rule-inadequate problems mirrored the one just described for standard base-rate problems. Stereotype-based and base-rate responses to conflict versions of rule-inadequate problems (where the base rates opposing the stereotype-based information were irrelevant) also took longer and were given with less confidence than responses to no-conflict versions of the same problems. In other words, participants' sensitivity to conflict, as reflected by these measures (response times and confidence ratings), does not seem to discriminate a real conflict between stereotype-based and valid base-rate information from a spurious conflict between stereotype-based information and irrelevant base rates.

This is not to say that participants simply did not distinguish between the two versions of standard and rule-inadequate base-rate problems. In fact, participants responded more often according to base rates when they were a valid source of statistical information (standard problems) than when base rates were irrelevant (rule-inadequate problems). In addition, manipulation of the order of presentation of the information (i.e., presenting the subsample actually used after the base rates and before participants were prompted to respond) further reduced the reliance on base rates on rule-inadequate problems.

However, a large positive correlation was consistently observed between base-rate responses to standard and rule-inadequate versions of the base-rate problems. This correlation suggests a tendency to use statistical information (base rates) even when it ceases to provide relevant statistical information. This overgeneralization bias is attenuated among high-reflective participants (as measured by the CRT), which is congruent with the notion that inaccurate rule-based reasoning can be a critical source of response bias.

6.1. *Inadequate Use of Decision Rules in Reasoning about Everyday Problems*

Rule-inadequate versions of standard base-rate problems were quite useful in disentangling adequate from inadequate use of base rates in responding to these problems.

As aforementioned, we are not the first to resort to this type of research strategy. The rationale for our approach was inspired by previous work by Fong et al. (1986). These authors developed so-called false-alarm versions of reasoning problems opposing statistical information (e.g., large but biased samples) to stereotype-based information, to test the effects of brief statistical training on the ability to use statistical principles such as the law of large numbers. By adding these false-alarm problems to their dependent measures, Fong et al. were able to show that statistical training increased the adequate use of the law of large numbers without leading to its over-application to situations where it was not called for.

Klauer and Singmann (2013) used rule-inadequate versions of syllogistic reasoning problems, which they referred to as "pseudo-syllogisms"[7], to further test Morsanyi and Handley's (2012) argument that logicality of syllogisms can be detected in an intuitive manner via changes in affective state (conceptual fluency). They showed that the surface features of the problems were confounded with their logical status and that these features, rather than an implicit intuitive access to the logically correct solution, drove participants' responses.

Likewise, the rule-inadequate versions of the base-rate problems used in the research here reported make the base-rate information irrelevant for the judgment, while leaving the surface features of the problems largely unchanged (e.g., information concerning group composition is still presented). In this sense, rule-inadequate problems may be useful for determining whether conflict-detection effects are driven by surface features that happen to covary with the normative implications of the problems (i.e., the spurious conflict between invalid base rates and stereotype-based information), or whether the normative implications themselves are causally responsible.

Analogous to the work of Klauer and Singmann (2013) in syllogistic reasoning, the use of what we referred to as rule-inadequate problem version (i.e., false-alarm problems) revealed that measures of conflict detection do not necessarily show participants' sensitivity to normative prescriptions. This may contribute to a better characterization of the interaction between different T1 intuitions and of T2 thinking.

6.2. *Contributions to the Debate among Dual-Process Models of Reasoning*

Two prevalent notions in the dual-process literature of reasoning are (a) the view that T2 thinking is responsible for the application of relevant rules of logic and probability that often correct T1 outputs (e.g., Kahneman 2003); and (b) that even when opting for the heuristic-based response in classic reasoning tasks, people have competing logical intuitions (i.e., intuitive access to logical and probabilistic principles) that point to the normatively correct response in classic reasoning tasks, such as base-rate problems (e.g., De Neys 2012). Previous research (Klauer and Singmann 2013) and our results suggest that both of these notions may need to be qualified.

Indeed, the present findings suggest that a sizeable fraction of those individuals who appear to respond in an adequate fashion (i.e., according to the normative implications of base rates) in standard base-rate problems may often also rely on statistically irrelevant information rather than making an adequate use of formal rules. In the same vein, processing measures such as (longer) response times and (decreased) confidence may point to the existence of faulty logical intuitions (opposing heuristic ones). This might occur if logical intuitions are at least sometimes driven by salient surface features (e.g., extreme base rates opposing stereotype-based information) that happen to co-vary with sound probabilistic judgment in the case of standard base-rate problems but not in the case of rule-inadequate versions of these problems.

In other words, what have been termed logical intuitions (e.g., De Neys 2012) may not always follow valid probabilistic or logic principles. These intuitions may sometimes merely conform to this principle in standard base-rate problems (though, to be sure, some studies offer compelling evidence for logical intuitions—e.g., Frey et al. 2018; Šrol and De Neys 2021).

Inferring the sound use of statistical rules from normatively correct responses to standard conflict problems might therefore be unwarranted when the tendency to use base rates, even when they are not a relevant source of statistical information (overgeneralization bias), is not considered.

Further research is certainly needed to clarify the conceptual implications of the current findings. The two-response procedure developed and validated by Thompson et al. (2011) could be used to better examine the time course of intuitive and deliberate processing in rule-inadequate problems. In this procedure, participants are asked to give two consecutive responses. At time 1, they are encouraged to respond intuitively and are put under a deadline to minimize the role of analytic thinking. They then provide feeling of rightness (FOR), followed by a second answer, during which reflection is encouraged. Two measures of T2 engagement are obtained: the length of the rethinking period and the probability of changing the initial answer. To what extent faulty logical intuitions would emerge at time 1, and to what extent associated FOR would predict T2 engagement and the confirmation or correction of the initial response to rule-inadequate problems, are interesting research questions waiting to be explored.

In conclusion, the present findings add to the growing body of evidence that challenges the traditional dual-process perspective (e.g., De Neys 2012; De Neys and Pennycook 2019; Newman et al. 2017; Thompson et al. 2018). However, they also beg for a more nuanced interpretation of findings suggesting that people have intuitive access to accurate logic and probability principles (e.g., De Neys 2012; De Neys and Pennycook 2019). It may be the case that sometimes people have intuitive access to faulty logical principles, which may eventually lead to biased rule-based responses.

Author Contributions: Conceptualization, M.B.F. and J.C.S.; methodology, M.B.F. and J.C.S.; software, J.C.S. and J.R.; validation, M.B.F., J.C.S. and A.M.; formal analysis, M.B.F. and J.C.S.; investigation, M.B.F., J.C.S., A.M. and J.R.; resources, M.B.F. and J.C.S.; data curation, J.C.S.; writing—original draft preparation, M.B.F. and J.C.S.; writing—review and editing, M.B.F., J.C.S., V.A.T. and A.M.; visualization, M.B.F., J.C.S., V.A.T. and A.M.; supervision, M.B.F. and J.C.S.; project administration, M.B.F.; funding acquisition, M.B.F. All authors have read and agreed to the published version of the manuscript.

Funding: This research was partially supported by a research fellowship for experienced researchers awarded by the Alexander von Humboldt-Stiftung (Germany) to the first author. This work also received national funding from FCT—Fundação para a Ciência e a Tecnologia, I.P., through the Research Center for Psychological Science of the Faculty of Psychology, University of Lisbon (UIDB/04527/2020; UIDP/04527/2020)—CICPSI, Faculdade de Psicologia, Universidade de Lisboa, Alameda da Universidade, 1649-013 Lisboa, Portugal.

Institutional Review Board Statement: The studies were approved by the ethics committee of Faculdade de Psicologia, Universidade de Lisboa.

Informed Consent Statement: Informed consent was obtained from all subjects involved in the study.

Data Availability Statement: The data presented in this study are available on request from the corresponding authors.

Conflicts of Interest: The authors declare no conflict of interest.

Appendix A

Problems from the cognitive reflection test presented in the experiments

In Experiment 2:

1- If 3 workers need 3 min to produce 3 toys, how long would it take 500 workers to produce 500 toys?

2- A TV and a DVD player cost, together, 110€. The TV costs 100€ more than the DVD player. How much costs the DVD player?

3- A virus is spreading in a computer. Every minute, the number of infected files double. If it takes 100 min to infect the whole system, how long does it take to infect half of the system?

Added in Experiment 4:

4- John had both the 15° best note and the 15° worse note on the class exam. How many students, in total, has the class?

Appendix B

Table A1. Means and standard errors of base-rate responses for different problems in each experiment.

	Subsample Last	Base Rates Last
	M (SE)	*M (SE)*
Experiment 1		
Standard no-conflict	.97 (.03)	-
Standard conflict	.46 (.03)	-
Rule-inadequate conflict	.26 (.03)	-
Experiment 2		
Standard no-conflict	-	.98 (<.01)
Standard conflict	-	.74 (.03)
Rule-inadequate no-conflict	-	.95 (.01)
Rule-inadequate conflict	-	.51 (.03)
Experiment 3		
Standard no-conflict	.95 (.01)	.98 (.01)
Standard conflict	.64 (.05)	.64 (.05)
Rule-inadequate no-conflict	.93 (.01)	.96 (.01)
Rule-inadequate conflict	.23 (.04)	.40 (.05)
Experiment 4		
Standard no-conflict	.97 (.01)	.97 (.01)
Standard conflict	.61 (.07)	.63 (.07)
Rule-inadequate no-conflict	.90 (.03)	.93 (.03)
Rule-inadequate conflict	.25 (.06)	.50 (.06)

Table A2. Response times (in ms) for each kind of response in each type of problem across experiments.

	No Conflict Correct	Conflict Base Rate	Conflict Stereotype	N Included in the Analysis
	M (SE)	M (SE)	M (SE)	
Experiment 1				
Standard	9615.04 (839.41) *	12,386.22 (839.41)	12,052.92 (839.41)	44
Rule-inadequate	10,117.63 (857.88) *	11,995.65 (857.88)	10,987.56 (857.88)	34
Experiment 2				
Standard	1045.85 (103.76)	1390.52 (103.76)	1847.12 (103.76)	84
Rule-inadequate	1037.08 (201.68)	2072.12 (201.68)	1229.29 (201.68)	94
Experiment 3				
Standard	1016.19 (99.99)	1199.60 (99.99)	1830.12 (99.99)	70
Rule-inadequate	1056.20 (90.83)	1584.46 (90.83)	1227.47 (90.83)	72
Experiment 4				
Standard	1051.35 (114.95)	1292.32 (114.95)	1651.77 (114.95)	47
Rule-inadequate	1007.27 (99.06)	1349.53 (99.06)	1188.64 (99.06)	48

* The response time associated to Rule-inadequate problems in Experiment 1 is from no-conflict standard problems, because there was no Rule-inadequate versions of no-conflict problems in this experiment. Because of different number of missed responses in standard and rule-inadequate conflict trials, different means were taken in account in each ANOVA.

Table A3. Examples for all variations of problems used in Experiment 1.

Problem Version	Problem Type: No-Conflict	Problem Type: Conflict
Standard problems	In a study 1000 people were tested. Among the participants there were 4 women and 996 men. *In the first week of the study, all men and women were interviewed. Jo was one of the people interviewed in the first week.* Jo is 23 years old and is finishing a degree in engineering. On Friday nights, Jo likes to go out cruising with friends while listening to loud music and drinking beer. What is most likely? a. Jo is a man b. Jo is a woman	In a study 1000 people were tested. Among the participants there were 996 women and 4 men. *In the first week of the study, all men and women were interviewed. Jo was one of the people interviewed in the first week.* Jo is 23 years old and is finishing a degree in engineering. On Friday nights, Jo likes to go out cruising with friends while listening to loud music and drinking beer. What is most likely? a. Jo is a man b. Jo is a woman
Rule inadequate		In a study 1000 people were tested. Among the participants there were 996 women and 4 men. *In the first week of the study, as many men as women were interviewed. Jo was one of the people interviewed in the first week.* Jo is 23 years old and is finishing a degree in engineering. On Friday nights, Jo likes to go out cruising with friends while listening to loud music and drinking beer. What is most likely? a. Jo is a man b. Jo is a woman

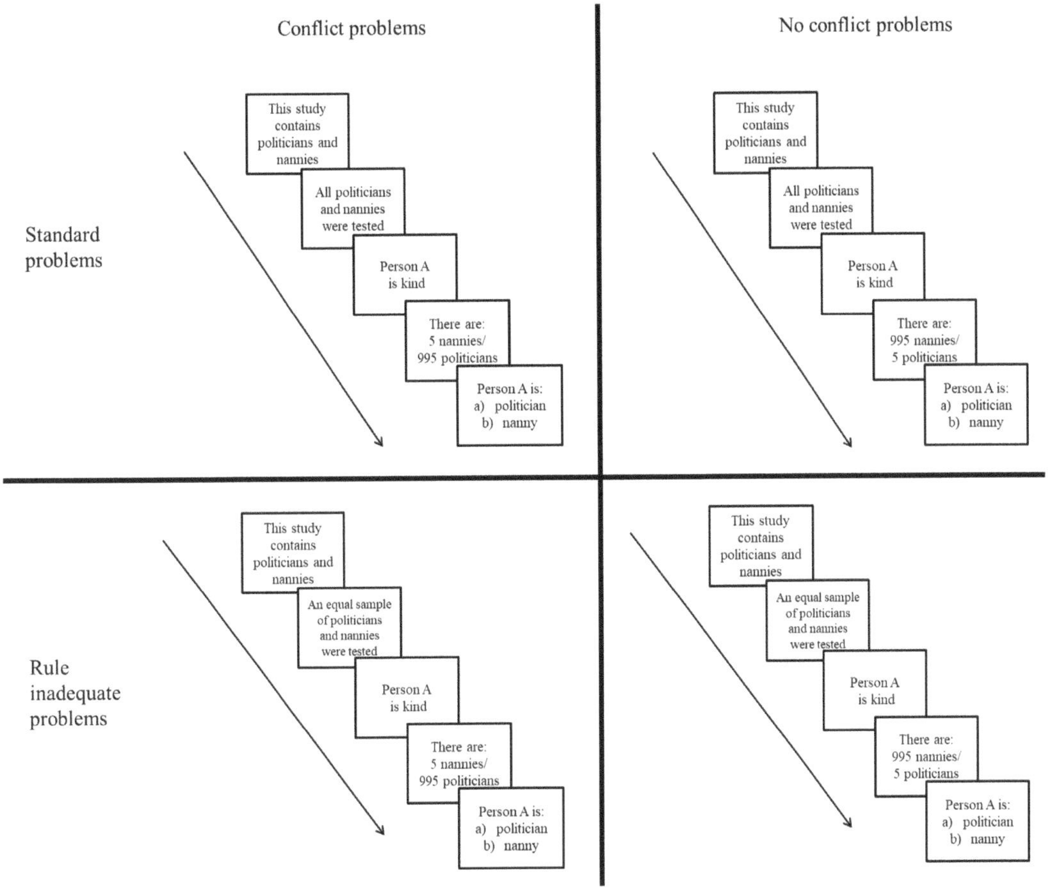

Figure A1. Examples for all variations of problems used in Experiment 2.

Notes

1 The base rates presented in the original version of the lawyer–engineer problem (70 lawyers and 30 engineers; Kahneman and Tversky 1972) were replaced in this example for more extreme base rates in order to make sure that when the base rates are taken into account, the likelier response option is "Dan is a lawyer").

2 Base rates and group composition coincide in the case of standard problems but not in the rule-inadequate problems, where base rates are 50/50. However, for ease of presentation, throughout the text we refer to "base-rate responses" for both standard and rule-inadequate problems. These are adequate responses in the case of standard problems and responses based on invalid base rates in the case of rule-inadequate problems.

3 The number of participants in Experiment 1 and in the remaining experiments here reported were not pre-defined based on power analysis. Experiments were included in different experimental waves, which were ran to completion. The obtained Ns are comparable to previous studies using the same experimental paradigms. Specifically, a total of 86 participants completed De Neys and Glumicic's (2008) Experiment 2 (the same number completed our Experiment 1). The number of participants who completed Pennycook et al.'s (2015) Studies varied between N = 60 and N = 88; whereas in the Experiments 2 to 4 here reported, they varied between N = 64 and N = 116). Tests of sensitivity were ran for all studies.

4 By "correct responses to no-conflict problems" we mean responses according to the base rates and stereotype-based information, which both converge in the same response option in these problems.

5 We thank Gordon Pennycook for making available to us the sample of traits used in the studies in Pennycook et al. (2015).

6 The asymmetry in the number of standard problems—twice as many as the rule-inadequate problems—was a mistake in the code of the E-prime program used to run this Experiment. In Experiment 3, this issue was corrected.

7 "pseudo-syllogisms" are false alarm versions of classical syllogisms in the sense that they modify the logical (valid or invalid) status of the syllogism while keeping the superficial structure of the problems unchanged.

References

Bago, Bence, and Wim De Neys. 2017. Fast logic?: Examining the time course assumption of dual process theory. *Cognition* 158: 90–109. [CrossRef] [PubMed]

Ball, Linden J., Valerie A. Thompson, and Edward J. N. Stupple. 2017. Conflict and Dual-Process Theory: The Case of Belief Bias. In *Dual Process Theory 2.0*, 1st ed. Edited by Wim De Neys. London: Routledge, Abingdon: Taylor and Francis Group, pp. 100–20.

De Neys, Wim, and Gordon Pennycook. 2019. Logic, Fast and Slow: Advances in Dual-Process Theorizing. *Current Directions in Psychological Science* 28: 503–9. [CrossRef]

De Neys, Wim, and Tamara Glumicic. 2008. Conflict monitoring in dual process theories of thinking. *Cognition* 106: 1248–99. [CrossRef] [PubMed]

De Neys, Wim. 2012. Bias and Conflict: A Case for Logical Intuitions. *Perspectives on Psychological Science* 7: 28–38. [CrossRef] [PubMed]

De Neys, Wim. 2014. Conflict detection, dual processes, and logical intuitions: Some clarifications. *Thinking & Reasoning* 20: 169–87. [CrossRef]

De Neys, Wim. 2018. Bias, conflict, and fast logic: Towards a hybrid dual process future? In *Dual Process Theory 2.0*, 1st ed. Edited by Wim De Neys. London: Routledge, Abingdon: Taylor and Francis Group, pp. 47–65.

Elqayam, Shira, and Jonathan St. B. T. Evans. 2011. Subtracting "ought" from "is": Descriptivism versus normativism in the study of human thinking. *Behavioral and Brain Sciences* 34: 233–48. [CrossRef]

Evans, Jonathan St. B. T. 2018. Dual process theory: Perspectives and problems. In *Dual Process Theory 2.0*, 1st ed. Edited by Wim De Neys. London: Routledge, Abingdon: Taylor and Francis Group, pp. 137–55.

Evans, Jonathan St. B. T., and Keith E. Stanovich. 2013. Dual-Process Theories of Higher Cognition: Advancing the Debate. *Perspectives on Psychological Science* 8: 223–41. [CrossRef] [PubMed]

Fong, Geoffrey T., David H. Krantz, and Richard E. Nisbett. 1986. The effects of statistical training on thinking about everyday problems. *Cognitive Psychology* 18: 253–92. [CrossRef]

Frederick, Shane. 2005. Cognitive Reflection and Decision Making. *Journal of Economic Perspectives* 19: 25–42. [CrossRef]

Frey, Darren, Eric D. Johnson, and Wim De Neys. 2018. Individual differences in conflict detection during reasoning. *Quarterly Journal of Experimental Psychology* 71: 1–52. [CrossRef]

Gilovich, Thomas, Dale Griffin, and Daniel Kahneman, eds. 2002. *Heuristics and Biases: The Psychology of Intuitive Judgment*, 1st ed. Cambridge: Cambridge University Press. [CrossRef]

Handley, Simon J., Stephen E. Newstead, and Dries Trippas. 2011. Logic, beliefs, and instruction: A test of the default interventionist account of belief bias. *Journal of Experimental Psychology: Learning, Memory, and Cognition* 37: 28–43. [CrossRef] [PubMed]

Just, Marcel A., Patricia A. Carpenter, and Jacqueline D. Woolley. 1982. Paradigms and processes in reading comprehension. *Journal of Experimental Psychology: General* 111: 228–38. [CrossRef]

Kahneman, Daniel, and Amos Tversky. 1972. Subjective probability: A judgment of representativeness. *Cognitive Psychology* 3: 430–54. [CrossRef]

Kahneman, Daniel, and Amos Tversky. 1973. On the psychology of prediction. *Psychological Review* 80: 237–51. [CrossRef]

Kahneman, Daniel, and Shane Frederick. 2002. Representativeness Revisited: Attribute Substitution in Intuitive Judgment. In *Heuristics and Biases*, 1st ed. Edited by Thomas Gilovich, Dale Griffin and Daniel Kahneman. Cambridge: Cambridge University Press, pp. 49–81. [CrossRef]

Kahneman, Daniel, Stewart Paul Slovic, Paul Slovic, and Amos Tversky, eds. 1982. *Judgment under Uncertainty: Heuristics and Biases*, 1st ed. Cambridge: Cambridge University Press. [CrossRef]

Kahneman, Daniel. 2003. A perspective on judgment and choice: Mapping bounded rationality. *American Psychologist* 58: 697–720. [CrossRef]

Klauer, Karl Christoph, and Henrik Singmann. 2013. Does logic feel good? Testing for intuitive detection of logicality in syllogistic reasoning. *Journal of Experimental Psychology: Learning, Memory, and Cognition* 39: 1265–73. [CrossRef]

Krosnick, Jon A., Fan Li, and Darrin R. Lehman. 1990. Conversational conventions, order of information acquisition, and the effect of base rates and individuating information on social judgments. *Journal of Personality and Social Psychology* 59: 1140–52. [CrossRef]

Morsanyi, Kinga, and Simon J. Handley. 2012. Logic feels so good—I like it! Evidence for intuitive detection of logicality in syllogistic reasoning. *Journal of Experimental Psychology: Learning, Memory, and Cognition* 38: 596–616. [CrossRef]

Newman, Ian R., Maia Gibb, and Valerie A. Thompson. 2017. Rule-based reasoning is fast and belief-based reasoning can be slow: Challenging current explanations of belief-bias and base-rate neglect. *Journal of Experimental Psychology: Learning, Memory, and Cognition* 43: 1154–70. [CrossRef]

Pennycook, Gordon, Jonathan A. Fugelsang, and Derek J. Koehler. 2015. What makes us think? A three-stage dual-process model of analytic engagement. *Cognitive Psychology* 80: 34–72. [CrossRef]

Šrol, Jakub, and Wim De Neys. 2021. Predicting individual differences in conflict detection and bias susceptibility during reasoning. *Thinking & Reasoning* 27: 38–68. [CrossRef]

Stanovich, Keith E. 1999. *Who Is Rational? Studies of Individual Differences in Reasoning*. Mahwah: Lawrence Erlbaum Associates.

Stanovich, Keith E. 2009. Distinguishing the reflective, algorithmic, and autonomous minds: Is it time for a tri-process theory? In *In Two Minds: Dual Processes and Beyond*, 1st ed. Edited by Jonathan St. B. T. Evans and Keith Ed Frankish. Oxford: Oxford University Press, pp. 55–88. [CrossRef]

Stanovich, Keith E., and Richard F. West. 1998. Individual differences in rational thought. *Journal of Experimental Psychology: General* 127: 161–88. [CrossRef]

Stanovich, Keith E., and Richard F. West. 2008. On the relative independence of thinking biases and cognitive ability. *Journal of Personality and Social Psychology* 94: 672–95. [CrossRef] [PubMed]

Thompson, Valerie A., Gordon Pennycook, Dries Trippas, and Jonathan St. B. T. Evans. 2018. Do smart people have better intuitions? *Journal of Experimental Psychology: General* 147: 945–61. [CrossRef] [PubMed]

Thompson, Valerie A., Jamie A. Prowse Turner, and Gordon Pennycook. 2011. Intuition, reason, and metacognition. *Cognitive Psychology* 63: 107–40. [CrossRef] [PubMed]

Toplak, Maggie E., Richard F. West, and Keith E. Stanovich. 2011. The Cognitive Reflection Test as a predictor of performance on heuristics-and-biases tasks. *Memory & Cognition* 39: 1275–89. [CrossRef]

Tversky, Amos, and Daniel Kahneman. 1974. Judgment under Uncertainty: Heuristics and Biases: Biases in judgments reveal some heuristics of thinking under uncertainty. *Science* 185: 1124–1131. [CrossRef]

Tversky, Amos, and Daniel Kahneman. 1983. Extensional versus intuitive reasoning: The conjunction fallacy in probability judgment. *Psychological Review* 90: 293–315. [CrossRef]

Journal of
Intelligence

MDPI

Article

Single-Heuristic Reasoning: Is It Still Dual-Process?

Pavle Valerjev [1,*] and Marin Dujmović [2]

1 Department of Psychology, University of Zadar, 23000 Zadar, Croatia
2 School of Psychological Science, University of Bristol, Bristol BS8 1QU, UK
* Correspondence: valerjev@unizd.hr

Abstract: Traditionally, paradigms used to study conflict in reasoning (and metacognition during reasoning) pit heuristic processes against analytical processes. Findings indicate that the presence of conflict between processes prolongs reasoning and decreases accuracy and confidence. In this study, we aimed to explore reasoning and metacognition when only one type of heuristic process is exploited to cue multiple responses. In two experiments, a novel modification of the Base Rate neglect task was used to create versions in which one belief-based heuristic competes, or works in concert, with another of the same type to provide a response. Experiment 1 results reveal that the presence of conflict between cued responses does not affect meta-reasoning, which indicates that reasoning defaulted to a single process. An alternative explanation was that the effect of conflict was masked due to an imbalance in the strength of the dominant response between conflicting and congruent versions. Experiment 2 was designed to test hypotheses based on these competing explanations. Findings show that when the strength of a response was no longer masking the effect, the conflict did result in longer reasoning times and lower confidence. The study provides more robust evidence in favor of the dual-process account of reasoning, introduces a new methodological approach, and discusses how conflict may be modulated during reasoning.

Keywords: dual-process theory; reasoning; meta-reasoning; heuristics; intuitive reasoning; single-process reasoning; conflict detection; logical intuitions

1. Introduction

Research into reasoning processes has been dominated by the dual-processing approach and the various theories it has spawned over the decades. Traditionally, System 1 or Type 1 processes were considered quick, automatic, cognitively undemanding, and based on intuitions. On the other hand, System 2 or Type 2 processes were considered slow, deliberate, cognitively demanding, and based on analytical thinking (Evans 2012). However, developments in the past two decades have led to the conclusion that responses previously ascribed to the slow System 2 are indeed available to System 1. Key findings leading to these conclusions come from research on reasoning under cognitive load and/or time pressure and research with the rethinking paradigm. When forcing participants to reason under cognitive load and under time pressure conditions, the frequency of analytical responses does decrease but only slightly, meaning that both intuitive and analytical responses are still being generated (Bago and De Neys 2017; Lawson et al. 2020). Similarly, when participants give quick initial responses and are then given the opportunity to rethink their choices, the vast majority of analytical responses after rethinking are not a result of switching from an intuitive response (Thompson et al. 2011; Dujmović et al. 2021). Rather, analytical responses after rethinking were already given as the initial response. These and other findings have led to a continuous development of models within the dual-process approach, which takes into account the fact that System 1 generates various types of responses, presumably accommodating different types of quick processes. An early idea about multiple intuitive processes is introduced by Glockner and Witteman (2010), while De Neys (2012) refers to logical intuitions which are part of System 1 and cue analytically correct responses in reasoning tasks. Logical intuitions would include mathematical, probabilistic, formal logic, and others not traditionally considered to be intuitions. The more traditionally used intuitions, like biases stemming from representativeness, matching, and other heuristics, will be referred to as *heuristic* intuitions.

Based on these more recent findings, a common framework is emerging, and the proposed working model is depicted in Figure 1. Intuitive processes generate initial intuitive responses (IR). Depending on task demands, and previous experience, one or more of these responses may be generated. Most tasks in the field are designed to pit a heuristic response against an analytical response. For example, take a simplified version of the Base Rate Neglect task (De Neys and Glumicic 2008; Dujmović and Valerjev 2018). Participants are told that a random person is chosen from a group of 1000 people. The random person is very tall. The group consists of 10 basketball players and 990 mathematicians. The task is to pick whether it is more probable that the randomly chosen person is a basketball player or a mathematician. In this example, one intuitive response is based on height being extremely typical of basketball players compared to mathematicians. Thus, it indicates that the randomly chosen person is most likely a basketball player. The second response is based on the information indicating there are many more mathematicians in the group, and thus indicates that the randomly chosen person is a mathematician.

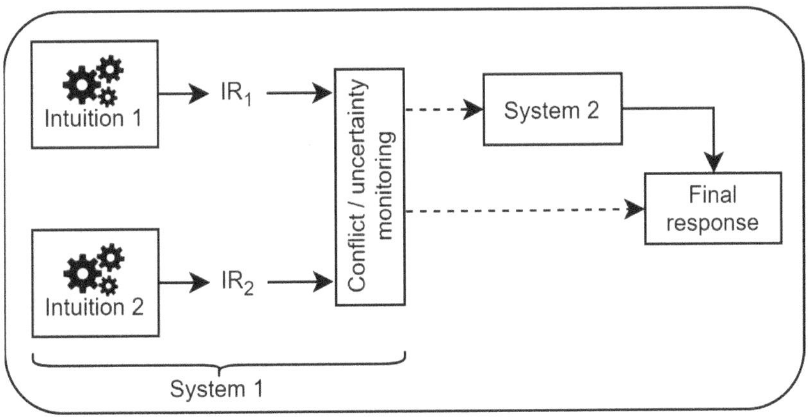

Figure 1. Schematic of a dual-process model of reasoning.

The two initial responses in this example are conflicting, but responses could point towards the same option congruently. The strength of these initial intuitive responses is going to depend on a number of factors. Stanovich (2018) refers to *mindware* as one determining factor. Mindware refers to knowledge and experiences of rules, procedures, and strategies which help detect and override intuitive, miserly, but incorrect responses. In traditional heuristic and bias tasks, the mindware necessary to perform well is usually tied to probabilistic, causal, and scientific reasoning as well as numeracy. The heuristic that basketball players are very tall has been built through experience and many examples over a lifetime. The same can be said about relying on the high proportion of mathematicians in the group. Knowledge and experiences handling various quantities, ratios, and probabilities build up mindware which can intuitively use that type of information. The strength of each intuitive response will therefore depend on the instantiation of the mindware. The average person might have a much stronger mindware generating the heuristic response (basketball player) when compared to a professional statistician who may have built up a strong mindware for handling probabilities, which favors what is deemed to be the analytical response (mathematician). De Neys (2022) further speculates that the strength of these intuitions might be variable over time. Therefore, the relative strength of intuitive responses may not simply be a function of peak magnitude but also the point in time at which the system is prompted to give a response. In our previous work, we have shown that the relative strength of responses can be influenced via instruction (Valerjev and Dujmović 2017) as well as a change in the modality of presentation (Dujmović and Valerjev 2017), implying attention and attentional resources play a mediating role between mindware and response strength. When the two responses are congruent, no further deliberation or activation of System 2 is required since there is no uncertainty, and a final response is given. If the responses are in conflict, a monitoring process may trigger System 2 activation. It is not clear how exactly conflict or the resulting uncertainty is computed, but many studies have shown that the presence of conflict increases the likelihood of System 2 activation (Dujmović and Valerjev 2018; Pennycook et al. 2014, 2015; Thompson and Johnson 2014). When the level of uncertainty/conflict is high enough to be detected, System 2 is activated to resolve the situation. This may include rationalization of one of the responses, considering additional evidence for one or both responses, or attempting to generate a novel response. If the level of uncertainty does not trigger System 2, the final response is simply the stronger of the two initial responses.

In recent years, more researchers have recognized the importance of metacognitive processes during reasoning. Indeed, the processes that detect conflict and/or compute uncertainty could be viewed as metacognitive monitoring. This has led to the emergence of meta-reasoning (Ackerman and Thompson 2015, 2017), in which measuring various metacognitive judgments such as perceived difficulty, solvability, and confidence has accompanied the usual measures of accuracy (usually defined as the proportion of analytical responses) and response time. In fact, metacognitive measures have proven to be very useful and sensitive to experimental manipulations even when accuracy and response times are not. For example, if all conditions in a task are completed at a very high level of accuracy, then there will be very little or no difference among them, but confidence judgments (for example) may be impacted by the experimental manipulation. Standard findings show participants are less confident in conflict when compared to congruent versions of reasoning tasks (Dujmović and Valerjev 2018; Dujmović et al. 2021; Mevel et al. 2015; Shynkaruk and Thompson 2006). Accuracy is often not a good predictor of confidence (Bajšanski et al. 2014; Markovits et al. 2015; Thompson et al. 2011), while response time as a proxy for response fluency oftentimes is (Ackerman and Zalmanov 2012; Dujmović and Valerjev 2018; Thompson et al. 2013; Thompson and Johnson 2014).

While the introduction of metacognitive measures to reasoning tasks has increased our understanding of both metacognitive and reasoning processes, it is becoming increasingly more difficult to probe deeper into these processes using current research paradigms. The tasks which have been used in reasoning research track their roots to Tversky and Kahneman (1982, 1983) and their conjunction fallacy (Linda problem) task. The task describes Linda, an outspoken, former philosophy student who was concerned with issues of discrimination and social injustice and participated in a number of demonstrations. Participants are then asked whether, years later, it is more probable that A—Linda is a bank teller or that B—Linda is a bank teller *and* active in the feminist movement. In many different variations of this task, Tversky and Kahneman found that the vast majority of participants judged the conjunction of two statements (bank teller *and* active feminist) as more likely than one of the two constitutive statements (bank teller). This is a fallacy; the conjunction of any two events is less likely than either of the two individually. The fallacy is committed because the description of Linda is more representative of someone who is to become active in the feminist movement, and thus, the conjunction is deemed more probable than the single statement even though it is, in fact, less probable (for a thorough investigation of the conjunction fallacy from a modern meta-reasoning perspective see Dujmović et al. 2021). Considering how prevalent the fallacy was and how difficult it was for participants to give the correct response, it naturally served as evidence for reasoning being dual-process. The representativeness heuristic provided one response (the conjunction) and correct probability assessment the other. This type of design then became prevalent in practically all reasoning tasks of note. This includes the various versions of the Base Rate Neglect task (De Neys and Glumicic 2008; Dujmović and Valerjev 2018), the Cognitive Reflection Test (Frederick 2005), the Covariation Detection task (Stanovich et al. 2016; Valerjev and Dujmović 2019), and syllogisms (Shynkaruk and Thompson 2006). In all of these tasks, the design pits one type of process (e.g., representativeness or believability) against another (e.g., probability computation or formal logic) in conflict versions or has the two processes aim at the same response in congruent versions. While these tasks have produced many important findings, there are inherent limitations in how they have been used. Namely, the tasks always pit qualitatively different processes of unknown strengths against each other. One can vary the group sizes in a base rate task (Dujmović and Valerjev 2018), but the strength of the heuristic response is almost always unknown or only roughly estimated as more or less extreme. Likewise, one can vary the believability when using syllogisms (Shynkaruk and Thompson 2006). However, with any of these manipulations, the relative strengths of the two processes are unknown. Indeed, the fact that the mismatch is quite extreme in most cases has perpetuated what De Neys (2022) deems as an unfounded assumption of exclusivity (that some responses are only available to System 2). The mismatch can be seen as a severe difference in mindware. In the Linda problem, representativeness as a heuristic is well developed for the vast majority of people, while properly combining probabilities or understanding rules of formal logic is tied to a much weaker mindware for most. Even if one could match (via pre-studies) the strengths of one process with another, these processes are still qualitatively different, and it is unknown if a match in peak estimated strength means a match when embedded in a reasoning task. Without more stringent matching, subtle manipulations are more difficult, as is testing more nuanced aspects of dual-process models.

The aims of this study were two-fold. First, we aimed to develop materials for reasoning tasks that focus on only one type of heuristic process with good measurements of relative heuristic strength. In order to accomplish this, we started from the simplified Base Rate task, but instead of pitting a heuristic intuitive process against a logical intuitive process, we chose to use the same type of intuitive heuristic process in order to cue both responses. In comparison to the example given above, instead of stating that a randomly chosen, very tall person came from a group of 10 basketball players and 990 mathematicians, we would state that a randomly chosen person is very tall and very intelligent while being selected from a group which is half basketball players and half mathematicians. In this version, being very tall is indicative of being a basketball player, while being very intelligent is indicative of being a mathematician (for details on all conditions and materials, see the Methods section of Experiment 1). Second, we aimed to test whether standard findings considering the influence of conflict detection and resolution hold for these tasks in which the same type of intuitive process cues different responses. Our pre-study (see Methods) results made fine control of the level of conflict/congruence possible. This allows for comparisons of congruent and conflict conditions in which the dominance of the final response can be held constant, making an evaluation of the influence of conflict easier. Given previous research, we hypothesized that conflict would manifest through longer reasoning times, lower accuracy (measured as a lower proportion of response in accordance with the dominant heuristic), and lower confidence. It was further hypothesized that the level of congruence/conflict would influence reasoning speed, accuracy, and confidence. Higher congruence should result in an increase in speed, accuracy, and confidence, while higher conflict should result in a decrease in all three dependent variables.

2. Experiment 1

2.1. Methods

In order to grasp the design of Experiment 1 and the materials created for this purpose, it is important to first explain the principles and the pre-study based on which these materials were created. In the first phase of the pre-study, participants ($N = 22$) generated typical and atypical characteristics for each of the 12 groups of people (e.g., mathematicians, basketball players, etc.). Based on the materials generated, 12 characteristics were chosen for the second phase in which participants (N = 309) judged the typicality of each characteristic for each of the groups on a −4 to 4 scale where high scores indicate very typical characteristics, while low scores indicate very atypical characteristics (e.g., being very tall is very typical of basketball players, but being very wealthy is atypical of cleaning ladies). These typicality judgments can then be used in various combinations in the manner depicted in Table 1. The heuristic response based on a characteristic will depend on the relative typicality of the characteristic for either of the two groups. In Table 1, the first characteristic (e.g., being highly educated) is quite typical of group A (e.g., mathematicians) but not group B (e.g., basketball players). The resulting heuristic response is that the person is more likely to be from group A, with a strength of 3.45. The reverse pattern is true for the second characteristic (e.g., being physically strong), which is typical of group B (basketball players) but not of group A (mathematicians). For this characteristic, the person is more likely to be from group B with a strength of 2.

Table 1. Schematic example of combining characteristics to form conditions.

	Typicality		Probable Heuristic Response (HR)	Nominal Strength of HR [1]	Probable Final Response (FR)	Nominal Strength of FR [2]
	Group A	Group B				
Characteristic 1	3.50	0.05	A	3.45	A	1.45
Characteristic 2	0.50	2.50	B	2.00		

[1] The difference in typicality scores; [2] The difference of HR strengths in conflict, and the sum of strengths in congruent conditions (in the direction of the dominant response if there is one).

These two heuristics could be combined in a single task if the objective was to select to which group a randomly chosen person high on both characteristics (highly educated and physically strong) belongs. Table 1 depicts an example of conflict, where one heuristic process results in one strong response, whereas the other heuristic process results in a competing response. These tasks can be designed for congruent as well as conflict conditions. In the congruent condition, both heuristic processes indicate the same response; therefore, the nominal strength of the probable final response is the sum of the two heuristic response strengths. Alternatively, in the conflict condition, the final response strength is the difference between the two heuristic response strengths. This also allows us to vary the level of congruence and conflict since the magnitude of the final response strength is a function of these levels. It is worth noting that only the typicality values from Table 1 were measured, while combining them to compute all other values served to produce the stimuli for our experimental conditions. If a similar computation does occur in the brain, it is likely to be considerably more complex and include additional factors.

2.1.1. Design

As mentioned, based on typicality judgments, tasks with multiple characteristics can be formed such that heuristic processes are congruent or in conflict, and the level of each can be varied. Experiment 1 follows a 2 (congruence) × 2 (level) design. Tasks could be congruent or in conflict, and the level of congruence/conflict had two magnitudes (low and high). In the low-level conditions, the final response strength was (on average) 1.50. Therefore, the low-congruent and low-conflict conditions were particularly interesting since the final response strength remained identical. In the high-level conditions—the high conflict was achieved by combining heuristic responses of equal nominal strengths but opposite directions, resulting in a final response strength of 0. This is the maximal level of conflict/uncertainty. In the high-congruence condition, two strong heuristics indicating the same response were combined to form tasks with an average nominal strength of 5.

2.1.2. Participants

A total of $N = 145$ participants were recruited amongst undergraduate students across a number of Croatian universities. After processing, the data from $N = 139$ participants (92.70% female, $M_{age} = 22.28$, $SD_{age} = 3.31$) was retained for statistical analyses. Sensitivity analysis conducted in G*Power v3.1.9.4 reveals that with a sample of $N = 139$, a desired power level of .80, and an alpha of .05, the smallest effect we could confidently detect was $\eta_p^2 = .054$.

2.1.3. Materials

A total of 16 tasks were generated (four per condition) by combining typicality scores into heuristic response strengths resulting in different combinations of congruence and level. Table 2 describes all four conditions with examples. Each task consisted of a scenario in which a person was selected at random from a group. Half of the group were members of one cohort (e.g., scientists) and half of another (e.g., gymnasts). The randomly chosen person is described with two characteristics (e.g., very intelligent and physically strong). The question posed to participants was whether it was more probable that the randomly chosen person was a member of one cohort or the other.

Table 2. Examples from all experimental conditions in Experiment 1.

	High		Low	
	Conflict	**Congruent**	**Conflict**	**Congruent**
Group 1	Scientists	Ballerinas	Actresses	Boxers
Group 2	Gymnasts	Cleaning ladies	Scientists	Basketball players
Characteristic 1 (strength of HR)	Intelligence (+2.40 group 1)	Gentleness (+2.50 group 1)	Sociability (+2.80 group 1)	Intelligence (+0.75 group 2)
Characteristic 2 (strength of HR)	Strength (+2.40 group 2)	Beauty (+2.50 group 1)	Diligence (+1.30 group 2)	Wealth (+0.75 group 2)
Strength of FR	0	+5 group 1	+1.50 group 1	+1.50 group 2

Note: HR—Heuristic response; FR—Final response.

In the pre-study, typicality judgments were made independently for each group rather than relatively across groups. Therefore, groups in all tasks were made up of cohorts of the same gender (e.g., ballerinas and cleaning ladies). This was done in order to control for the fact that typicality judgments were likely given in relation to other same-gender individuals. For example, physical strength was judged as highly typical for both male basketball players and female gymnasts. If these groups were part of the same trial, there is no doubt that the characteristic would be associated with basketball players to a much higher degree than gymnasts, even though the typicality judgments would not suggest it.

2.1.4. Procedure and Data Processing

Participants first provided demographic information and were given instruction in the form of a walkthrough of an unused example of the task. They then completed the 16 tasks (a single trial procedure for each task can be seen in Figure 2). The task order was randomized for each participant. The order in which responses appeared was balanced so that in half of the trials, the response according to the dominant heuristic was the first response option, and in half, it was the second option. The experiment was designed in PsychoPy (Peirce et al. 2019) and conducted via the Pavlovia online platform. Participants were required to complete the study on a desktop or laptop computer with a mouse and asked to isolate themselves in a quiet environment during the study.

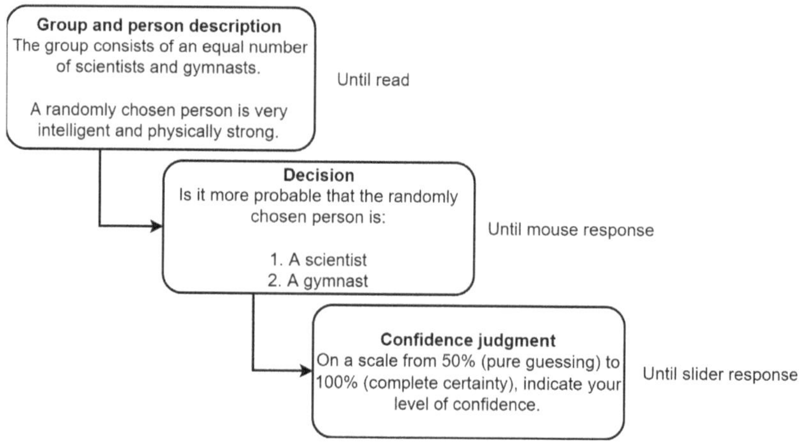

Figure 2. Single-trial procedure for tasks in Experiment 1.

Reading times in the Group and personal information, response times in the Decision phase, and accuracy and confidence judgments were recorded for each trial. Responses were considered as correct if they were in line with the more dominant heuristic (according to Table 2). The high-conflict condition was an exception since there was no dominant response; as such, the top response to appear in the Decision phase was coded as correct for convenience. Considering that the reasoning process included both the first and second phases (Figure 2), the total time for each trial was computed as a sum of reading and decision time.

Only trials with correct responses were selected for further analyses of reasoning times and confidence judgments (but all trials in the high-conflict condition). This was done because correct trials made up the vast majority of all trials and because a response in line with the probable final response according to our nominal strengths indicates that it did indeed fit within the ascribed experimental condition for the particular participant. Additionally, we excluded 5% of all trials based on total time (2.5% of the fastest and 2.5% of the slowest), considering this was an online study, and these more extreme times deviated considerably from the median. After this filter was applied, a median for each of the dependent variables was computed for each participant in each experimental condition to form the final dataset for analysis.

2.2. Results and Discussion

Total times were considerably positively skewed; therefore, analyses were conducted after computing the natural logarithm, and no other data processing was required prior to running further analyses. In order to determine whether congruence and level had an effect on accuracy, reasoning times, and confidence judgments, three 2 × 2 repeated measures ANOVAs were conducted (results can be seen in Table 3).

Table 3. Congruence by level ANOVA results for accuracy, reasoning times, and confidence in Experiment 1.

Effect	Accuracy			Reasoning Times			Confidence		
	$F(1, 138)$	η_p^2	p	$F(1, 138)$	η_p^2	p	$F(1, 138)$	η_p^2	p
Congruence	93.82	.41	<.001	48.77	.26	<.001	112.51	.45	<.001
Level	0.72	<.01	.399	0.62	<.01	.431	18.46	.12	<.001
Interaction	193.38	.58	<.001	36.91	.21	<.001	182.39	.57	<.001

The ANOVA of accuracy rates revealed that participants were more accurate in congruent conditions. However, this is qualified by a large interaction effect. Participants were equally accurate in low-conflict and low-congruent conditions (as tested by Tukey HSD post hoc tests) while being significantly more accurate in the high-congruence condition and significantly less accurate in the high-conflict condition (Figure 3). Note that accuracy in the high-conflict condition refers to the percentage of times participants chose the first response which appeared in the trial. If our tasks were perfectly calibrated for every single participant at a nominal strength of 0, we would expect the ratio to be exactly 50%. As can be seen in Figure 3, accuracy was only slightly above 50%, meaning that when the conflict was very high, participants chose randomly most of the time with a slight preference for the first response presented in the Decision phase (Figure 2). This may be an interesting effect of primacy. Both of the responses are cued strongly by opposing processes, and the first one may have a slight advantage simply due to primacy. The more important result in the context of our aims is that participants were not significantly more accurate in both of the congruent conditions when compared to conflict conditions. In fact, participants were slightly (though not significantly) more accurate in the low-conflict when compared to the low-congruence condition. This implies that the nominal strength of the final response itself was the determining factor rather than whether conflict was present or not.

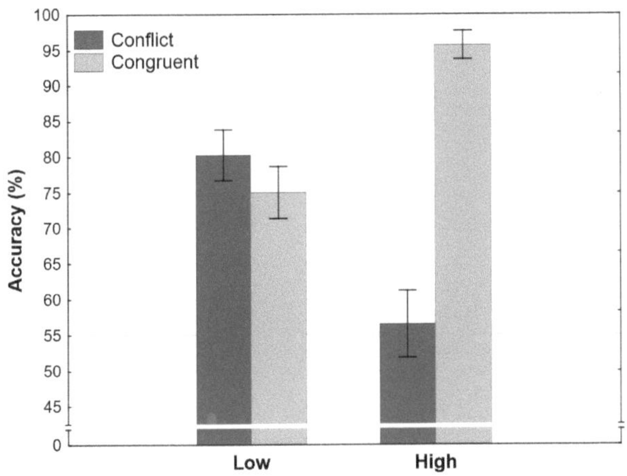

Figure 3. Accuracy as a function of congruence and level in Experiment 1 (error bars represent 95% CI).

The ANOVA of reasoning times revealed the same pattern of results. Participants were generally faster in the congruent conditions, but this is qualified by a large interaction effect. Participants completed tasks in the low-conflict condition just as fast as in the low-congruence condition but significantly faster in the high-congruence and significantly slower in the high-conflict condition (Figure 4, note that the figure depicts raw response times for convenience, the ANOVA was conducted on logarithms). The implications are the same as for accuracy rates; it seems that final response strength is the key determining factor in how long reasoning is going to take, not whether conflict was present or not.

Finally, the ANOVA of confidence judgments resulted in the same pattern. The large interaction effect was significant because there was no difference between the low-level conditions, while participants were much more confident in the high-congruence and much less confident in the high-conflict condition (Figure 5).

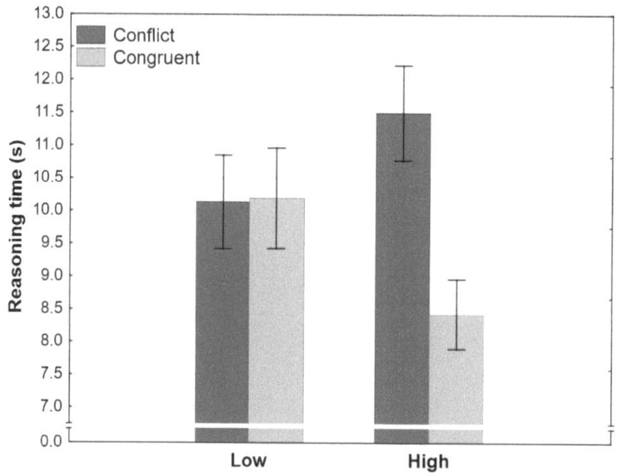

Figure 4. Reasoning time as a function of congruence and level in Experiment 1 (error bars represent 95% CI).

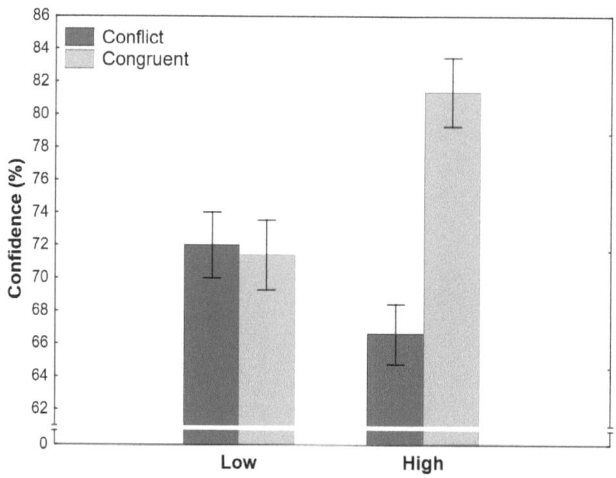

Figure 5. Confidence judgments as a function of congruence and level in Experiment 1 (error bars represent 95% CI).

These results would indicate, contrary to modern dual-processing theories, when two heuristic processes of the same type are pitted against each other, the presence of conflict does not impact reasoning. The key comparison which leads to this conclusion is between the low-conflict and low-congruence conditions. If both conflict and nominal final response strength were influencing reasoning, given that strength is equal for these two conditions, participants should have been more accurate, faster, and more confident in the low-congruence condition. On the other hand, if only strength influenced reasoning, then we would expect exactly the pattern of results we observed in Experiment 1. Note that the remaining two conditions would result in the exact same pattern of results regardless of the explanation. The result seems to imply that when a single type of heuristic process is required to cue responses, only one heuristic process actually *is* activated. This heuristic process would then compute all of the information of the same type and cue one of the responses with some level of certainty.

However, the low-congruent and conflict conditions are not entirely balanced, nor can complete balance be achieved due to the fact that one combines the strengths of two responses while the other is a difference between them. Table 4 highlights this difference.

Table 4. Schematic of differences between low congruent and conflict conditions.

	Congruent	**Conflict**
Strength of HR_1	+0.75—Group A	+3.00—Group A
Strength of HR_2	+0.75—Group A	+1.50—Group B
Strength of FR	+1.50—Group A	+1.50—Group A

Note: HR—Heuristic response; FR—Final response.

Apart from final response strength and the presence of conflict, there is a third difference between the conditions which may influence reasoning. Note that the strength of heuristic responses in the congruent condition is quite low, even though they both cue the same response. On the other hand, even though competing responses are cued in the conflict condition, the dominant response is very strong. This high level of strength, not present for either heuristic response in the congruent condition, may mitigate the effect of conflict on reasoning. The relative dominance of one response might make conflict detection less probable, or it may make it easier to resolve conflict in favor of the more dominant heuristic response. The final response then may seem more salient to participants. These effects may contribute to the low-conflict condition resulting in the same accuracy, speed, and confidence as the low-congruent condition. It has to be noted again that the nominal strengths of final responses and how they are computed are surely an oversimplification of any such computation during reasoning. Regardless of whether and how this computation is done—it may be modulated by the strength of cued heuristic responses. While the final response strength is nominally +1.50 in both situations, that strength is computed based on independent typicality judgments and then combined with the estimated strengths of heuristic responses. During reasoning, the presence of conflict (and how large the conflict is) would modulate final response strength to a lower score, but the relative dominance of one heuristic response would increase the score. In the congruent condition from Experiment 1, the reverse may be true, congruence (or absence of conflict) would strengthen the final response strength, but lack of a strong heuristic response would decrease (or at least not contribute as much) to the score. Thus, the effect of conflict, indirectly measured by looking at the differences between these two conditions, may be masked.

3. Experiment 2

It is not possible to completely balance congruent and conflict conditions. Nominal strengths of final responses can be kept identical, as in the low-conflict and low-congruence conditions in Experiment 1, but simply due to the nature of the conflict condition having competing heuristics pulling in opposite directions, it, by definition, means that the stronger heuristic in the conflict condition will be stronger than either of the heuristic responses in the congruent condition. However, nominal heuristic response strength likely would not have a linear, unbound effect. The logic behind Experiment 2 was to increase the strength of heuristic responses in both the congruent and conflict conditions sufficiently so that the difference between them would not mask the effect of conflict if it does influence reasoning. Therefore, the aim of Experiment 2 was to resolve which of the two explanations led to the pattern of results in Experiment 1.

3.1. Methods

3.1.1. Design

The design of Experiment 2 was quite similar to Experiment 1. It was again a 2(congruence) × 2(level) design. The main difference is that instead of the *level* independent variable relating to low or high congruence, and low or high conflict, it now directly relates to the nominal strength of the final response. This strength in the low conditions was again +1.50, which allows for the replication of the main, unexpected finding from Experiment 1. The strength in the high conditions was, on average, +3.00 for both congruent and conflict conditions. If strength was the only influence on reasoning, then we expected just a main effect of strength level when analyzing accuracy, reasoning times, and confidence. Increasing strength should lead to higher accuracy, faster reasoning, and higher confidence, irrespective of congruence. If, on the other hand, the presence of conflict does influence reasoning, then we would expect a different pattern. Given the results from Experiment 1, we would expect a congruence by strength-level interaction effect. The increase in accuracy, speed up of reasoning, and increase of confidence at higher nominal strength should be significantly larger in the congruent when compared to the conflict condition.

3.1.2. Participants

A total of $N = 117$ participants were recruited amongst undergraduate students across a number of Croatian universities. After data processing, a total of $N = 114$ participants (86.73% female, $M_{age} = 22.78$, $SD_{age} = 6.84$) were retained for statistical analyses. Sensitivity analysis conducted in G*Power v3.1.9.4 reveals that with a sample of $N = 114$, a desired power level of .80, and an alpha of .05, the smallest effect we could confidently detect was $\eta_p^2 = .065$.

3.1.3. Materials

The materials for Experiment 2 were constructed using similar logic. In fact, the same types of tasks from the low-conflict and congruent conditions from Experiment 1 (Table 2) were used in Experiment 2. New types of tasks were created for high-strength conflict and congruent conditions.

As can be seen in Table 5, the strength of the final response was equalized across these two conditions, but unlike the low-congruence condition from Experiment 1, the high-strength congruent condition here had an increased level of strength for both heuristic responses, as discussed in the Design section. The precautions taken when generating the tasks and the number of tasks per condition were the same as in Experiment 1.

Table 5. High-strength conflict and congruent conditions in Experiment 2.

	Conflict	**Congruent**
Group 1	Actresses	Scientists
Group 2	Cleaning ladies	Ballerinas
Characteristic 1 (strength of HR)	Wealth (+4.00 group 1)	Strength (+1.50 group 2)
Characteristic 2 (strength of HR)	Strength (+1.00 group 2)	Dexterity (+1.50 group 2)
Strength of FR	+3.00 group 1	+3.00 group 2

Note: HR—Heuristic response; FR—Final response.

3.1.4. Procedure and Data Processing

The procedure and all data pre-processing mirrored Experiment 1.

3.2. *Results and Discussion*

Natural logarithms of response times were again computed for further analyses, and no other data processing was required. In order to test whether congruence and strength level had an effect on reasoning, three 2 × 2 ANOVAs were conducted on the accuracy, reasoning times, and confidence judgments (results in Table 6).

The analysis of accuracy revealed a strong effect of strength level and no other significant effects. Participants were more accurate (chose the dominant response at a higher rate) in high-strength conditions irrespective of congruence (Figure 6). There was a slight trend towards higher accuracy in congruent conditions in general, but this did not reach significance.

Reasoning time analysis, however, did result in a significant interaction effect. The increase in nominal strength led to a significant decrease in reasoning times in the congruent but not in the conflict condition (Figure 7). The main effect of the level was also significant but inconsequential considering the interaction effect and the fact that effect size was under what our sensitivity analysis indicates would be the smallest confidently detectable effect.

Table 6. Congruence by response strength level ANOVA results for accuracy, reasoning times, and confidence in Experiment 2.

Effect	Accuracy			Reasoning Times			Confidence		
	$F(1, 113)$	η_p^2	p	$F(1, 113)$	η_p^2	p	$F(1, 113)$	η_p^2	p
Congruence	3.86	.03	.052	0.16	<.01	.687	0.10	<.01	.752
Level	45.53	.29	<.001	4.55	.04	.035	64.83	.36	<.001
Interaction	0.51	<.01	.477	10.35	.08	.002	16.81	.13	<.001

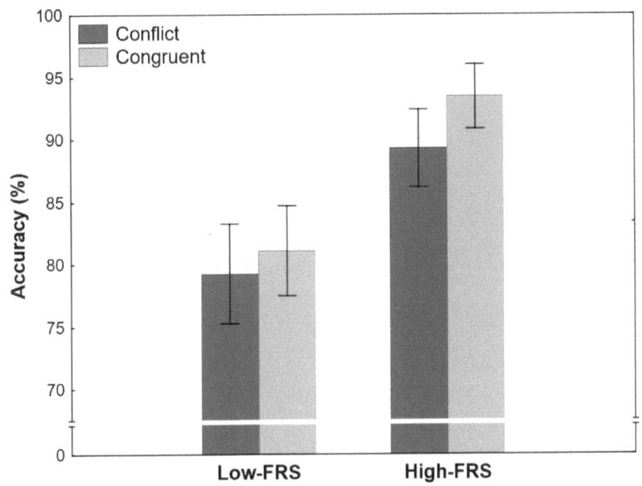

Figure 6. Accuracy as a function of congruence and final response strength level in Experiment 2 (error bars represent 95% CI). FRS refers to the final response strength level (see Table 1).

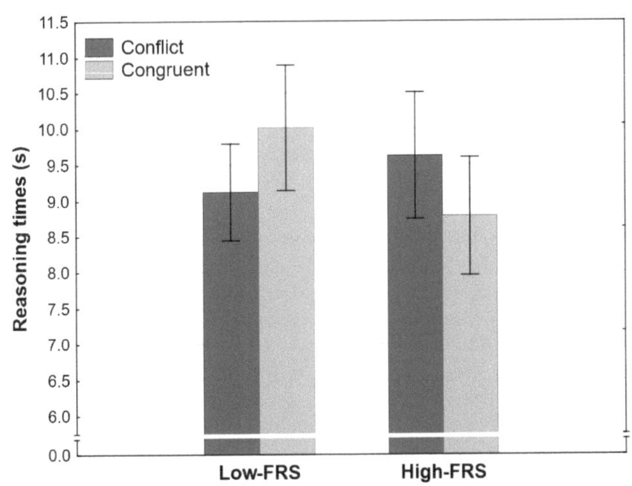

Figure 7. Reasoning time as a function of congruence and final response strength level in Experiment 2 (error bars represent 95% CI). FRS refers to the final response strength level (see Table 1).

Finally, the analysis of confidence judgments revealed both a strong effect of strength level and a strong interaction effect. Participants were generally more confident when reasoning in high-strength conditions (Figure 8). However, the increase of confidence between high and low strength levels was significantly larger in the congruent than it was in the conflict condition.

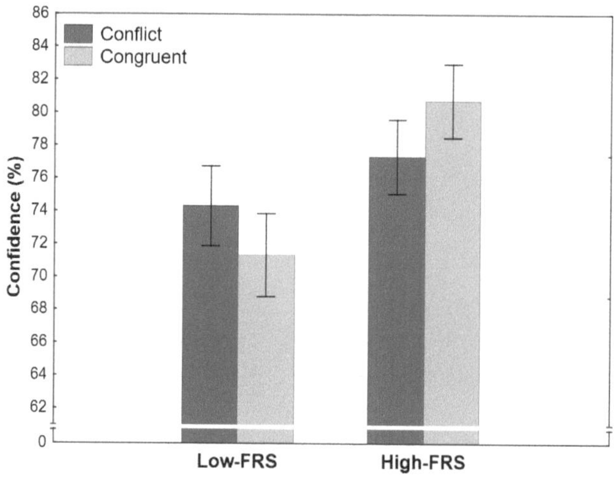

Figure 8. Confidence as a function of congruence and final response strength level in Experiment 2 (error bars represent 95% CI). FRS refers to the final response strength level (see Table 1).

Overall, results from Experiment 2 replicate the finding of Experiment 1 in the low-level conditions and provide evidence that conflict does affect reasoning performance. This effect seems to have been masked in Experiment 1 due to the strength of the dominant heuristic, but the masking was reduced in Experiment 2, revealing the effect of conflict. The increase of the prevailing response strength had a larger impact on reasoning in the congruent than it did in the conflict condition (a decrease in reasoning times and a larger increase in confidence).

4. General Discussion

We set out to test whether conflict detection and resolution retained a strong influence on meta-reasoning when pitting responses cued by the same type of heuristic process rather than two vastly different processes. Thus far, the influence of conflict in reasoning has always been tested by inducing processes relying on different types of mindware. Further, the strength of responses cued by these processes is not strongly controlled or calibrated. Usually, the only manipulation is whether a conflict is present or not; at the most, the level of conflict is lightly manipulated when it is simple to do so (e.g., ratios in the Base Rate task as in Dujmović and Valerjev 2018). Therefore, it is difficult to disambiguate the influence of the presence and magnitude of the conflict itself from other differences between the processes cueing these responses. The key finding from Experiment 1 was that when nominal strengths of the prevailing responses were matched in congruent and conflict conditions, there was no difference in accuracy, reasoning times, or confidence. These findings are in sharp contrast with most research within the modern dual-process framework (Dujmović and Valerjev 2018; Dujmović et al. 2021; Mevel et al. 2015; Shynkaruk and Thompson 2006). Indeed, the most parsimonious explanation of the results was a single-process account in which one heuristic process simply computes all of the information in the task and cues a response with some level of certainty. However, as pointed out in the discussion of Experiment 1, the presence or absence of conflict was not the only difference between the two key conditions. The strength of the dominant heuristic response was also different, which is unavoidable due to the nature of conflict vs. congruent tasks if one aims to match the nominal strength of final responses. This may have masked the impact of conflict in Experiment 1. In order to account for this, Experiment 2 was designed to test whether an increased nominal strength level would have more of an impact on congruent conditions than conflict conditions. We hypothesized that an increase in heuristic response strength would, even if the two conditions remained unbalanced in that regard, unmask the effect of conflict if there was one. Results indeed revealed that increased strength had a higher impact in the congruent condition. Accuracy was equally increased between strength levels irrespective of congruence, but analyses of the remaining two dependent variables revealed strong interaction effects. Participants reasoned significantly faster in the high-strength congruent when compared to the low-strength congruent condition, while the difference between the two conflict conditions was not significant. Further, confidence was higher with the increase of strength level in both congruent and conflict conditions, but the increase was significantly larger in the congruent condition. The results indicate that even when a single type of heuristic process is responsible for cueing multiple responses, there is an effect of conflict. This provides further evidence that two heuristic processes cue the responses; conflict can then be detected and does influence meta-reasoning.

The results revealed an interesting modulation effect of dominant response strength on the impact of conflict in our tasks. A higher strength of the dominant response can mitigate the negative impact conflict has on reasoning times and confidence levels. This may be due to factors that are not mutually exclusive. First, there may be a lower probability of conflict being detected due to the high strength of one of the responses. Second, it may be due to an easier resolution of the conflict when it is detected. It is likely that this effect is bound and non-linear. In Experiment 2, the high-strength conflict condition boasts a higher final response strength (which can be seen as lower conflict), a higher heuristic strength of the dominant, and a lower heuristic strength of the subordinate heuristic response than the low-strength conflict condition. However, these conditions do not differ when looking at reasoning times. Additionally, the increase in confidence was much smaller than in the congruent conditions—despite both heuristic responses in the high-strength congruent condition being much weaker than the dominant response in the high-strength conflict condition.

The findings presented here provide a promising foundation as the approach taken allows for more control and a more detailed analysis of the targeted processes. As far as we are aware, this is the first study to investigate processing that gives rise to System 1 responses rather than just the outcomes of those processes. In the future, materials can be generated using a similar approach in order to match the strength of qualitatively different heuristic processes and explore differences due to other factors. The recent review of dual-process theory by De Neys (2022) poses questions for future research, many of which we feel our approach is uniquely positioned to explore. That being said, there are clear limitations to this approach. There is a clear disconnect between independent typicality judgments and solving reasoning tasks with combinations based on those judgments. Indeed, additional factors that occur during reasoning led to Experiment 1 findings fitting a misleading single-process account. Furthermore, it is entirely unknown whether computations similar to the ones used here to generate experimental conditions mirror anything happening in the brain during reasoning. If they do, they are likely to be an oversimplification of the actual computations. However, the behavioral results presented here match expectations based on our computations rather well (e.g., participants basically guessing in the high-conflict condition of Experiment 1), and this approach may be an avenue to better explore the dynamics of reasoning in the future.

Author Contributions: Conceptualization, P.V. and M.D.; Methodology, P.V. and M.D.; Software, M.D.; Formal analysis, M.D.; Data curation, M.D.; Writing—original draft, P.V. and M.D.; Visualization, M.D. All authors have read and agreed to the published version of the manuscript.

Funding: This research was published with the financial support of the University of Zadar.

Institutional Review Board Statement: The study was conducted in accordance with the Declaration of Helsinki, and approved by the Ethic Committee of the Department of Psychology, University of Zadar (Approval Code: n/a Approval Date: 1 May 2022).

Informed Consent Statement: Participants provided explicit consent in order to take part in the study. The identity of the participant was not required or recorded, and they were informed that data was anonymized and analyzed on a group level.

Data Availability Statement: The data presented in this study are openly available on the Open Science Framework at https://doi.org/10.17605/OSF.IO/J8XTN.

Conflicts of Interest: The authors declare no conflict of interest.

References

Ackerman, Rakefet, and Hagar Zalmanov. 2012. The persistence of the fluency-confidence association in problem solving. *Psychonomic Bulletin & Review* 19: 1187–92. [CrossRef]

Ackerman, Rakefet, and Valerie A. Thompson. 2015. Meta-reasoning: What can we learn from meta-memory? In *Reasoning as Memory*. Edited by Aidan Feeney and Valerie A. Thompson. London: Psychology Press, pp. 164–82.

Ackerman, Rakefet, and Valerie A. Thompson. 2017. Meta-reasoning: Monitoring and control of thinking and reasoning. *Trends in Cognitive Sciences* 21: 607–17. [CrossRef] [PubMed]

Bago, Bence, and Wim De Neys. 2017. Fast logic?: Examining the time course assumption of dual process theory. *Cognition* 158: 90–109. [CrossRef] [PubMed]

Bajšanski, Igor, Maja Močibob, and Pavle Valerjev. 2014. Metacognitive judgments and syllogistic reasoning. *Psychological Topics* 23: 143–65.

De Neys, Wim. 2012. Bias and conflict a case for logical intuitions. *Perspectives on Psychological Science* 7: 28–38. [CrossRef]

De Neys, Wim. 2022. Advancing theorizing about fast-and-slow thinking. *Behavioral and Brain Sciences, in press*. [CrossRef]

De Neys, Wim, and Tamara Glumicic. 2008. Conflict monitoring in dual process theories of thinking. *Cognition* 106: 1248–99. [CrossRef]

Dujmović, Marin, and Pavle Valerjev. 2017. An image is worth a thousand words, but what of numbers? The impact of multi-modal processing on response times and judgments of confidence in base-rate tasks. In *Proceedings of the XXIII Science Conference Empirical Studies in Psychology*. Edited by Oliver Tošković, Kaja Damnjanović and Ljiljana Lazarević. Belgrade: University of Belgrade, pp. 30–36.

Dujmović, Marin, and Pavle Valerjev. 2018. The influence of conflict monitoring on meta-reasoning and response times in a base rate task. *Quarterly Journal of Experimental Psychology* 71: 2548–61. [CrossRef]

Dujmović, Marin, Pavle Valerjev, and Igor Bajšanski. 2021. The role of representativeness in reasoning and metacognitive processes: An in-depth analysis of the Linda problem. *Thinking & Reasoning* 27: 161–86. [CrossRef]

Evans, Jonathan St. B. T. 2012. Questions and challenges for the new psychology of reasoning. *Thinking & Reasoning* 18: 5–31. [CrossRef]

Frederick, Shane. 2005. Cognitive reflection and decision making. *Journal of Economic Perspectives* 19: 25–42. [CrossRef]

Glockner, Andreas, and Cilia Witteman. 2010. Beyond dual-process models: A categorization of processes underlying intuitive judgment and decision making. *Thinking & Reasoning* 16: 1–25. [CrossRef]

Lawson, M. Asher, Richard P. Larrick, and Jack B. Soll. 2020. Comparing fast thinking and slow thinking: The relative benefits of interventions, individual differences, and inferential rules. *Judgment and Decision Making* 15: 660–84. [CrossRef]

Markovits, Henry, Valerie A. Thompson, and Janie Brisson. 2015. Metacognition and abstract reasoning. *Memory & Cognition* 43: 681–93. [CrossRef]

Mevel, Katell, Nicolas Poirel, Sandrine Rossi, Mathieu Cassotti, Gregory Simon, Olivier Houde, and Wim De Neys. 2015. Bias detection: Response confidence evidence for conflict sensitivity in the ratio bias task. *Journal of Cognitive Psychology* 27: 227–37. [CrossRef]

Peirce, Jonathan, Jeremy R. Gray, Sol Simpson, Michael MacAskill, Richard Höchenberger, Hiroyuki Sogo, Erik Kastman, and Jonas K. Lindeløv. 2019. PsychoPy2: Experiments in behavior made easy. *Behavior Research Methods* 41: 195–203. [CrossRef]

Pennycook, Gordon, Dries Trippas, Simon J. Handley, and Valerie A. Thompson. 2014. Base rates: Both neglected and intuitive. *Journal of Experimental Psychology: Learning, Memory, and Cognition* 40: 544–54. [CrossRef]

Pennycook, Gordon, Jonathan A. Fugelsang, and Derek J. Koehler. 2015. What makes us think? A three-stage dual-process model of analytic engagement. *Cognitive Psychology* 80: 34–72. [CrossRef]

Shynkaruk, Jody M., and Valerie A. Thompson. 2006. Confidence and accuracy in deductive reasoning. *Memory & Cognition* 34: 619–32. [CrossRef]

Stanovich, Kieth E. 2018. Miserliness in human cognition: The interaction of detection, override and mindware. *Thinking & Reasoning* 24: 423–44. [CrossRef]

Stanovich, Kieth E., Richard F. West, and Maggie E. Toplak. 2016. *The Rationality Quotient. Toward a Test of Rational Thinking*. Cambridge: The MIT Press.

Thompson, Valerie A., and Stephen C. Johnson. 2014. Conflict, metacognition, and analytic thinking. *Thinking & Reasoning* 20: 215–44. [CrossRef]

Thompson, Valerie A., Jamie A. Prowse Turner, and Gordon Pennycook. 2011. Intuition, reason, and metacognition. *Cognitive Psychology* 63: 107–40. [CrossRef] [PubMed]

Thompson, Valerie A., Jamie A. Prowse Turner, Gordon Pennycook, Linden J. Ball, Hannah Brack, Yael Ophir, and Rakefet Ackerman. 2013. The role of answer fluency and perceptual fluency as metacognitive cues for initiating analytic thinking. *Cognition* 128: 237–51. [CrossRef] [PubMed]

Tversky, Amos, and Daniel Kahneman. 1982. Judgments of and by representativeness. In *Judgment under Uncertainty: Heuristics and Biases*. Edited by Daniel Kahneman, Paul Slovic and Amos Tversky. Cambridge: Cambridge University Press, pp. 84–98.

Tversky, Amos, and Daniel Kahneman. 1983. Extensional versus intuitive reasoning: The conjunction fallacy in probability judgment. *Psychological Review* 90: 293–315. [CrossRef]

Valerjev, Pavle, and Marin Dujmović. 2017. Instruction type and believability influence on metareasoning in a base rate task. In *Proceedings of the 39th Annual Meeting of the Cognitive Science Society*. Edited by Glenn Gunzelmann, Andrew Howes, Thora Tenbrink and Eddy J. Davelaar. London: Cognitive Science Society, pp. 3429–34.

Valerjev, Pavle, and Marin Dujmović. 2019. Performance and Metacognition in Scientific Reasoning: The Covariation Detection Task. *Psychological Topics* 28: 93–113. [CrossRef]

Journal of
Intelligence

MDPI

Article

I Choose to Opt-Out of Answering: Individual Differences in Giving Up Behaviour on Cognitive Tests

Marvin K. H. Law *, Lazar Stankov and Sabina Kleitman

School of Psychology, University of Sydney, Sydney, NSW 2006, Australia
* Correspondence: marvin.law@sydney.edu.au

Abstract: Under the Meta-reasoning model, the process of giving up when a solution may not be feasible reflects an adaptive metacognitive strategy, where individuals opt-out of responding to mitigate error and resource costs. However, research is still needed to determine whether individuals systematically vary in this behaviour and if so, which variables it meaningfully relates with. The current study (N = 176) is the first to examine factorial stability in giving up tendencies and its relationships with on-task confidence, cognitive ability, decision-making predispositions, and academic performance. To measure giving up tendencies, participants completed three cognitive tests which allowed for opting out, thereby capturing giving up frequency within each test and its consistency across tests. Participants also completed five other cognitive tasks embedded with confidence ratings, and a decision-making styles questionnaire. Confirmatory factor analyses were conducted on all giving up, confidence, and accuracy variables, with a three-factor solution having the best fit (containing a giving up factor, confidence factor, and cognitive ability factor). Supporting the proposed adaptive nature of giving up tendencies, the giving up factor correlated positively with cognitive ability, rational decision making, and academic performance. This research establishes factorial stability in giving up tendencies and provides a foundation for further investigation into its role within Meta-reasoning theory.

Keywords: giving up; metacognition; cognition; meta-reasoning; individual differences

1. Introduction

When faced with problems to solve, we balance our feelings of certainty before making the decision to continue solving or to give up. Depending on the strength of these feelings, the decision-making process can be quick and intuitive or slow and reflective (Ackerman et al. 2020). Ackerman and Thompson's (2017) Meta-reasoning framework combines cognitive and metacognitive processes to explain how individuals solve problems and make decisions. Under this model, the decision to give up can occur at any point of the problem-solving process and has been posited to be an adaptive strategy to mitigate potential errors and resource costs (Payne and Duggan 2011). However, is there a trait-like construct surrounding giving up behaviour? Namely, do individuals vary systematically in these tendencies across different tasks? Furthermore, if so, what are the variables that share meaningful relationships with giving up tendencies? Using the Meta-reasoning model to establish hypotheses, this research examines these questions, contributing to further development of the model and the broader decision-making field of research.

According to the Meta-reasoning model, metacognitive processes both monitor and regulate reasoning during the problem-solving and decision-making process (Ackerman and Thompson 2017). These metacognitive processes allow individuals to gauge the effectiveness of their cognitive input whilst adjusting their cognitive resources based on the outcome and demand requirements (full description is available in Ackerman and Thompson 2017). One line of research has focussed on individual differences in cognitive and metacognitive traits. Findings from this research suggest that cognitive and metacognitive traits vary between individuals and can predict better decision making as well as academic achievement and life success (de Bruin et al. 2007; Parker and Fischhoff 2005; Toplak et al. 2017; West et al. 2008). For example, feelings of certainty, measured through on-task post-response confidence ratings, have been established as a robust metacognitive trait, confidence, which relates with life outcomes such as academic performance (Stankov 2013; Stankov et al. 2014; Stankov et al. 2015). However, one caveat of this research is that confidence ratings are based on chosen responses, where individuals must firstly decide on an answer to a problem before responding with a confidence rating that their chosen response is correct (Stankov et al. 2015). This caveat neglects problem-solving contexts where individuals may feel uncertain about a problem to the extent that they instead decide to give up on answering the question, choosing "I don't know". The current study seeks to determine whether there are robust individual differences in this type of giving up behaviour where the individual chooses to opt-out from the problem at hand.

Although there has been some investigation into opt-out decision making within experimental psychology (Ackerman 2014; Ackerman and Beller 2017; Goldsmith et al. 2014; Koriat and Goldsmith 1996; Payne and Duggan 2011; Weber and Brewer 2008; Weber and Perfect 2011) and animal psychology (Coutinho et al. 2015; Smith et al. 1997; Washburn et al. 2005), no research has yet been conducted on how individuals systematically vary in these decisions. Similarly, to the best of our knowledge, there is no research on how giving up tendencies relate to metacognitive traits, cognitive ability, and decision-making predispositions. Furthermore, the role of these tendencies on real-life performance outcomes have also not been investigated. Thus, this paper is a first proof of concept which aims to extend the current individual differences framework on decision making by including measures of opting out or giving up behaviour as adapted mainly from the animal metacognition literature. More specifically, we aim to examine the factorial structure of giving up tendencies, measured through the consistency in "giving up" behaviour within a task and across different tasks, and then compare it with the traits confidence, cognitive ability, decision-making predispositions, and academic performance.

1.1. Cognition and Metacognition

Although varying in some aspects, most theories of metacognition distinguish two groups of processes: (a) metacognitive monitoring, which monitors the allocation of cognitive resources; and (b) metacognitive control, which allocates cognitive resources (Ackerman and Thompson 2017, Nelson and Narens 1990). On-task, post-response confidence ratings (i.e., how confident are you that your answer is correct? Stankov et al. 2015) have been proposed to capture a key component of metacognitive self-monitoring, and have been vigorously investigated for decades, including their relationship with cognitive abilities and academic achievements. These feelings of confidence are theorised to accumulate throughout the problem-solving process, ultimately reaching a threshold, resulting in an individual to respond with a chosen answer (Koriat and Goldsmith 1996). Furthermore, this confidence can also determine a person's willingness to commit their chosen answer for further scrutiny, such as submitting it for marking (Ackerman 2014; Koriat and Goldsmith 1996; Undorf et al. 2021).

Within the literature of individual differences, confidence has been shown to be a trait-like disposition, stable across differing contexts (see Stankov et al. 2015, for a review). Stemming from past decision-making research (Lichtenstein and Fischhoff 1977), a common paradigm is to measure confidence alongside cognitive ability within reasoning tasks using post-response confidence ratings (Ackerman 2014; Stankov and Lee 2008). In these studies, confidence and cognitive ability are measured by calculating overall mean confidence and accuracy, respectively, with both constructs being independent but positively correlated around the .3–.6 mark (Stankov and Lee 2008). The two constructs have been shown to predict real-life outcomes such as academic performance, with both contributing unique and significant variance (Neisser et al. 1996; Rohde and Thompson 2007;

Stankov et al. 2012; Stankov et al. 2014). These findings highlight the combined role of cognitive and metacognitive capacities within real-life contexts.

However, this approach imposes the need for a forced-choice paradigm, where individuals must respond with an answer before rating their confidence. This paradigm ignores the possibility where an individual realises that they do not know the answer and thus prefers to opt-out from answering the item. The decision to give up, however, is proposed to be a key decision-making pathway which is driven by metacognitive monitoring. Importantly, by not providing an answer, the individual may be making a better choice for minimising error and resource costs.

The importance of examining strategically opting out of providing an answer to avoid errors ("giving up behaviour") becomes particularly apparent when comparing research investigating metacognitive miscalibrations such as overconfidence and meta-ignorance (Blanchard et al. 2020; Buratti et al. 2013; Händel et al. 2020; Kleitman et al. 2019; Moore and Healy 2008; Schlösser et al. 2013). An individual who never gives up on any problem is bound to waste unnecessary time and cognitive resources. Similarly, within self-regulated learning, individuals who cannot give up on any learning goals may become overwhelmed (Boekaerts 1999). Acknowledging a lack of knowledge is particularly crucial in settings where ignorance can lead to further risk or harm (e.g., medical risk perception and health behaviours; Waters et al. 2013).

Despite its relevance to real-life problem solving and its inclusion within metacognitive models such as Ackerman and Thompson (2017)'s Meta-reasoning model, little research has investigated how people vary in this giving up behaviour and its relationship with other theoretically relevant variables. To address this gap in the literature, the current study examines individual differences in giving up behaviour in two important ways. Firstly, using a factor analytic approach, the current study aims to determine whether individuals give up systematically and in a generalisable manner across different cognitive tasks. If so, we will determine whether this behaviour defines a potential novel trait, which we will label as giving up. We will then relate this construct with the well-established trait confidence (Stankov et al. 2015), cognitive ability, decision-making predispositions, and academic performance.

1.2. Opting Out or Giving Up Behaviour

Although slight differences may exist in "opting out" and "giving up" terms, for the purpose of this research, we use both terms interchangeably to mean avoiding responding to a cognitive test item by indicating that they feel uncertain to continue. Existing research on "giving up" behaviour offers several different approaches to explain why an individual would choose to opt-out from solving a cognitive problem. Within this literature, some research has examined how individuals may give up on answering based on the perceived solvability of a problem and expended resources (Ackerman 2014; Payne and Duggan 2011). To explain giving up behaviour, Payne and Duggan (2011) proposed that individuals may give up by weighing the perceived probability of a question being solvable (P), the perceived gains of successfully answering the question (G), and finally the perceived costs of attempting the question (C) with the final guiding rule being PG-C. As PG-C approaches zero, participants become increasingly incentivised to give up on the question, rather than to solve it. The theory, however, does not stipulate whether the participants will apply this guiding rule consistently across different cognitive stimuli, thus, the factorial stability of this construct needs to be determined.

Following this guiding rule, we propose that the trait confidence may relate with individual differences in giving up tendencies, as confidence judgements would reflect the perceived capacity to solve a problem (P). Individuals with lower trait confidence would have lower values of PG-C, resulting in greater frequency of giving up. This dynamic between confidence and giving up would reflect the role of confidence ratings in driving decisions to withhold answers, a result which has been well established within experimental psychology (Koriat and Goldsmith 1996). However, as giving up tendencies are also determined by costs (C) and gains (G) associated with a problem, we do not expect giving up scores and confidence ratings to converge onto a single giving up/confidence factor. In contrast, we hypothesise that giving up scores and confidence ratings would converge onto two independent but negatively correlated giving up and confidence factors respectively. The negative correlation between the two would reflect the role of monitoring processes in driving control decisions, as presented within metacognitive models such as the Meta-reasoning framework (Ackerman and Thompson 2017).

The theoretical relationship between the proposed giving up factor and the cognitive ability factor is less clear and has never been investigated. Individuals with greater cognitive ability may give up less frequently simply due to their capacity to answer more problems correctly. However, we propose that individuals with greater cognitive ability should exhibit greater metacognitive awareness, by giving up on problems which are too difficult or resource intensive. Based on this latter premise, individuals with greater cognitive ability would be able to accurately follow the *PG-C* guideline to strategically give up, thereby minimising errors and resource costs. Conversely, individuals with low cognitive ability are more likely to deny or fail to follow *PG-C*, e.g., continuing solving despite the great cost of continued deliberation or responding with a likely incorrect response, despite having the option to opt-out. Thus, we hypothesise that giving up tendencies should relate positively with cognitive ability.

In addition to confidence and cognitive ability, decision-making styles may also relate with differences in giving up tendencies. Decision-making research has found that individuals vary in their approaches surrounding problem solving and making decisions, with five independent but interrelated styles of decision making (Scott and Bruce 1995). These styles include (1) rational, reflecting logical and methodical evaluations, (2) avoidant, reflecting delaying and avoiding decisions, (3) dependent, reflecting a need for advice and thoughts from others, (4) intuitive, making decisions based on feelings and instincts, and (5) spontaneous, reflecting making decisions quickly and impulsively without thinking deeply. Although some styles have been found to relate with metacognitive awareness, these decision-making predispositions have not been examined with giving up tendencies, nor with metacognitive trait confidence (Basu and Dixit 2022). Under the *PG-C* guideline, decision-making styles which prioritise strategic decision making or response avoidance should be associated with greater giving up tendencies. Based on this premise, we hypothesise that rational and avoidant decision-making styles should relate with greater giving up tendencies. The positive relationship with rational decision making will support the adaptive, strategic nature of giving up, while the positive relationship with avoidant decision making will suggest that giving up tendencies are reflective of response avoidance. It is unclear, however, how decision-making styles based on external dependence, minimal strategic thinking, and low risk consideration relate to giving up tendencies. As such, the remaining dependent, intuitive, and spontaneous decision-making styles are hypothesised to relate negatively or share no relationships with giving up tendencies. Ultimately, the primary aims for this study are to determine whether individual differences in giving up tendencies are systematic across tasks, and if so, whether these systematic tendencies would relate to confidence, cognitive ability as well as varying decision-making styles.

1.3. Academic Performance

The current study also aims to examine the role of giving up tendencies on real-world performance outcomes, i.e., academic performance. Within the individual differences literature, the role of both metacognition and cognition on academic performance is clear, with both confidence and cognitive ability factors positively relating to academic performance (Stankov et al. 2012, 2014). The error and resource cost mitigation strategy surrounding giving up tendencies can also be beneficial within academic settings, reflecting a capacity to know when to opt-out of answering. Thus, it is hypothesised that giving up tendencies would relate positively with academic performance.

1.4. Uncertainty–Monitoring Paradigm from Animal Metacognition

Existing experimental research has examined giving up behaviour within mostly animal participants, but also occasionally with humans, using a perceptual decision-making task known as the sparse–uncertain–dense (SUD) task (Beran et al. 2009; Coutinho et al. 2015; Smith et al. 1997, 2006; Washburn et al. 2005). The task was developed as an alternative approach to examining metacognitive monitoring, since traditional measures of self-report ratings (e.g., confidence ratings) were not feasible with animal participants (Smith et al. 1997). In this paradigm, participants are provided with boxes of varying degrees of pixel density and asked to determine whether they judged the box was sparse (S), dense (D) or whether they were uncertain (U) (example items shown below in Figure 1). To opt-out of responding, the participant must indicate that they feel uncertain, thus giving up on responding for the item and moving on to the next. Using 42 varying box items, participants were provided feedback for their S and D responses with either a 0.5 s whoop sound for correct responses or an 8 s buzz sound for incorrect responses. Uncertain responses were not given any feedback.

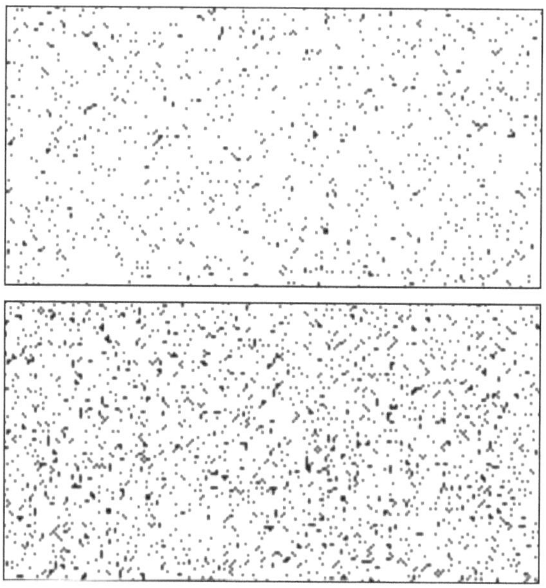

Figure 1. Example items from the SUD task. The box on the top is the least dense item, and the box on the bottom is the densest.

For animal participants, they would also be given food as a reward for correct responses. By running these items over multiple iterations, such animal research found the uncertain response to be used as a means for error monitoring, with animals (e.g., rhesus monkeys) and humans adaptively choosing the uncertain response for more difficult items (Smith et al. 1997). For both animals and humans, some individual differences emerged in the optimality of uncertainty responding both as the frequency as well as in the range of difficulty of choosing uncertain (Smith et al. 1997; Washburn et al. 2005).

Although the uncertainty response is a well-validated finding in animal research (Coutinho et al. 2015; Smith et al. 1997), it has yet to be compared systematically with confidence, cognitive ability, and decision-making predispositions as measured within the individual differences paradigm[1]. Adapting the SUD task to the current study, which uses human participants, aggregate uncertainty scores (i.e., frequency of indicating uncertainty and thus giving up) for each participant were computed for each test and for the relationships between these cumulative instances of "giving up" behaviour on different tests examined.

1.5. Aims and Hypotheses

In summary, we have two broad exploratory aims in the current study. The first aim is to examine whether there are robust individual differences in giving up behaviour across tasks. Secondly, we aim to investigate the relationships between giving up tendencies, and other theoretically important constructs, namely, the confidence trait, cognitive ability, and decision-making predispositions, as well as with academic performance. To the best of our knowledge, the current study is the first to examine the factorial structure of giving up tendencies, as well as its relationship with other key variables. We hypothesise that:

1. We postulate and compare four models: (1) a one-factor model with all performance accuracy, on-task confidence and giving up metrics defining one factor; (2) a two-factor model with performance accuracy defining one factor with confidence and giving up scores converging to define the second factor; (3) a three-factor model with a giving up factor, defined by all giving up scores across the three tasks, a confidence factor, defined by all confidence ratings, and a cognitive ability factor, defined by all performance accuracy variables; and finally, (4) a four-factor model with the broad accuracy factor splitting into two possible factors, accuracy for the tests with the embedded giving up scores and accuracy for the tests with the embedded confidence ratings. We expect that a three-factor solution would fit the data best.
2. The giving up factor will (a) correlate negatively with the confidence and (b) positively with the cognitive ability factors. It will also correlate positively with (c) rational and (d) avoidant decision-making styles, and negatively or not at all with (e) dependent, (f) intuitive, and (g) spontaneous decision-making styles. Finally, the giving up factor will (h) correlate positively with academic performance.

To determine the factorial structure of giving up tendencies, a Confirmatory Factor Analysis (CFA) was conducted on all giving up, confidence, and accuracy scores. Correlations between the subsequent factors and other variables (as specified above) were then examined.

2. Materials and Methods

2.1. Participants

The current study was conducted on 181 participants, 156 first year psychology students from the University of Sydney who participated for partial course credit using SONA recruitment, and 25 paid participants recruited through paid SONA at the University of Sydney. The paid sample were from a similar undergraduate cohort compared with those unpaid participants[2].

Data cleaning: Data from five (one paid) participants were removed from any further analyses due to either non-attempts on most of the battery (item response times lower than one second for several measures) or non-completion of a substantial (>35%) number of measures. After filtering non-attempts for specific measures, except for academic performance (see below), some variables contained random missing values (together up to 6.8%). Expectation maximisation imputation was conducted on all missing values aside from academic performance using SPSS (IBM Corp 2020), as it provides reliable and unbiased parameter estimates (Graham et al. 2013). For the final analyses, there were 176 participants (131 Females, mean age = 20.31, SD = 3.90, age range = 17–59), which was sufficient to conduct factor analyses which required approximately 10–15 participants per variable within the model (Tabachnick and Fidell 2013).

2.2. Measures

To measure giving up, confidence, and cognitive ability, the current battery included eight cognitive tasks: three perception-based cognitive tasks (SUD, Cube Comparison and Visual Search tests), three traditional reasoning tasks (Esoteric Analogies, Advanced

Raven's Progressive Matrices, and Berlin Numeracy tests) and two heuristics and biases tasks (Cognitive Reflection and Applying Decision Rules tests). The battery also included the General Decision-Making Styles questionnaire, a measure of varying decision-making approaches (Scott and Bruce 1995). Other measures were included (e.g., personality, executive function tasks, additional heuristics and biases tasks), in conjunction with the measures for the current study, but were omitted as either being outside the scope of the focus of this study and/or having no systematic relationship with the variables of interest. Descriptive statistics, reliabilities, and correlation results are shown in Table A2 in Appendix A. Furthermore, the full data set and their descriptions are available on request from the corresponding author.

Giving up scores were measured within each of the three perception-based cognitive tasks, captured as the percentage of opt-out/giving up responses. To mitigate the possibility that giving up tendencies may converge simply due to being an artefact of response methodology, we varied the three tasks based on (1) punishment/reward mechanism, (2) the complexity/difficulty surrounding the cognitive processes involved, and (3) method of responding. *The punishment/reward system* for the SUD task was taken from the animal psychology literature and was based on time delays, i.e., participants were punished for incorrect responses by having to wait longer for the next item. As such, giving up decisions on this task would allow for avoiding punishment. In contrast, this type of punishment was not utilised for the Cube Comparison and Visual Search tasks. We also varied *the difficulty between tasks*, with the SUD task being the least difficult due to being adapted from an animal psychology paradigm. In comparison, the Cube Comparison Test was the most difficult, as it was from the Educational Testing Service (ETS) Kit of Factor Referenced Cognitive Tests and has been assumed to capture more complex visual processes associated with fluid reasoning (Nusbaum and Silvia 2011). Finally, the Visual Search task was judged to be in between the other two tasks in complexity and difficulty due to requiring emotional visual perception processes (further details available in Table 1). Finally, we varied *the response method* for the Visual Search Task compared to two other measures (see Table 1 in the Method section).

Confidence ratings were measured in the remaining five cognitive tasks, three traditional reasoning tasks and two heuristics and biases tasks. Confidence for each task was calculated by averaging post-response confidence ratings for each item, "how certain are you that your answer is correct?": either from 0% to 100% or from 25% to 100%). These ratings were then averaged to determine each participants' average confidence. Finally, performance accuracy was also captured for all cognitive tasks, measured by the percentage of correct responses.

2.3. Procedure

The study was conducted via computers within an experimental university lab, with participants completing all tasks and a demographic questionnaire measuring variables including age, sex, and English as a first language in a 2 h block. To account for counterbalancing, participants completed one of three different test blocks with differing task orders. Due to the long protocol, participants were encouraged and were able to take short breaks during the testing session.

Table 1. Summary of measures used in this study.

Measure	Example Item	Reliability from Previous Studies
Cognitive Tasks with embedded giving up options		
Sparse-Uncertainty-Dense (SUD) Task—Beran et al. (2009). A visual processing task which was developed within animal psychology research to examine metacognition in animals and humans (Beran et al. 2009; Smith et al. 2006). For this task, participants are shown pixel density boxes and were asked to decide whether the number of pixels in the box constitutes as sparse or dense with correct responses eliciting rewards and incorrect responses eliciting punishments. The most sparse and dense items were given to participants as practise prior to the task. For the current study, a point system and item time delays were used as reward and punishment incentives with correct responses leading to a point increase and time delays between items of one second whilst incorrect responses had a point reduction and a time delay of eight seconds [a]. To measure giving up, participants were given a third option to select that they are uncertain about the answer, which will skip over the current item, eliciting no point change and a time delay of two seconds. Following prior studies, pixel density plots were 200 × 100 pixels in area with pixel densities ranging from 1085 to 2255 with a 1.8% increase in pixel density per level (Beran et al. 2009). Ultimately, 42 levels were constructed with the first 21 defined as sparse and the latter 21 defined as dense. These pixel boxes were constructed using a pixel density randomisation script in JavaScript (source code available by request). Participants were shown each level three times over the task. Overall accuracy and giving up scores were measured for this task, with accuracy being the percentage of correct responses out of all attempted items and giving up scores, being the percentage of uncertain responses out of all items (126—three repetitions of the 42 items).	see Figure 1	-
Cube Comparisons—Ekstrom et al. (1976). The Cube Comparisons Task is a measure of visual processing from the ETS's Kit of Factor Referenced Cognitive Tests with 42 items. For each item, participants are shown three faces of two cubes and must decide if they think that the cubes could potentially be the same cube or are definitely different. Similar to the SUD task, a third "uncertain" option was added for when participants were uncertain about answering. Prior to the task, participants were shown some practise items with feedback. Overall accuracy and giving up scores were calculated in the same way as that in the SUD task. For this task, there was no feedback nor reward/punishment for responding.	See Ekstrom et al. (1976) and Figure A1 in the Appendix A.	Original measure: .84 (Kozhevnikov and Hegarty 2001).
Visual Search for Faces with Corresponding Emotion Expressions of Different Intensity—Short Form—Wilhelm et al. (2014). This task is a short 20-item measure of emotion perception. For each of the twenty items, participants are shown 9 different faces of the same individual in a 3 × 3 grid, with each face displaying a certain emotion. Participants must discern which of the faces do not express the same emotion as the majority of the faces. Adapting to measure giving up, an "uncertain" option was provided for each of the faces in the items, with participants being allowed to choose uncertain rather than make a decision on whether the face was part of the majority emotion or not. As a result, participants were not allowed to select uncertain and make a decision on the same face. Participants were shown an example item before the task. Following prior research, accuracy on the task was measured using Unbiased Hit Rates which control for both Hit and False rates. Akin to the other two tasks measuring giving up, overall giving up scores were measured as the percentage of uncertain responses out of all possible face selections (180).	See Wilhelm et al. (2014) and Figure A2 in the Appendix A.	.89 (Law et al. 2018)

Table 1. *Cont.*

Measure	Example Item	Reliability from Previous Studies
Cognitive tasks with embedded confidence ratings		
Esoteric Analogies Test (EAT)—Stankov (1997). The EAT is a measure of both fluid and crystallised intelligence with the current study using a twenty-item version. For each item, participants are shown a pair of words and an additional word. They are then given four multiple choice options and must decide which option relates to the additional word in the same way that the pair of words relate to each other. Accuracy on this score is measured by the percentage of items correct.	SPACE is to POINT as TIME is to: CLOCK, ETERNITY, MOMENT*, POSITION	.70 to .72 (Kleitman et al. 2018; Stankov 1997)
Berlin Numeracy Test (BNT)—Cokely et al. (2012). The BNT is a measure of numerical ability, more specifically, statistical and risk literacy. We used the four-item version of the BNT, with every item being a mathematical question which asks for an open-ended numeric answer. Accuracy was measured through percentage of items correct. Although the measure only consists of four items, it has been shown to predict risk understanding more effectively than other longer numeric and intelligence measures.	Imagine we are throwing a five-sided die 50 times. On average, out of these 50 throws how many times would this five-sided die show an odd number (1, 3 or 5)? ____ out of 50 throws	Test-retest reliability: .91 (Cokely et al. 2012).
Raven's Advanced Progressive Matrices (RAPM)—Raven (1938) and Waschl et al. (2016). RAPM is a measure of fluid intelligence which has been used frequently within the individual differences literature (McGrew 2009; Waschl et al. 2016). For each item, participants are shown a 3 × 3 grid of abstract figures with the bottom right figure being left blank. The participants are shown eight different options and must decide which option best fits the pattern shown in the grid. Based on previous research, the current study used a 15-item version which has been validated in prior research (Waschl et al. 2016). Accuracy is measured as the percentage of correct items.	See Raven (1938) and Figure A3 in the Appendix A for example.	.80 (Jackson et al. 2017)
Cognitive Reflection Test (CRT)—Frederick (2005) and Toplak et al. (2014). The CRT is a reasoning measure of susceptibility to heuristics and biases where participants must answer open-ended questions which have been constructed to prompt incorrect intuitive answers. The current study used the extended 7-item CRT (Toplak et al. 2014). Accuracy on the CRT is measured by percentage correct.	A bat and a ball cost $1.10 in total. The bat costs a dollar more than the ball. How much does the ball cost? ____ cents.	.70 (Jackson et al. 2016)
Applying Decision Rules (ADR)—(de Bruin et al. 2007). The ADR is a reasoning measure testing heuristics and biases susceptibility where participants are shown attribute information on five different DVD players, with four different attributes (e.g., sound quality) each with varying levels of performance ratings from 1 (very low) to 5 (very high). For each of the ten items, participants are given a scenario and must then select the DVD player/s which best fit the scenario. An accuracy score is computed by the percentage of answers correct.	LaToya only wants a DVD player that has got a "Very High" rating on Sound Quality	.60 (Jackson et al. 2016)

Table 1. *Cont.*

Measure	Example Item	Reliability from Previous Studies
Decision-Making Styles		
General Decision-Making Styles (GDMS) Scale—(Scott and Bruce 1995). The GDMS scale measures people's decision-making tendencies with 25 self-report items with 5 items each relating to 5 different decision-making styles: rational, avoidant, dependent, intuitive, and spontaneous. Participants must respond from 1 (strongly disagree) to 5 (strongly agree) for each item. Average scores for each decision-making style are calculated, with higher scores indicating higher style propensity.	Rational: I make decisions in a logical and systematic way. Avoidant: I avoid making important decisions until the pressure is on. Dependent: I rarely make important decisions without consulting other people. Intuitive: When making decisions, I rely upon my instincts. Spontaneous: I often make impulsive decisions	.67 to .87 (Scott and Bruce 1995; Spicer and Sadler-Smith 2005)
Real-Life Outcomes		
Academic performance. To measure academic performance, overall weighted average marks (WAM) from consenting participants were collected (n = 130).	-	-

[a] Initial testing used four-second time periods as delays for incorrect response of (n = 25). This was changed to eight seconds following participant feedback, suggesting that four-second delays were insufficient/not noticeable. No significant differences were found between participants with the different levels of delays however, and thus, results were examined together for analyses, $t(174) = -1.78$, $p = .08$ for accuracy and $t(174) = -1.04$, $p = .31$ for giving up scores.

3. Results

To determine the factorial structure of giving up tendencies, after examining the pattern of correlations, we conducted a confirmatory factor analysis (CFA; maximum likelihood with geomin rotation) on giving up variables (SUD, Cube Comparisons, and Visual Search tasks), confidence variables (Esoteric Analogies, Berlin Numeracy, Advanced Raven's Progressive Matrices, Cognitive Reflection Test and Applying Decision Rules tasks), and performance accuracy scores from all eight tasks.

3.1. Descriptive Statistics, Reliabilities, and Correlations between Giving Up, Confidence, Performance Accuracy and Life Outcome Measures

Descriptive statistics, reliabilities, and correlations between all metacognitive and cognitive variables are provided in Table 2 below. The descriptive statistics and reliabilities for all remaining variables can be found in Appendix A (Table A1).

Internal consistency ranged from acceptable to excellent (.64 to .94) for all metacognitive and cognitive variables (Cronbach 1951), except for BNT accuracy (.55). Average giving up scores were generally low and varied substantially between the three tasks, ranging from 3.73% to 14.02%. The average frequency of giving up decisions for all tasks ranged between five and seven times. Despite high positive skew on giving up scores (see Figure 2 below), only five participants (less than 3%) had never given up on any task. Furthermore, 35%, 43% and 54% of participants had given up at least five times on the SUD, Cube Comparisons, and Visual Search tasks. respectively. These results suggest that despite having low average giving up scores, the current study captured a reasonable distribution of opt-out behaviour, and these tendencies were systematic with the internal consistency estimates for giving up tendencies ranging between .78 and .94. Despite having an added opt-out option, performance accuracy for the Visual Search task in the current study (~71%) was similar compared with the same task in other studies where no opt-out option was provided (~72%, Law et al. 2018). Accuracy scores between the three tasks embedded with opt-out options also varied. Confirming our expectations, accuracy was highest for the SUD task (87%), being 12% to 16% higher than the other two tasks. Although we had expected the Cube Comparisons to be more difficult than the Visual Search task, their accuracy scores were similar, being 75% and 71% respectively.

To examine potential item order effects, giving up rates were examined based on the presentation order of each item for the three tasks (randomised for the SUD and non-randomised for the other tasks). Although rates of giving up appeared to be relatively stable across item order, early to middle items for the SUD and Visual Search tasks appeared to have greater marked differences in giving up rate between proximal items (e.g., giving up rates for the Visual Search task was 8% for item 4 but 3% for item 5). In comparison, items shown later in the task exhibited more stable giving up rates (e.g., items 11 to 20 of the Visual Search task have giving up rates between 2% and 4%). This behavioural pattern was not observed for the Cube Comparison task.

Between the giving up variables, two of the three correlations were positive and significant: .21 and .32. Furthermore, there was a positive manifold between all confidence ratings (correlations range: .29 to .61). Finally, most performance accuracy variables significantly and positively correlated with each other aside from SUD Task accuracy and Visual Search Task accuracy. SUD Task accuracy did not correlate significantly with performance accuracy from any other task. Visual Search accuracy correlated positively, albeit weakly with all other performance accuracy variables, with only two correlations being significant: EAT accuracy ($r = .15, p < .05$) and BNT accuracy ($r = .18, p < .05$).

Table 2. Descriptive statistics, reliabilities, and correlations between metacognitive and cognitive variables.

	Mean	SD	1	2	3	4	5	6	7	8	9	10	11	12	13	14	15	16
Giving Up Score																		
1. SUD	4.61	7.01	(.94)															
2. CUBE	14.02	15.35	.32 **	(.91)														
3. VS	3.73	3.45	.21 **	.03	(.78)													
Confidence																		
4. EAT	68.96	11.22	.02	−.21 **	.06	(.83)												
5. BNT	63.38	24.93	.07	−.17 *	.06	.29 **	(.83)											
6. RAPM	65.60	19.93	−.01	−.16 *	.05	.46 **	.46 **	(.91)										
7. CRT	76.70	17.77	.00	−.29 **	.03	.47 **	.54 **	.52 **	(.79)									
8. ADR	83.65	14.69	.02	−.22 **	.02	.53 **	.41 **	.57 **	.61 **	(.87)								
Performance Accuracy																		
9. SUD	87.17	3.80	.20 **	.07	.16 *	−.01	−.02	−.02	−.04	−.02	(.79)							
10. CUBE	74.55	13.39	.14	.10	.07	.24 **	.36 **	.36 **	.17 *	.21 **	.02	(.85)						
11. VS	70.61	12.10	−.01	−.05	.00	.01	.11	.05	−.04	−.10	.07	.12	(.86)					
12. EAT	64.13	16.06	.09	−.05	.11	.43 **	.25 **	.30 **	.30 **	.29 **	.02	.36 **	.15 *	(.69)				
13. BNT	42.61	30.39	.16 *	−.07	.12	.17 *	.52 **	.29 **	.31 **	.19 *	.09	.30 **	.18 *	.39 **	(.55)			
14. RAPM	60.68	22.75	.05	.02	.12	.23 **	.38 **	.69 **	.30 **	.30 **	.06	.52 **	.10	.37 **	.39 **	(.79)		
15. CRT	46.43	32.16	.11	−.04	.15 *	.27 **	.50 **	.40 **	.45 **	.31 **	−.02	.44 **	.13	.39 **	.53 **	.49 **	(.78)	
16. ADR	62.69	21.55	.11	−.09	.05	.21 **	.36 **	.23 **	.19 *	.37 **	.01	.44 **	.12	.43 **	.38 **	.38 **	.50 **	(.64)

Note: * $p < .05$, ** $p < .01$; SUD—Sparse Uncertainty Dense, CUBE—Cube Comparisons, VS—Visual Search, EAT—Esoteric Analogies Test, BNT—Berlin Numeracy Test, RAPM—Raven's Advanced Progressive Matrices, CRT—Cognitive Reflection Test, ADR—Applying Decision Rules. (Cronbach's alpha estimates) are shown in the diagonal for all variables. Descriptive statistics (mean and SD) are given as %s.

Giving up scores only correlated significantly with eight variables based on confidence ratings and performance accuracy. Cube Comparisons giving up score shared significant negative correlation with all confidence variables (ranging between −.29 and −.16). Furthermore, SUD Task giving up score shared significant positive correlations with BNT and SUD Task accuracy (.16 and .20 respectively). Finally, Visual Search giving up score correlated significantly and positively with both CRT and SUD Task accuracy. All performance accuracy variables besides SUD Task and Visual Search Task accuracies had significant and positive correlations with all confidence variables (correlations range: .18 to .69). In contrast, SUD Task and Visual Search Task performance accuracy did not correlate significantly with any of the confidence variables. These relationships are summarised by the results of the Confirmatory Factor Analysis below.

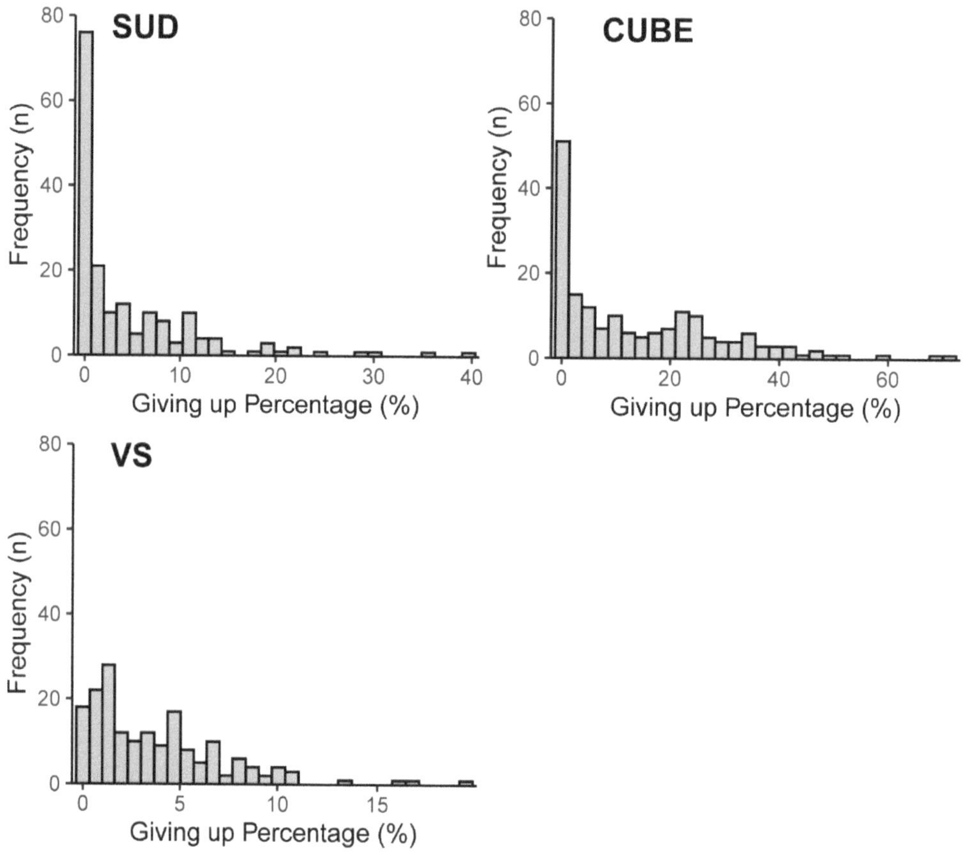

Figure 2. Frequency histograms for the giving up scores in the SUD, CUBE, and Visual Search tasks.

3.2. *Confirmatory Factor Analysis (CFA) of Cognitive and Metacognitive Variables Using Maximum-Likelihood Method*

To test the factorial structure underlying giving up, confidence, and performance accuracy variables, five CFAs solutions were conducted with the Maximum Likelihood method and geomin rotation using the lavaan package within the R software (R Core Team 2022; Rosseel 2012) (descriptions and fit indices of all models can be found in Table 3 below). Using fit indices cut-off conventions from the literature, model fit was defined to be acceptable if Standardized Root Mean Squared Residual (SRMR) $\leq$ .08, Root Mean Square Error of Approximation (RMSEA) ≤ .06, Comparative Fit Index (CFI) ≥ .95, and Tucker–Lewis Index (TLI) ≥ .95 (Brown 2015; Hu and Bentler 1999).

Table 3. Model description and fit statistics for CFA with all giving up, confidence, and accuracy variables.

	χ^2	df	χ^2/df	TLI	CFI	RMSEA (90% CI)	AIC	SRMR
One-factor model	387.502	104	3.726	.618	.669	.124 (.111–.138)	22,370.5	.092
Two-factor model	309.666	103	3.006	.719	.759	.107 (.093–.121)	22,294.7	.084
Three-factor model	284.124	101	2.813	.746	.786	.101 (.088–.116)	22,273.1	.080
Four-factor model	277.246	98	2.829	.744	.791	.102 (.088–.116)	22,272.2	.079
Three-factor model—Modified	123.273	95	1.298	.958	.967	.041 (.015–.060)	22,124.3	.060

χ^2—Chi-squared, df—degrees of freedom, TLI—Tucker-Lewis Index, CFI—Comparative Fit Index, RMSEA—Root Mean Square Error of Approximation, CI—Confidence Interval, AIC—Akaike Information Criterion, SRMR—Standardized Root Mean Squared Residual.

Firstly, an initial one-factor model was conducted with all variables loading onto a single factor with the model fit being poor (Brown 2015; Hu and Bentler 1999). Subsequently, a two-factor model (capturing a giving up/confidence factor and a cognitive ability factor) and a three-factor model (capturing a giving up factor, a confidence factor, and a cognitive ability factor) were conducted to determine the factorial independence of giving up tendencies. For the two-factor model, all giving up variables and confidence ratings defined a single giving up/confidence factor whilst all accuracy scores defined a cognitive ability factor, with both factors being allowed to covary. In contrast, the three-factor model had all giving up variables define a giving up factor, all confidence ratings define a confidence factor, and all accuracy scores define a cognitive ability factor, with all factors covarying freely. Comparing differences in χ^2 statistics indicated that the two-factor solution was an improvement compared to the one-factor solution (χ^2 diff = 77.84, $p < .001$); however, the fit indices remained poor (Brown 2015; Hu and Bentler 1999). Subsequently, the three-factor solution demonstrated a significant improvement over the two-factor solution (χ^2 diff = 25.54, $p < .001$) with better fit indices; however, the overall model fit was still poor. Thus, a four-factor solution was conducted with the broad accuracy factor being split into two possible factors: accuracy for the tests with the embedded giving up scores and accuracy for the tests with the embedded confidence scores. This model, however, showed no significant improvement over the three-factor-model (χ^2 diff = 6.88, $p = .08$), suggesting that the three-factor model fit the data best; however, the low Tucker–Lewis Index suggested that some important paths may have been missing. Thus, we examined and incorporated several minor modifications.

Firstly, following prior literature, given the strong relationship between confidence ratings and accuracy, residuals between accuracy and confidence from the same measures were allowed to covary (Jackson et al. 2016; Kleitman et al. 2019). Although there are concerns about allowing residuals to correlate due to violating the assumption of independence of residuals, the correlations between the residuals for the different scores within the same test are acceptable, as they capture non-random disturbances due to the nature of the test (Brown 2015). Cube Comparisons giving up score was also allowed to load onto the confidence factor, as it correlated negatively with all confidence ratings. Finally, SUD accuracy was allowed to load on the giving up factor instead of the cognitive ability factor given that its only significant (albeit weak) correlations were with giving up scores. The final modified model had acceptable fit based on cut-off fit conventions (Brown 2015; Hu and Bentler 1999). Although loadings varied in sizes, they all were statistically significant ($p < .05$). A path diagram summarising the final model is shown in Figure 3 below.

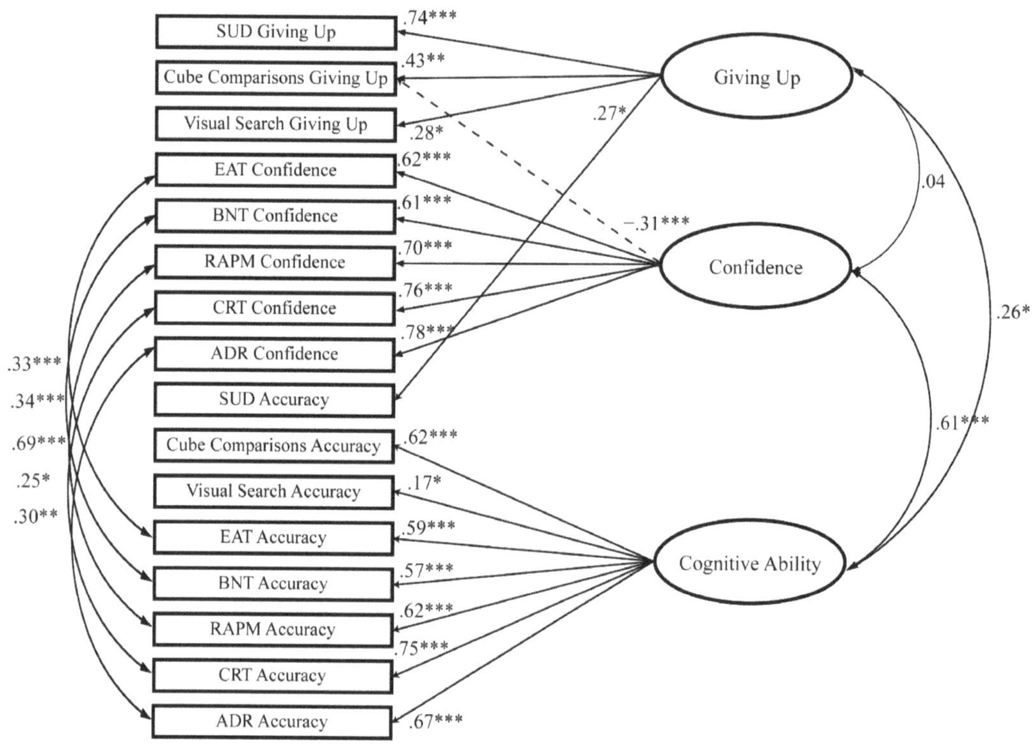

Figure 3. Summary path model of the final three-factor CFA model with standardised regression and correlation coefficients * $p < .05$, ** $p < .01$, *** $p < .001$.

The three factors in the final model were interpreted as:

Factor 1: Giving up—this factor was defined by all giving up variables as well as with a positive loading from SUD accuracy. This factor is indicative of robust individual differences in giving up tendencies.

Factor 2: Confidence—this factor was defined by all confidence variables, as well as negatively defined by giving up score on the Cube Comparisons task. This factor is clearly indicative of the metacognitive confidence trait. The negative loading from Cube Comparisons giving up scores indicates that high-confidence individuals may be related to giving up tendencies in some contexts.

Factor 3: Cognitive ability—this factor was defined by all accuracy variables except for SUD accuracy. This factor is indicative of cognitive ability. The low loading from Visual Search accuracy on this cognitive ability factor indicates that this factor is more aligned with fluid and crystallised intelligence, rather than perceptual ability. However, Cube Comparisons accuracy had a strong positive loading on this factor, supporting the general nature of this factor which included a test embedded with giving up response options.

The giving up factor did not correlate significantly with the confidence factor (r = .04). However, supporting Hypothesis 2b, the giving up factor correlated significantly and positively with the cognitive ability factor (r = .26). Finally, replicating numerous studies (e.g., Stankov and Lee 2008; Stankov et al. 2015), the confidence factor correlated significantly, positively, and strongly with the cognitive ability factor (r = .61).

3.3. Comparisons between Giving Up, Confidence, and Cognitive Ability Factors with All Other Variables

To compare the resultant factors with other variables, factor scores were estimated using the Bartlett method within the lavaan package (Rosseel 2012). Subsequently, *t* tests and correlations were conducted to compare the three extracted factors (giving up, confidence, cognitive ability) with decision-making styles and academic performance. Pairwise comparisons between sex and English as a first language indicated no significant differences in giving up factor scores for either variable. However, females had significantly lower confidence (mean difference = −4.54) and cognitive ability (mean difference = −4.79) than males, $t(174) = -3.60$, $p < .001$ and $t(174) = -3.07$, $p < .001$, respectively. Correlations between the three factors resulted from the CFA, and other variables are shown in Table 4 below.

Table 4. Correlations between the giving up, confidence, and cognitive ability factors with decision-making styles and academic performance.

	Giving Up	Confidence	Cognitive Ability
GDMS Avoidant	.05	**−.15** *	−.03
GDMS Dependent	.08	−.01	.00
GDMS Intuitive	−.11	−.08	**−.16** *
GDMS Rational	**.18** *	**.20** **	.10
GDMS Spontaneous	.01	.05	.02
Academic Performance	**.19** *	**.18** *	**.35** **

GDMS—General Decision-Making Style, ** $p < .01$, * $p < .05$.

Seven significant correlations were found between the three factors and other variables. Supporting Hypothesis 2c and 2h, the giving up factor correlated positively with both rational decision-making style and academic performance, albeit weakly. The confidence factor also correlated positively and weakly with rational decision-making style and academic performance, and negatively and weakly with an avoidant decision-making style. Finally, the cognitive ability factor correlated negatively and weakly with intuitive decision-making style, and positively and moderately with academic performance.

4. Discussion

The current research is a novel study which is a proof of concept into systematic individual differences in giving up tendencies, and their relationships with confidence, cognitive ability traits, decision-making predispositions, and academic performance. The decision to give up has been posited to be an adaptive strategy to mitigate potential errors and resource costs (Ackerman 2014; Koriat and Goldsmith 1996; Payne and Duggan 2011). However, to the best of our knowledge, previous research has not examined whether individuals vary systematically in the use of this strategy, and if so, what other psychological constructs might hold meaningful relationships with these tendencies. The current study aimed to investigate these two questions. Following Payne and Duggan's (2011) *PG-C* model, we proposed that confidence, cognitive ability traits, decision-making predispositions, and academic performance should all relate with systematic giving up tendencies.

To examine these premises, the current study employed eight cognitive tasks, three embedded with an opt-out option, and five embedded with post-response confidence ratings. CFA were conducted on all giving up, confidence, and accuracy variables to determine the factorial structure of giving up tendencies. Finally, *t* tests and correlations were conducted between the resultant factor scores with all other variables.

4.1. Are Giving Up Tendencies Systematic?

Our CFA results support the premise that giving up tendencies are systematic across tasks, with the best fitting solution having three factors: all giving up scores converging onto a novel giving up factor, and confidence ratings and performance accuracies converging onto two separate confidence and cognitive ability factors, respectively. However, to improve the model's fit, we incorporated two post hoc cross-loadings. Although post hoc, these cross-loadings make sense and can potentially inform future studies. Firstly, SUD accuracy loaded positively and weakly onto the giving up factor. This cross-loading can be explained by the strategic role of giving up towards minimising potential errors embedded in the test's scoring system. Individuals who give up more often are likely to have higher accuracy scores within the same task due to opting out of more difficult questions. Thus, this cross-loading demonstrates the error mitigation strategy of giving up, with participants having increased accuracy by opting out and reducing potential errors. Secondly, giving up score on the Cube Comparisons task loaded negatively onto the confidence factor. Although the giving up factor did not correlate significantly with the confidence factor, this cross-loading suggests that giving up tendencies may be negatively related to confidence within certain task contexts. As such, we suggest further studies comparing giving up tendencies and confidence from similar or the same tasks. Despite these two significant cross-loadings, the CFA solution has robust fit, establishing the novel giving up factor, which is representative of systematic giving up tendencies. Within the tests, participants also exhibit high internal consistency in their giving up responses, with Cronbach's alpha estimates ranging between .78 and .94. Altogether, although largely exploratory, these results provide strong support that giving up tendencies are systematic within and across tasks.

4.2. Correlations with Other Variables

We had expected that the giving up factor would correlate significantly with confidence, cognitive ability, decision-making predispositions, and academic performance. However, there were only three significant and positive, though weak, correlations, namely, with cognitive ability, rational decision making, and academic performance. Normative expectations suggest that individuals who give up more often would likely be those who are less capable. However, our findings support an alternative, and possibly strategic nature of giving up, such that individuals who give up are those with greater cognitive ability and are doing so based on a rational and logical approach. Decisions to give up on responding are often viewed negatively, particularly when frequent (Grice 1975; Goldsmith 2016). Thus, rational decision makers and those who are more intelligent may be able to judge the benefits of giving up as an error mitigation strategy, resulting in more frequent opt-out decisions. The adaptive benefits of this approach can ultimately allow individuals to perform more effectively within academic settings. Although promising, we acknowledge that these correlations are weak and need to be replicated in future studies.

Despite the strong theoretical relationship, there was no significant correlation between the confidence and giving up factors, suggesting that those with low confidence did not opt to give up more often. One explanation for this non-significant relationship may be due to the strategic nature of giving up. As giving up is posed as a strategy to avoid potential errors and costs, individuals who give up more may be those who are error and cost avoidant, rather than those with low confidence. Namely, some individuals may still prefer to avoid incorrect responses despite having high trait confidence. In contrast, some individuals with low trait confidence may prefer to respond with an answer, despite having a lower perceived likelihood of answering correctly. This preference may explain participants choosing low-confidence responses even when given the opportunity to opt-out (Ackerman 2014). However, we note that in experimental research, opt-out decisions have been found to be associated with low-confidence judgements even in conditions where alternatives to opting-out are permitted such as seeking help or providing less informative answers (Undorf et al. 2021; Weber and Brewer 2008). Thus, our current findings should be replicated and extended within experimental paradigms, to determine the nature of its relationship, however, focusing between trait confidence and the novel giving up factor.

Instead of trait confidence, individual differences reflective of other metacognitive judgements may drive giving up tendencies. One potential avenue for investigation is the initial judgements of solvability (iJOS), the perceived judgment of being able to solve a problem after initial evaluation (Ackerman and Thompson 2017). Unlike confidence ratings, IJOS are made early on in the problem-solving process and have been shown to predict later decision-making behaviour, e.g., solving time and post-response confidence (Lauterman and Ackerman 2019). They have also been suggested to control decisions to give up and time until giving up (Ackerman and Thompson 2017; Lauterman and Ackerman 2019; Payne and Duggan 2011). Thus, we suggest future studies to examine other metacognitive processes, e.g., solvability judgments, which may relate to giving up behaviour.

4.3. Limitations and Future Direction

There were several limitations within the current study. Firstly, our results were limited by low rates of giving up decisions. These low rates of opt-out decision have been seen in previous studies, suggesting that decisions to opt-out are inherently low even within other contexts (Ackerman 2014; Hanczakowski et al. 2013; Washburn et al. 2005). Furthermore, as mentioned before, these low rates may be reflective of the large number of items used in our tasks, with the average giving up frequency ranging from five to seven for all three tasks. Thus, we propose that the current distribution of giving up rates, although low, is reasonable in examining individual differences in a behaviour that is generally avoided. Nevertheless, increasing rates of giving up behaviours may provide greater spread and individual differences in the behaviour. The current study measured giving up rates using reasoning tasks which were easier than standard reasoning tasks used to measure cognitive ability. Thus, future studies can increase giving up rates by implementing tasks with greater task difficulty. Additionally, giving up rates may also increase with greater costs associated with incorrect responses. As such, tasks embedded with greater punishments for incorrect answers are also more likely to incentivise giving up decisions. Furthermore, we also suggest that future studies should investigate potential item order effects within tasks using the opt-out procedure.

Another limitation was that only frequency of giving up was examined within our study. Existing models of giving up (Ackerman and Thompson 2017; Payne and Duggan 2011) suggest that there are varying stages during problem solving where an individual may decide to give up, with differing motivations. For example, the decision to give up may come immediately after seeing a problem if it is judged to be unsolvable, or it may come after deliberation if solving efforts are unfruitful. Predispositions towards these divergent giving up decisions may also be a source of individual differences worth investigation. Thus, we suggest future studies to examine giving up at different stages of problem solving, as well as individual differences in varying giving up approaches.

As the current study was a preliminary examination into giving up behaviour, its investigation into task motivation was also limited, with only the SUD task having a reward/punishment system. With no incentives for high task performance and no punishment for choosing to give up, participants may have chosen to give up due to low task motivation. However, this is an unlikely explanation, as giving up scores were greatest for the Cube Comparisons task where participants were not punished for responding incorrectly. Furthermore, frequencies of giving up were similar across all giving up tasks. Nevertheless, we recommend that future studies should incorporate more substantial reward/punishment paradigms as well as control for task motivation.

The study was also limited by differences in the cognitive tasks measured. There were substantial differences in average accuracy scores across the eight employed cognitive tasks (42% to 87%), and this could have affected the results given the known hard–easy effect, where individuals overestimate their success towards hard tasks and underestimate success towards easy tasks (Suantak et al. 1996). In particular, the tasks embedded with giving up had higher accuracy when compared with the tasks embedded with confidence ratings. These differences may have been due to the role of giving up in increasing output-bound accuracy, where participants can choose to give up for more difficult items, leaving only the easier items to be calculated for accuracy (Koriat and Goldsmith 1996). To address these limitations, we suggest further studies to compare giving up and confidence from similar tasks with the same range of difficulty. We must note, however, that further studies in our laboratory have utilised the mentioned modifications to the research design and have replicated these findings.

Our current study was further limited by the length of the protocol. Due to the extensive battery required to investigate our aims, testing sessions lasted approximately two hours, undoubtedly causing fatigue. Although our protocol was counter-balanced to control for fatigue, future research would benefit by using these findings to develop more concise studies. As a further limitation, we acknowledge that behaviour can vary based on environmental factors such as time constraints and task demands outside of the experiment. These factors are difficult to control for within single-session experiments such as the current study, and we recommend that future studies investigate the stability of individual differences in opt-out behaviour using longitudinal experiment protocols.

5. Conclusions

The current study is a novel investigation into individual differences in giving up behaviour, more broadly examining the existence of the metacognitive control trait, giving up. CFA revealed a best fitting three-factor solution, with giving up, confidence, and performance accuracy variables loading respectively onto three separate and independent factors. These findings support the premise that individuals give up systematically across different tasks as a metacognitive strategy to minimise errors and resource costs during problem solving. This is a premise that serves as a much-needed foundation for future investigations. Importantly, the study also found significant positive, albeit weak, relationships between this newly established giving up factor with cognitive ability, rational decision making, and academic performance, suggesting that more intelligent and rational decision makers may be more predisposed to give up as an adaptive strategy. These tendencies may also benefit academic outcomes. Future research should investigate both the characteristics of this novel giving up factor and its potential underlying processes to further understand its metacognitive nature and its role in learning, cognitive processing, and decision making.

Author Contributions: Conceptualization, M.K.H.L. and S.K.; methodology, M.K.H.L. and S.K.; formal analysis, M.K.H.L. and S.K.; investigation, M.K.H.L.; data curation, M.K.H.L.; writing—original draft preparation, M.K.H.L.; writing—review and editing, M.K.H.L., L.S. and S.K.; visualization, M.K.H.L.; supervision, S.K. (main) and L.S. (associate). All authors have read and agreed to the published version of the manuscript.

Funding: This research received no external funding.

Institutional Review Board Statement: Ethics approval was received from the Human Research Ethics Committee at the University of [redacted], project number: 2019/185.

Informed Consent Statement: Informed consent was obtained from all subjects involved in the study.

Data Availability Statement: The data presented in this study are openly available from the OSF repository at DOI: 10.17605/OSF.IO/5NK8R.

Conflicts of Interest: The authors declare no conflict of interest.

Appendix A

Table A1. Descriptive statistics and reliabilities for decision-making variables and academic performance.

	Mean	SD	α
GDMS Avoidant	3.15	0.89	.87
GDMS Dependent	3.82	0.65	.75
GDMS Intuitive	3.64	0.61	.76
GDMS Rational	3.86	0.59	.75
GDMS Spontaneous	2.90	0.69	.80
Academic Performance	70.77	9.88	-

GDMS—General Decision-Making Styles.

Table A2. Descriptive statistics, reliabilities, and correlations with factor scores of all other measured variables.

	Mean	SD	α	Correlations		
				Giving Up	Confidence	Cognitive Ability
Executive Function						
Flanker Difference Score	0.10	0.09	-	−.05	−.15	**−.16 ***
Task Switching Difference Score	0.08	0.15	-	−.10	−.01	−.05
Running Letters Accuracy	0.47	0.19	.71	**.15 ***	**.26 ****	**.38 ****
Heuristics and Biases						
Risky Gambles	0.42	0.17	.66	.09	.08	.04
Resistance to Framing	36.18	4.85	.62	.10	.06	.09
Social Norms Rank Score	0.51	0.22	.37	.00	.07	**.16 ***
Consistency in Risk Perception	0.87	0.12	.51	.00	**.18 ***	.12
Personality						
Mini-IPIP Extraversion	3.05	1.01	.87	−.01	−.07	**−.15 ***
Mini-IPIP Agreeableness	4.00	0.72	.72	.03	−.13	−.14
Mini-IPIP Conscientiousness	3.22	0.81	.68	.10	.04	−.06
Mini-IPIP Neuroticism	3.09	0.81	.65	−.06	**−.17 ***	−.03
Mini-IPIP Intellect	3.86	0.69	.69	.03	.06	.11
BAS Drive	2.72	0.59	.80	−.01	.10	−.13
BAS Fun Seeking	3.00	0.54	.70	−.05	.03	−.10
BAS Reward Responsiveness	3.39	0.42	.72	−.10	−.02	**−.25 ****
BIS	3.15	0.51	.79	−.01	−.10	−.07
Need for Closure	3.83	0.64	.82	−.05	.03	−.02
Other Variables						
Medical Decision-Making Test Accuracy	0.49	0.24	.78	.01	**.15 ***	**.34 ****
Adapted Decision Outcomes Inventory Score	−0.15	0.08	-	−.03	.14	.13

Note: * $p < .05$, ** $p < .01$.

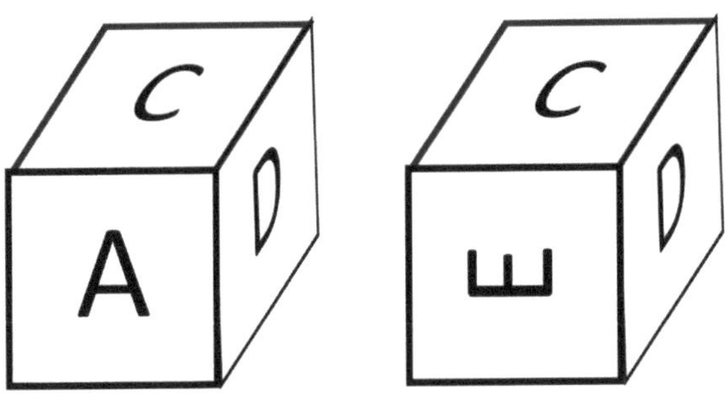

Figure A1. Pseudo-example item for the Cube Comparisons Task. In this example, the two cubes must be different, as the A and E occupy the same position (below the C, and to the left of the D), and any letter can only appear once on each cube.

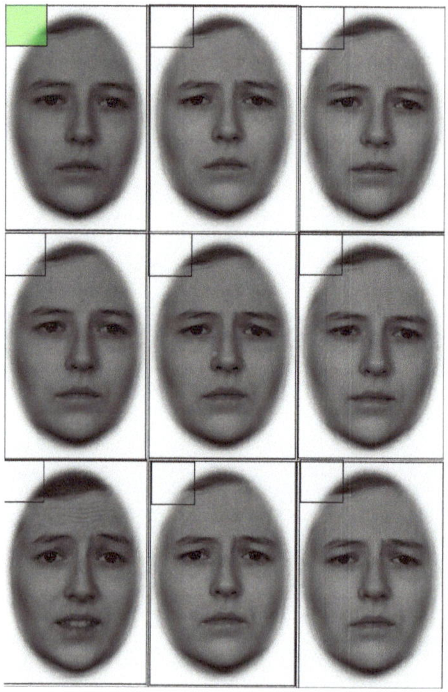

Figure A2. Example item for the Visual Search task. In this example, there are eight faces indicating sadness and one indicating disgust. The top left corner of the top left face was selected, indicating uncertainty towards whether the face displayed the majority emotion or not.

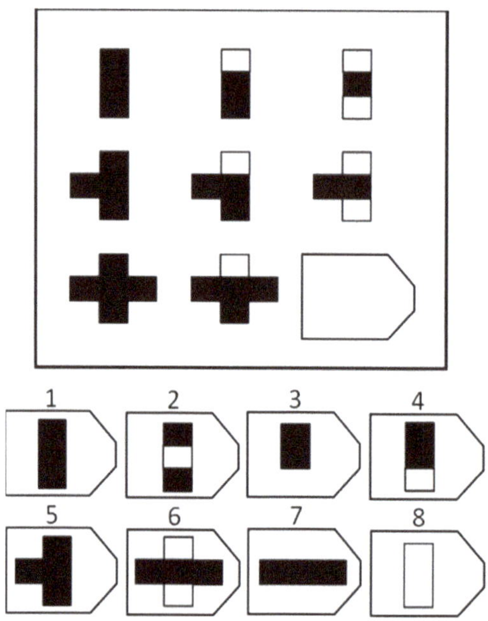

Figure A3. Pseudo-example item for the Raven's Advanced Progressive Matrices. In the example, the current answer is pattern 6, as it fits both the vertical and horizontal patterns.

Notes

1 Whilst the animal psychology literature labels this paradigm as "uncertainty monitoring", this terminology may be confusing due to the metacognitive and decision-making literatures' use of certainty judgements (i.e., confidence). In such fields, uncertainty is treated synonymously as the antithesis of certainty (e.g., 0% confidence on a response being correct). Thus, to increase clarity, we refer to the "uncertainty monitoring" variables measured in the current study as "giving up" behaviour to align with terminology which is more commonly used in the cognitive and decision-making literature.

2 When factor analyses were conducted with the paid participants removed, the pattern of results was analogous to the full sample analyses. Thus, both samples were combined to improve power (analyses conducted without paid participants can be provided upon request from the corresponding author).

References

Ackerman, Rakefet. 2014. The Diminishing Criterion Model for Metacognitive Regulation of Time Investment. *Journal of Experimental Psychology: General* 143: 1349–68. [CrossRef]

Ackerman, Rakefet, and Yael Beller. 2017. Shared and Distinct Cue Utilization for Metacognitive Judgements during Reasoning and Memorisation. *Thinking & Reasoning* 23: 376–408. [CrossRef]

Ackerman, Rakefet, and Valerie A. Thompson. 2017. Meta-Reasoning: Monitoring and Control of Thinking and Reasoning. *Trends in Cognitive Sciences* 21: 607–17. [CrossRef]

Ackerman, Rakefet, Elad Yom-Tov, and Ilan Torgovitsky. 2020. Using Confidence and Consensuality to Predict Time Invested in Problem Solving and in Real-Life Web Searching. *Cognition* 199: 104248. [CrossRef] [PubMed]

Basu, Sabnam, and Shikha Dixit. 2022. Role of Metacognition in Explaining Decision-Making Styles: A Study of Knowledge about Cognition and Regulation of Cognition. *Personality and Individual Differences* 185: 111318. [CrossRef]

Beran, Michael J., J. David Smith, Mariana V. C. Coutinho, Justin J. Couchman, and Joseph Boomer. 2009. The Psychological Organization of 'Uncertainty' Responses and 'Middle' Responses: A Dissociation in Capuchin Monkeys (Cebus Apella). *Journal of Experimental Psychology. Animal Behavior Processes* 35: 371–81. [CrossRef]

Blanchard, Matthew D., Simon A. Jackson, and Sabina Kleitman. 2020. Collective Decision Making Reduces Metacognitive Control and Increases Error Rates, Particularly for Overconfident Individuals. *Journal of Behavioral Decision Making* 33: 348–75. [CrossRef]

Boekaerts, Monique. 1999. Self-Regulated Learning: Where We Are Today. *International Journal of Educational Research* 31: 445–57. [CrossRef]

Brown, Timothy A. 2015. *Confirmatory Factor Analysis for Applied Research, Second Edition*. New York: Guilford Publications.

Buratti, Sandra, Carl Martin Allwood, and Sabina Kleitman. 2013. First- and Second-Order Metacognitive Judgments of Semantic Memory Reports: The Influence of Personality Traits and Cognitive Styles. *Metacognition and Learning* 8: 79–102. [CrossRef]

Cokely, Edward T., Mirta Galesic, Eric Schulz, Rocio Garcia-Retamero, and Saima Ghazal. 2012. Measuring Risk Literacy: The Berlin Numeracy Test. *Judgment and Decision Making* 7: 23.

Coutinho, Mariana V. C., Joshua S. Redford, Barbara A. Church, Alexandria C. Zakrzewski, Justin J. Couchman, and J. David Smith. 2015. The Interplay between Uncertainty Monitoring and Working Memory: Can Metacognition Become Automatic? *Memory & Cognition* 43: 990–1006. [CrossRef]

Cronbach, Lee J. 1951. Coefficient Alpha and the Internal Structure of Tests. *Psychometrika* 16: 297–334. [CrossRef]

de Bruin, Wandi Bruine, Andrew M. Parker, and Baruch Fischhoff. 2007. Individual Differences in Adult Decision-Making Competence. *Journal of Personality and Social Psychology* 92: 938–56. [CrossRef] [PubMed]

Ekstrom, Ruth B., John W. French, Harry H. Harmon, and Diran Dermen. 1976. *Manual for Kit of Factor-Referenced Cognitive Tests*. Princeton: Educational Testing Service.

Frederick, Shane. 2005. Cognitive Reflection and Decision Making. *Journal of Economic Perspectives* 19: 25–42. [CrossRef]

Goldsmith, Morris. 2016. Metacognitive Quality-Control Processes in Memory Retrieval and Reporting. In *The Oxford Handbook of Metamemory*. Oxford: Oxford Academic. [CrossRef]

Goldsmith, Morris, Ainat Pansky, and Asher Koriat. 2014. Metacognitive Control of Memory Reporting. In *The SAGE Handbook of Applied Memory*. Edited by Timothy J. Perfect and D. Stephen Lindsay. London: SAGE Publications, pp. 481–500.

Graham, John W., Patricio E. Cumsille, and Allison E. Shevock. 2013. Methods for Handling Missing Data. In *Handbook of Psychology: Research Methods in Psychology*, 2nd ed. Edited by Irving B. Weiner, John A. Schinka and Wayne F. Velicer. Hoboken: John Wiley & Sons, Inc., vol. 2, pp. 109–41.

Grice, Herbert. P. 1975. *Logic and Conversation*. Leiden: Brill. [CrossRef]

Hanczakowski, Maciej, Tomasz Pasek, Katarzyna Zawadzka, and Giuliana Mazzoni. 2013. Cue Familiarity and 'Don't Know' Responding in Episodic Memory Tasks. *Journal of Memory and Language* 69: 368–83. [CrossRef]

Händel, Marion, Anique B. H. de Bruin, and Markus Dresel. 2020. Individual Differences in Local and Global Metacognitive Judgments. *Metacognition and Learning* 15: 51–75. [CrossRef]

Hu, Li-tze, and Peter M. Bentler. 1999. Cutoff Criteria for Fit Indexes in Covariance Structure Analysis: Conventional Criteria versus New Alternatives. *Structural Equation Modeling: A Multidisciplinary Journal* 6: 1–55. [CrossRef]

IBM Corp. 2020. *IBM SPSS Statistics for Windows*. version 27.0. Armonk: IBM Corp.

Jackson, Simon A., Sabina Kleitman, Pauline Howie, and Lazar Stankov. 2016. Cognitive Abilities, Monitoring Confidence, and Control Thresholds Explain Individual Differences in Heuristics and Biases. *Frontiers in Psychology* 7: 1559. [CrossRef]

Jackson, Simon A., Sabina Kleitman, Lazar Stankov, and Pauline Howie. 2017. Individual Differences in Decision Making Depend on Cognitive Abilities, Monitoring and Control: Individual Differences in Decision Making. *Journal of Behavioral Decision Making* 30: 209–23. [CrossRef]

Kleitman, Sabina, Marvin K. H. Law, and Judy Kay. 2018. It's the Deceiver and the Receiver: Individual Differences in Phishing Susceptibility and False Positives with Item Profiling. *PLoS ONE* 13: e0205089. [CrossRef]

Kleitman, Sabina, Jessica Sik-Wai Hui, and Yixin Jiang. 2019. Confidence to Spare: Individual Differences in Cognitive and Metacognitive Arrogance and Competence. *Metacognition and Learning* 14: 479–508. [CrossRef]

Koriat, Asher, and Morris Goldsmith. 1996. Monitoring and Control Processes in the Strategic Regulation of Memory Accuracy. *Psychological Review* 103: 497–517. [CrossRef] [PubMed]

Kozhevnikov, Maria, and Mary Hegarty. 2001. A Dissociation between Object Manipulation Spatial Ability and Spatial Orientation Ability. *Memory & Cognition* 29: 745–56. [CrossRef]

Lauterman, Tirza, and Rakefet Ackerman. 2019. Initial Judgment of Solvability in Non-Verbal Problems—A Predictor of Solving Processes. *Metacognition and Learning* 14: 365–83. [CrossRef]

Law, Marvin K. H., Simon A. Jackson, Eugene Aidman, Mattis Geiger, Sally Olderbak, and Sabina Kleitman. 2018. It's the Deceiver, Not the Receiver: No Individual Differences When Detecting Deception in a Foreign and a Native Language. *PLoS ONE* 13: e0196384. [CrossRef] [PubMed]

Lichtenstein, Sarah, and Baruch Fischhoff. 1977. Do Those Who Know More Also Know More about How Much They Know? *Organizational Behavior and Human Performance* 20: 159–83. [CrossRef]

McGrew, Kevin S. 2009. CHC Theory and the Human Cognitive Abilities Project: Standing on the Shoulders of the Giants of Psychometric Intelligence Research. *Intelligence* 37: 1–10. [CrossRef]

Moore, Don A., and Paul J. Healy. 2008. The Trouble with Overconfidence. *Psychological Review* 115: 502–17. [CrossRef] [PubMed]

Neisser, Ulric, Gwyneth Boodoo, Thomas J. Bouchard, A. Wade Boykin, Nathan Brody, Stephen J. Ceci, Diane E. Halpern, John C. Loehlin, Robert Perloff, Robert Sternberg, and et al. 1996. Intelligence: Knowns and Unknowns. *American Psychologist* 51: 77–101. [CrossRef]

Nelson, Thomas O., and Louis Narens. 1990. Metamemory: A Theoretical Framework and New Findings. *Psychology of Learning and Motivation* 26: 125–73. [CrossRef]

Nusbaum, Emily C., and Paul J. Silvia. 2011. Are Intelligence and Creativity Really so Different?: Fluid Intelligence, Executive Processes, and Strategy Use in Divergent Thinking. *Intelligence* 39: 36–45. [CrossRef]

Parker, Andrew M., and Baruch Fischhoff. 2005. Decision-Making Competence: External Validation through an Individual-Differences Approach. *Journal of Behavioral Decision Making* 18: 1–27. [CrossRef]

Payne, Stephen J., and Geoffrey B. Duggan. 2011. Giving Up Problem Solving. *Memory & Cognition* 39: 902–13. [CrossRef]

R Core Team. 2022. *R: A Language and Environment for Statistical Computing*. Vienna: R Foundation for Statistical Computing.

Raven, John C. 1938. *Progressive Matrices*. New York: The Psychological Corporation.

Rohde, Treena Eileen, and Lee Anne Thompson. 2007. Predicting Academic Achievement with Cognitive Ability. *Intelligence* 35: 83–92. [CrossRef]

Rosseel, Yves. 2012. Lavaan: An R Package for Structural Equation Modeling. *Journal of Statistical Software* 48: 1–36. [CrossRef]

Schlösser, Thomas, David Dunning, Kerri L. Johnson, and Justin Kruger. 2013. How Unaware Are the Unskilled? Empirical Tests of the 'Signal Extraction' Counterexplanation for the Dunning–Kruger Effect in Self-Evaluation of Performance. *Journal of Economic Psychology* 39: 85–100. [CrossRef]

Scott, Susanne G., and Reginald A. Bruce. 1995. Decision-Making Style: The Development and Assessment of a New Measure. *Educational and Psychological Measurement* 55: 818–31. [CrossRef]

Smith, J. David, Wendy E. Shields, Jonathan Schull, and David A. Washburn. 1997. The Uncertain Response in Humans and Animals. *Cognition* 62: 75–97. [CrossRef]

Smith, J. David, Michael J. Beran, Joshua S. Redford, and David A. Washburn. 2006. Dissociating Uncertainty Responses and Reinforcement Signals in the Comparative Study of Uncertainty Monitoring. *Journal of Experimental Psychology: General* 135: 282–97. [CrossRef]

Spicer, David P., and Eugene Sadler-Smith. 2005. An Examination of the General Decision Making Style Questionnaire in Two UK Samples. *Journal of Managerial Psychology* 20: 137–49. [CrossRef]

Stankov, Lazar. 1997. *Gf–Gc Quickie Test Battery*. Unpublished Test Battery Available from the School of Psychology, University of Sydney. Sydney: University of Sydney.

Stankov, Lazar. 2013. Noncognitive Predictors of Intelligence and Academic Achievement: An Important Role of Confidence. *Personality and Individual Differences* 55: 727–32. [CrossRef]

Stankov, Lazar, and Jihyun Lee. 2008. Confidence and Cognitive Test Performance. *Journal of Educational Psychology* 100: 961–76. [CrossRef]

Stankov, Lazar, Jihyun Lee, Wenshu Luo, and David J. Hogan. 2012. Confidence: A Better Predictor of Academic Achievement than Self-Efficacy, Self-Concept and Anxiety? *Learning and Individual Differences* 22: 747–58. [CrossRef]

Stankov, Lazar, Suzanne Morony, and Yim Ping Lee. 2014. Confidence: The Best Non-Cognitive Predictor of Academic Achievement? *Educational Psychology* 34: 9–28. [CrossRef]
Stankov, Lazar, Sabina Kleitman, and Simon A. Jackson. 2015. Measures of the Trait of Confidence. In *Measures of Personality and Social Psychological Constructs*. Edited by Gregory J. Boyle, Donald H. Saklofske and Gerald Matthews. Cambridge: Academic Press, pp. 158–89. [CrossRef]
Suantak, Liana, Fergus Bolger, and William R. Ferrell. 1996. The Hard–Easy Effect in Subjective Probability Calibration. *Organizational Behavior and Human Decision Processes* 67: 201–21. [CrossRef]
Tabachnick, Barbara, and Linda Fidell. 2013. *Using Multivariate Statistics: Pearson New International Edition PDF EBook*, 6th ed. Harlow: Pearson Education, Limited.
Toplak, Maggie E., Richard F. West, and Keith E. Stanovich. 2017. Real-World Correlates of Performance on Heuristics and Biases Tasks in a Community Sample. *Journal of Behavioral Decision Making* 30: 541–54. [CrossRef]
Toplak, Maggie E., Richard F. West, and Keith E. Stanovich. 2014. Assessing Miserly Information Processing: An Expansion of the Cognitive Reflection Test. *Thinking & Reasoning* 20: 147–68. [CrossRef]
Undorf, Monika, Iris Livneh, and Rakefet Ackerman. 2021. Metacognitive Control Processes in Question Answering: Help Seeking and Withholding Answers. *Metacognition and Learning* 16: 431–58. [CrossRef]
Waschl, Nicolette A., Ted Nettelbeck, Simon A. Jackson, and Nicholas R. Burns. 2016. Dimensionality of the Raven's Advanced Progressive Matrices: Sex Differences and Visuospatial Ability. *Personality and Individual Differences* 100: 157–66. [CrossRef]
Washburn, David A., J. David Smith, and Lauren A. Taglialatela. 2005. Individual Differences in Metacognitive Responsiveness: Cognitive and Personality Correlates. *The Journal of General Psychology* 132: 446–61. [CrossRef]
Waters, Erika A., Jennifer L. Hay, Heather Orom, Marc T. Kiviniemi, and Bettina F. Drake. 2013. 'Don't Know' Responses to Risk Perception Measures: Implications for Underserved Populations. *Medical Decision Making: An International Journal of the Society for Medical Decision Making* 33: 271–81. [CrossRef] [PubMed]
Weber, Nathan, and Neil Brewer. 2008. Eyewitness Recall: Regulation of Grain Size and the Role of Confidence. *Journal of Experimental Psychology: Applied* 14: 50–60. [CrossRef] [PubMed]
Weber, Nathan, and Timothy J. Perfect. 2011. Improving Eyewitness Identification Accuracy by Screening Out Those Who Say They Don't Know. *Law and Human Behavior* 36: 28–36. [CrossRef]
West, Richard F., Maggie E. Toplak, and Keith E. Stanovich. 2008. Heuristics and Biases as Measures of Critical Thinking: Associations with Cognitive Ability and Thinking Dispositions. *Journal of Educational Psychology* 100: 930–41. [CrossRef]
Wilhelm, Oliver, Andrea Hildebrandt, Karsten Manske, Annekathrin Schacht, and Werner Sommer. 2014. Test Battery for Measuring the Perception and Recognition of Facial Expressions of Emotion. *Frontiers in Psychology* 5: 404. [CrossRef]

Journal of
Intelligence

Article

The Development of Intuitive and Analytic Thinking in Autism: The Case of Cognitive Reflection

Kinga Morsanyi [1,2,*] and Jayne Hamilton [2]

[1] Department of Mathematics Education, Loughborough University, Loughborough LE11 3TU, UK
[2] School of Psychology, Queen's University Belfast, Belfast BT7 1NN, UK; jayne@studyseed.co.uk
* Correspondence: k.e.morsanyi@lboro.ac.uk

Abstract: The cognitive reflection test (CRT) is a short measure of a person's ability to resist intuitive response tendencies, and to produce normatively correct responses that are assumed to be based on effortful, analytic thinking. A remarkable characteristic of the CRT is that although the questions are open-ended, for each item, the vast majority of people either produce a correct, analytic response or a typical incorrect (i.e., intuitive) response. This unique feature of the CRT makes it possible to investigate the question of whether autistic and neurotypical people share the same intuitions. We report a study that included adolescents and young adults. In both age groups, autistic and neurotypical participants were matched on age, gender, cognitive ability, and educational background. In line with previous findings, the results showed an age-related increase in analytic responding on the CRT, and a decrease in intuitive responding. Crucially, the proportion of both intuitive and analytic responses across autistic and neurotypical participants was identical in both age groups. The current results are in contrast with claims that autistic individuals have an increased tendency toward an analytic/rational type of processing, which is commonly attributed to an impairment within their intuitive reasoning mechanisms.

Keywords: analytic thinking; attention to detail; autism; cognitive reflection; developmental change; intuition; reasoning

1. Introduction

Autism Spectrum Disorder (ASD) is characterized by a combination of persistent deficits in social communication and social interaction, and restricted, repetitive patterns of behavior, interests, and activities. For an individual to obtain a diagnosis of ASD, these symptoms must be present in early childhood, should cause clinically significant impairments in social, occupational, or other areas of functioning, and they should not be better explained by intellectual disability (APA 2013). It is estimated that the prevalence of ASD is about 1 in 36 individuals (Maenner et al. 2023), although diagnosis rates are lower, especially in the case of women (e.g., Hull et al. 2020).

Individuals with ASD are often described as having a strong focus on literal meaning and explicitly presented details (e.g., Happé 1997; Jolliffe and Baron-Cohen 1999), and, at the same time, they might miss the "big picture" (e.g., Frith 1989) or interpret information in a different way than other people. This may be evident in the verbal domain, for example in understanding figurative language, such as metaphors (see Kalandadze et al. 2018 for a systematic review and meta-analysis), although such differences may be smaller or may not even be present when autistic and neurotypical participants are carefully matched on intelligence or verbal ability, or when samples involve high-ability participants (Morsanyi et al. 2020, 2022).

Studies on thinking, reasoning, and judgment in autism suggested that there might be a tendency in individuals with autism to be hyper-systematic in their thinking (e.g., Baron-Cohen and Belmonte 2005), or to exhibit enhanced logical consistency (De Martino et al. 2008). This has been contrasted with a tendency to lack some basic intuitions, such as the ability to read others' emotions (e.g., Baron-Cohen et al. 2003). Previous studies showed that autistic individuals were less susceptible to framing effects (i.e., framing gambles of equal expected values as loss/gain; De Martino et al. 2008; Shah et al. 2016, although see Levin et al. 2015), and to the conjunction fallacy (Morsanyi et al. 2010). Additionally, using a self-report questionnaire, the rational-experiential inventory (Pacini and Epstein 1999), several studies (Brosnan et al. 2016; Levin et al. 2015; Taylor et al. 2022) showed lower levels of self-reported engagement in experiential/intuitive thinking in the ASD group, although the groups did not differ in their self-reported rational/analytic thinking styles. Additionally, Brosnan et al. (2014) found that autistic adolescents showed a reduced tendency to "jump to conclusions" (i.e., they requested more information before making a judgment about probabilistic sampling distributions), which suggests that they are less likely to follow their gut instincts than neurotypical adolescents.

Although the above studies suggest that there may be differences between autistic and neurotypical individuals in terms of their reasoning, decision-making and judgments, there is some difficulty with interpreting the results from these studies. Specifically, most reasoning and judgment problems that contrast intuitive and normative/analytic reasoning do not make it possible to distinguish between the hypotheses that autistic participants tend to perform better on some of these problems because they have a stronger tendency to rely on analytic reasoning or because they do not share the typical intuitions of other individuals. (Of course, a third possibility is that both of these hypotheses are correct.)

The studies based on self-report measures (e.g., Levin et al. 2015) suggest that the difference may only lie in intuitive reasoning and there may be no difference between autistic and neurotypical people in their analytic thinking abilities. In line with these studies, Morsanyi et al. (2010) also found that although autistic adolescents were less susceptible to the conjunction fallacy (which is based on intuitive thinking), they were not any more likely to rely on the conjunction rule (i.e., analytic processing) in their judgments than neurotypical participants. Nevertheless, the evidence for a reduced tendency for intuitive thinking in autism remains scarce.

In the current study, we investigated intuitive and analytic processing in autism, using the cognitive reflection test (CRT; Frederick 2005). The CRT measures a person's ability to resist intuitive response tendencies, and to produce normatively correct responses that are assumed to be based on effortful, analytic reasoning. For example, consider the following item: *"If it takes 5 min for five machines to make five widgets, how long would it take for 100 machines to make 100 widgets?"* The correct answer is 5 min, but the majority of participants (about 60%—see Primi et al. 2016), give the incorrect, but intuitively compelling response of 100 min. Indeed, it is a very remarkable property of the CRT that, although the questions are open-ended, for each item the majority of people either produce a normatively correct/analytic response or a typical incorrect (i.e., intuitive response). In other words, using the CRT makes it possible to investigate the tendency to give intuitive vs. analytic responses relatively independently, as participants who do not give the analytic response could still produce a response that is different from the typical intuitive response.

Performance on the CRT in autism has already been investigated in previous studies, although the results were somewhat mixed. Brosnan et al. (2016) administered the CRT to autistic and neurotypical late adolescents who were matched on age and gender (all participants in this study were males). Autistic participants gave significantly fewer intuitive and significantly more analytic responses to the CRT problems.

In another study, Brosnan et al. (2017) administered the CRT to autistic and neurotypical late adolescents and young adults. The groups were matched on gender and non-verbal intelligence, but the autistic group was slightly older. The results showed no significant difference in analytic responses to the CRT problems, but autistic participants gave significantly fewer intuitive responses (with a medium effect size). Additionally, Brosnan et al. (2017) reported that autistic traits (as measured by the Autism Spectrum Quotient—Baron-Cohen et al. 2001) showed a small, but significant positive correlation with giving analytic responses to the CRT. In this correlational analysis, the autistic and neurotypical samples were pooled, and the effects of age, gender and diagnostic group membership were controlled.

Brosnan and Ashwin (2023) also investigated performance on the CRT among autistic and neurotypical late adolescents. The groups were matched on age, but they were not matched on cognitive ability. Participants performed the CRT either under the typical, untimed condition or under time pressure. The results showed that (in both conditions) autistic participants produced more analytic and fewer intuitive responses than the neurotypical participants. Additionally, the effect of time pressure was similar in both groups, with a higher proportion of intuitive and lower proportion of analytic responses under time pressure. Similar to Brosnan et al. (2017), Brosnan and Ashwin (2023) also reported that there was a small in size but significant positive correlation between autistic traits and providing correct responses to the CRT. In this analysis, they pooled the autistic and neurotypical samples, and controlled for the effect of condition (time pressure/untimed), gender, age and diagnostic status. Autistic traits were measured using the Subthreshold Autism Trait Questionnaire (SATQ; Kanne et al. 2012).

Overall, these results seem to paint a relatively clear picture regarding the performance of autistic participants on the CRT. In all of these studies, autistic participants gave fewer intuitive responses to the CRT, and in two out of the three studies (Brosnan et al. 2016; Brosnan and Ashwin 2023), they also gave a higher proportion of analytic responses. Two of these studies (Brosnan et al. 2017; Brosnan and Ashwin 2023) also showed a significant positive correlation between autistic traits and the proportion of analytic responses on the CRT, although this relationship was weak.

Nevertheless, there are some methodological concerns regarding these findings. Specifically, the studies that showed a significant difference in analytic responding between autistic and neurotypical participants did not match the samples on cognitive ability. This is problematic, because Frederick (2005, p. 26) described the CRT as a 'simple measure of one type of cognitive ability' and reported reliable relationships between cognitive reflection and measures of intelligence (see also Otero et al. 2022 for a meta-analysis). A recent study (Taylor et al. 2022; Study 4) overcame this methodological issue by comparing the performance of large samples of autistic and neurotypical participants, who were closely matched on age, gender, and cognitive ability. The results showed no group difference in either analytic or intuitive responding on the CRT, which was in contract with Brosnan et al. (2017) who reported a significant difference in intuitive responding on the CRT even when the autistic and neurotypical groups were matched on cognitive ability.

Taylor et al. (2022; Studies 1–3) also investigated the relations between autistic traits (as measured by the AQ scale; Baron-Cohen et al. 2001) and performance on different versions of the CRT in large online samples that were drawn from the general population. These studies showed no relationship between autistic traits in the general population and either analytic or intuitive responding on the CRT. Although in Study 3, AQ scores showed a significant (weak, positive) correlation with analytic responding on the CRT and a significant (weak, negative) correlation with intuitive responding, these correlations became non-significant when the effects of age, gender and cognitive ability were controlled for. Overall, Taylor et al.'s (2022) findings suggest that performance on the CRT may not be related to either a diagnosis of autism or autistic traits in the general population when the effects of cognitive ability are taken into account.

In addition to ignoring the effects of cognitive ability, another potential issue with the studies of Brosnan and colleagues is that the autistic participants were recruited from a university summer school for students on the autism spectrum, which was aimed at providing insight into university life, whereas the neurotypical group was recruited from either a general summer school (Brosnan et al. 2017; Brosnan and Ashwin 2023) or from first year student populations (Brosnan et al. 2016). The CRT has numerical content, and, although it is more than just a mathematics test (see Campitelli and Gerrans 2014; Liberali et al. 2012; Otero et al. 2022), it relies heavily on mathematics skills and mathematics-related feelings (e.g., Morsanyi et al. 2014; Primi et al. 2018). In fact, it has even been used as part of a numeracy test (Weller et al. 2013). Baron-Cohen et al. (2007) reported that an autism diagnosis was between three and seven times more likely among mathematics undergraduates than undergraduate students in medicine, law, and social sciences. Thus, a potential explanation for any group differences on the CRT between autistic and neurotypical participants is that the autistic participants simply had better mathematics knowledge/more background in mathematics.

The aim of the current study was to replicate and extend previous work regarding cognitive reflection in autism, while also offering some methodological improvements. Specifically, we made sure that participants were well-matched on age, gender, cognitive ability, and educational background. We also wanted to investigate the question of whether a potential reason for autistic participants to produce fewer intuitive responses could be that although they were not any more analytic in their thinking than their neurotypical peers, they were less susceptible to giving the typical intuitive response. To answer this question, we analyzed the responses given to each CRT item in more detail. In particular, we were interested in whether autistic participants produced more atypical (neither analytic nor intuitive) responses to these open-ended problems.

Brosnan et al. (2017) and Brosnan and Ashwin (2023) reported that autistic traits were related to analytic responding on the CRT, but Taylor et al. (2022) found no such relationship. Thus, we also revisited this question in our study. Additionally, Brosnan and colleagues did not give any details regarding the nature of this relationship. In particular, the AQ (Baron-Cohen et al. 2001) and the SATQ (Kanne et al. 2012) have several subscales relating to social skills, communication skills, rigidity, etc. It would be of interest to better understand which of these traits may be responsible for the tendency to provide analytic responses to the CRT.

A final, novel aim was to investigate developmental changes in cognitive reflection (and in the tendency to produce intuitive vs. analytic responses). So far, in all existing studies on cognitive reflection in autism, the participants were either late adolescents or young adults. This raises the question of whether these results would generalize to other age groups. A typical finding in the developmental literature on reasoning biases is that intuitive responding tends to decrease and analytic reasoning tends to increase throughout adolescence and into young adulthood (e.g., Chiesi et al. 2011; Klaczynski 2001; Klaczynski and Cottrell 2004; Morsanyi and Handley 2013), and a similar tendency has been identified in the case of the CRT as well (Primi et al. 2016). Thus, we were interested in whether any differences in intuitive and analytic responding between the autistic and neurotypical samples (if they exist) are stable across development, or whether these gaps increase or decrease with age. If these gaps change with age, this may signal a developmental advantage or delay in the case of the autistic sample.

2. Method

2.1. Participants

The participants were adolescents and adults from Northern Ireland. In the adolescent sample, there were 28 adolescents with ASD (23 males; between the ages of 11 years 9 months and 17 years 1 month; $M = 12.80$, $SD = 1.32$) and 50 typically developing adolescents (37 males; between the ages of 11 years 7 months and 14 years 4 months; $M = 12.97$, $SD = 0.57$). A chi-square test indicated that the gender distribution of the two samples was not significantly different ($p = .413$). The participants in the autistic and neurotypical samples were recruited from the same schools. In the adult sample, there were 26 participants with ASD (23 males; between the ages of 18 and 41 years; $M = 22.81$, $SD = 6.41$) and 27 neurotypical adults (22 males; between the ages of 18 and 52 years; $M = 22.26$, $SD = 7.02$). A chi-square test indicated that the gender distribution of the two samples was not significantly different ($p = .478$). All participants were recruited from university courses either at Queen's University Belfast or Ulster University. The participants with ASD were attending various courses, including math, physics, engineering, computing, law, arts, and biological and life sciences. The participants in the neurotypical group were recruited from the same university courses to control for the potential effects of educational background and, in particular, interest and experience in mathematics. In the adolescent sample, the participants' school shared any official diagnosis (autism, dyslexia, etc.) that the participants had as well as further details (e.g., date of diagnosis, and the person providing the diagnosis) when available. In the adult sample, the participants confirmed their own diagnosis and the details that they knew. In both samples, the participants had no diagnosis other than ASD. Table 1 presents more details regarding the characteristics of our participants. Blocked Operation Span scores were missing for two autistic participants due to a software error.

Table 1. Sample characteristics and matching (*t*-tests were run to compare the groups on each measure).

		ASD *M* (SD)	Neurotypical *M* (SD)	*p* Value
Adolescents	Age (years)	12.80 *(1.32)*	12.97 *(0.56)*	.525
	WISC-IV Vocabulary	37.68 *(7.37)*	39.16 *(5.60)*	.321
	WISC-IV Block Design	48.64 *(8.17)*	48.18 *(8.70)*	.818
	Raven's APM	7.43 *(2.49)*	7.78 *(2.08)*	.507
	Estimated IQ	102.29 *(11.84)*	104.00 *(11.61)*	.536
Adults	Age (years)	22.81 *(6.41)*	22.26 *(5.39)*	.768
	WASI-II Vocabulary	41.08 *(8.67)*	39.19 *(8.59)*	.429
	Operation Span	14.54 *(7.37)*	14.11 *(9.09)*	.854
	AQ	30.11 *(7.77)*	16.04 *(5.37)*	<.001

2.2. Materials

The *CRT-Long* (Primi et al. 2016), an extended, 6-item version of the CRT (Frederick 2005) was used to measure cognitive reflection (the full list of items and their scoring can be found in Appendix A). The CRT-Long includes 3 new items beside the original items, and it has been developed with the aim of making it possible to measure cognitive reflection in lower ability samples as well (such as developmental samples), who often score zero on the original CRT. All CRT items are open-ended and there is no time limit for the participants to complete the questions. Participants' responses were marked as analytic (i.e., correct), intuitive (which corresponds to the typical incorrect response) or other. In the current sample, Cronbach's alpha for the scale (based on the total number of analytic responses) was 0.78. Performance on the original CRT items and the CRT-Long showed a very strong correlation ($r(129) = 0.91$ $p < .001$), confirming that the CRT-Long measured the same construct as the original CRT.

2.2.1. Ability Measures Used for Matching the Adolescent Samples

To match the groups on *general intelligence*, we used a short form of the *Wechsler Intelligence Scale for Children* (Wechsler 2004), consisting of the Block Design and Vocabulary subtests. This short form of the WISC is reported to have the highest reliability and validity compared to any other two-subtest short forms (see Sattler and Dumont 2004). Using these scores, it was also possible to calculate the participants' estimated IQ (see Sattler 2008, p. 307, Table A-7).

As a measure of *fluid intelligence*, we used Set 1 of Raven's Advanced Progressive Matrices (Raven and Court 2003) consisting of 12 items. The test was administered together with 3 practice items taken from the Raven Standard Progressive Matrices (Raven 1998). This set is usually used as a practice and screening set for the full Raven's test as it covers all the intellectual processes assessed by the full test. However, it does not extend to the highest complexity levels. Chiesi et al. (2012) reported that Set-1 of the APM was a reliable test to be used in a short time frame as it cuts around 60% of testing time compared to the full Raven's test, thus helping to prevent the effects of fatigue and boredom on results.

2.2.2. Ability Measures Used for Matching the Adult Samples

As a measure of *verbal intelligence*, we used the Vocabulary Subtest of the Wechsler Abbreviated Scale of Intelligence (Version Two; WASI-II 2011). In the vocabulary subtest, participants have to define words which are presented to them verbally, one by one. This subtest is considered to measure expressive vocabulary, verbal knowledge, and crystallized and general intelligence. Performance also draws heavily on memory, learning ability, and concept and language development (Sattler 2008).

As a measure of *verbal working memory*, we used the Operation Span test (Foster et al. 2015). In this computer-based task, participants are asked to memorize a series of letters, while being distracted by math sums. The participant first completes a simple mental arithmetic problem, then they see a letter, then another arithmetic problem and another letter. This process is repeated between three and seven times, although the trial length is random and cannot be predicted. At the end of each trial, the participant has to recall the letters in the correct order. Scores are calculated by adding up the number of correct letters that were recalled in the correct order (Turner and Engle 1989). The absolute score is based on the total number of correctly recalled letters (based on only those sequences for which all letters in the sequence were recalled in the correct order).

Autism-Spectrum Quotient: The 50-item AQ scale (Baron-Cohen et al. 2001) measures a variety of autistic traits, using a four-point scale that ranges from *definitely agree* to *definitely disagree.* We used the dichotomous scoring system of Baron-Cohen and colleagues, where participants were given 1 point if they indicated that they agreed/definitely agreed with an item measuring autistic characteristics, and they were given 0 point if they indicated that they (definitely) disagreed. The scale assesses 5 different areas (with 10 questions each): *social skill (e.g., "I prefer to do things with others rather than on my own"); attention switching ("I frequently get so strongly absorbed in one thing that I lose sight of other things"); attention to detail ("I often notice small sounds when others do not"); communication ("I enjoy social chit-chat")* and *imagination ("I find making up stories easy").* Approximately half the items are worded to produce a 'disagree' response, and half an 'agree' response, in a high scoring autistic person. Cronbach's alpha for the full scale was 0.92 (0.83, 0.73, 0.55, 0.84, and 0.70 for the social skills, attention switching, attention to detail, communication, and imagination subscales, respectively). Higher scores on subscales indicate higher levels of autistic traits (i.e., worse social skills, attention switching, communication skills, and imagination, but better attention to detail).

2.3. *Procedure*

The study received ethics approval from the Faculty's Human Ethics Committee. In the case of the adolescent sample, after receiving permission from the schools to carry out the study, parents/caregivers provided informed consent and the participants provided assent. In the young adult sample, the participants were recruited from local universities. Adolescents were involved in two testing sessions. In the individual session, the participants completed the vocabulary and block design tests, which were administered by one of the experimenters. The CRT-Long and Raven's Matrices were completed in a paper-and-pencil format in a group session (which took place on a separate day), with groups of 3–4 children, but with all participants working individually. Adults completed a paper-based version of the CRT-Long and the AQ scale, and a computer-based verbal working memory test. As part of the same session, they also completed the vocabulary test, which was administered by one of the experimenters.

3. Results

The data that support the findings of this study are available at (https://osf.io/t7yn4/?view_only=d7be657e7a56445897a8ccc336e44c5f) Our first analysis investigated whether participants in each sample mostly gave either the intuitive or the analytic response to the CRT-L items (the exact proportion of intuitive and analytic responses given to each item by each group is presented in Table 2). The overwhelming majority of responses to each item were either intuitive or analytic. When we looked at the autistic and neurotypical groups separately within each age group, the proportion of intuitive/analytic responses to each CRT item ranged from 72% to 100%. We also checked whether there were any other typical responses apart from the intuitive and analytic responses given to any item by either group. For most items, the responses that were neither intuitive nor analytic were only given by a very small number of participants (i.e., no other typical responses were identified). The only exception was item 5, where 12 neurotypical adolescents (i.e., 24% of the sample) gave the response "15". Nevertheless, even in this case, a much larger proportion of the sample (58%) gave the typical intuitive response. Thus, from these analyses we can conclude that the dominant responses given by each group to each question were either the intuitive or the analytic response. In fact, overall, 88% of the responses given by the autistic adolescents were either intuitive or analytic, and the corresponding proportions were 88% for the neurotypical adolescents, 96% for the autistic adults, and 95% for the neurotypical adult group.

Table 2. The proportion of analytic and intuitive responses given to each item split by age and diagnostic group.

	Analytic Responses				Intuitive Responses			
	Adolescents		Adults		Adolescents		Adults	
	ASD	NT	ASD	NT	ASD	NT	ASD	NT
Item 1	0.18	0.20	0.39	0.37	0.75	0.76	0.58	59
Item 2	0.32	0.28	0.61	0.44	0.50	0.60	0.39	0.52
Item 3	0.11	0.18	0.62	0.63	0.82	0.76	0.35	0.30
Item 4	0.61	0.36	0.81	0.93	0.32	0.60	0.19	0.07
Item 5	0.07	0.14	0.58	0.37	0.77	0.58	0.35	0.48
Item 6	0.18	0.12	0.69	0.81	0.61	0.68	0.23	0.19
Total	0.24	0.21	0.62	0.59	0.61	0.66	0.35	0.36

In an additional analysis we also checked if the proportion of atypical (i.e., neither intuitive nor analytic) responses decreased with age. Such patterns were reported for other judgment and reasoning tasks (e.g., Morsanyi and Handley 2008, 2013; Klaczynski 2001). A 2 × 2 ANOVA with age group (adolescent/adult) and diagnostic group (ASD/neurotypical) on the overall number of atypical responses showed a significant effect of age ($F(3,127) = 9.06$, $p = .003$, $\eta_p{}^2 = 0.07$), with a lower number of atypical responses in the case of adult participants. There was no effect of diagnostic group and no interaction between age and diagnostic group ($ps > .81$).

Our next analysis was aimed at investigating the effects of age and diagnosis on analytic performance on the CRT-L (Figure 1). A 2 × 2 ANOVA with age group (adolescent/adult) and diagnosis (autistic/neurotypical) as between-subject factors on the number of analytic responses given to the CRT-L indicated a significant main effect of age ($F(1,127) = 53.75$, $p < .001$, $\eta_p{}^2 = 0.30$), but no effect of diagnosis and no age by diagnosis interaction ($ps > 0.60$).

Given that our analyses indicated no significant difference in analytic responding in either the adolescent or the adult sample, we followed up the above analyses with Bayesian *t* tests, comparing analytic responding separately in the adolescent and young adult samples. In the case of non-significant results, using a Bayesian analysis makes it possible to distinguish between inconclusive evidence, as opposed to evidence that supports the null hypothesis. A Bayes factor of above 3 indicates moderate evidence in favor of the null hypothesis over the alternative hypothesis (i.e., that there was a significant difference between the autistic and neurotypical samples in analytic responding), a Bayes factor of above 10 indicates strong evidence, and a Bayes factor of between 1 and 3 indicates anecdotal evidence. A Bayes factor of 1 indicates that the evidence in favor of the null hypothesis and the alternative hypothesis is equally strong (see, e.g., Lee and Wagenmakers 2014).

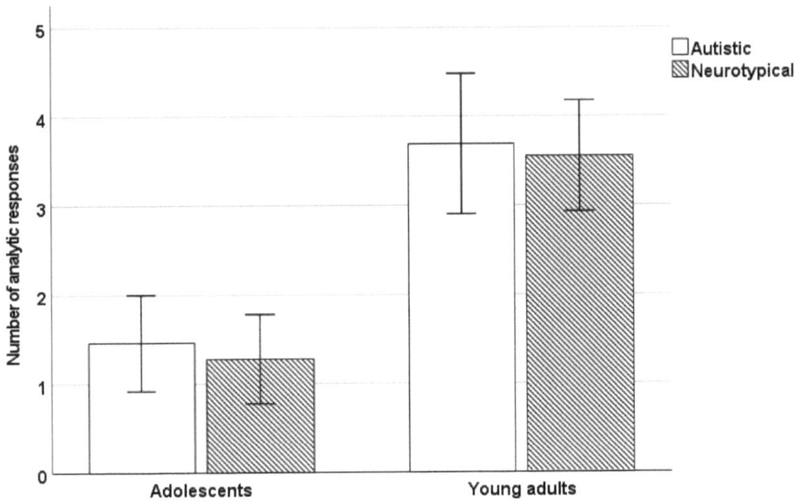

Figure 1. The number of analytic responses given by the autistic and neurotypical groups within each age group (error bars represent 95% CI).

A Bayesian t test in the case of the adolescent group indicated moderate evidence for no difference between the autistic and neurotypical groups in analytic responding ($BF = 3.73$). In the case of the adult sample, the results also showed moderate evidence in favor of no group differences ($BF = 3.51$).

Although our results regarding intuitive responding on the CRT-L very much mirrored the results obtained in the case of analytic responses, given the theoretical significance of these results, we also report these analyses (Figure 2). A 2 × 2 ANOVA with age group (adolescent/adult) and diagnosis (autistic/neurotypical) as between-subject factors on the number of intuitive responses given to the CRT-L indicated a significant main effect of age ($F(1,127) = 34.38$, $p < .001$, $\eta_p^2 = 0.21$), but no effect of diagnosis and no age by diagnosis interaction ($ps > .65$).

A Bayesian t test in the case of the adolescent group indicated moderate evidence for no difference between the autistic and neurotypical groups in intuitive responding ($BF = 3.70$). In the case of the adult sample, the results also showed moderate evidence in favor of no group differences ($BF = 3.59$).

Our next analyses investigated the correlations between participants' scores on the AQ scale and its subscales, and their tendency to give intuitive/analytic responses to the CRT-Long (Table 3). These analyses are only presented for the adult participants, as adolescents were not administered the AQ scale. We present these correlations separately for the autistic and neurotypical samples, as we did not necessarily expect that these correlations would be the same across the samples. Overall scores on the AQ scale were not correlated with either intuitive or analytic responding on the CRT in either the autistic or the neurotypical sample. However, a very strong correlation emerged between attention to detail and analytic responding on the CRT-L in the autistic group. The correlation was in the same direction in the neurotypical group, but it was much weaker and non-significant. The other subscales of the AQ were unrelated to performance on the CRT-L, although there was some indication of a relationship between social skills and imagination and performance on the CRT-L (i.e., autistic participants who displayed better social skills and imagination—i.e., who were less "autistic" in these respects, tended to score higher on the CRT-L).

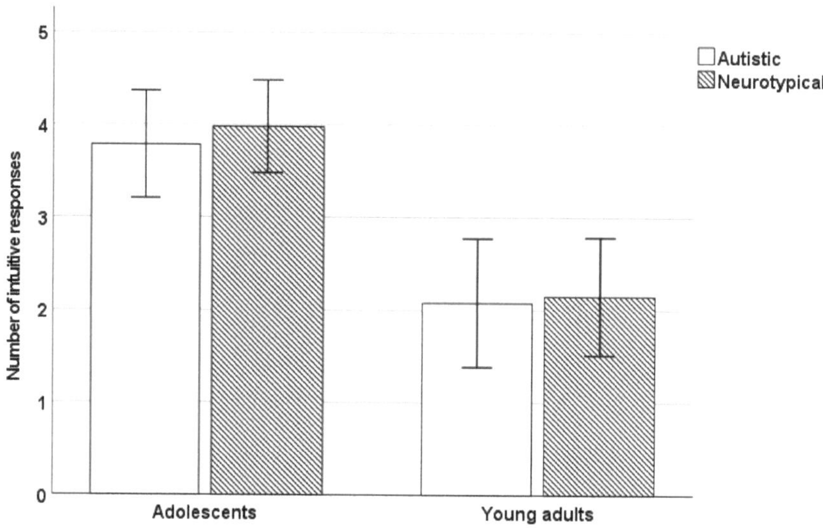

Figure 2. The number of intuitive responses given by the autistic and neurotypical groups within each age group (error bars represent 95% CI).

Table 3. Correlations between performance on the CRT-L and autistic traits across diagnostic groups (the correlations for neurotypical participants are presented in brackets).

	CRT-L Intuitive	CRT-L Analytic
Social skills	0.22 *(0.13)*	−.25 *(−0.13)*
Attention switching	−0.10 *(0.22)*	−0.002 *(−0.14)*
Attention to detail	−0.72 ** *(−0.29)*	0.72 ** *(0.21)*
Communication	0.09 *(0.001)*	−0.16 *(0.13)*
Imagination	0.29 *(−0.01)*	−0.31 *(0.02)*
AQ total	0.01 *(−0.005)*	−0.09 *(0.04)*

** $p < .001$.

Brosnan et al. (2017) reported that scores on the AQ scale showed a small, but significant positive correlation with giving analytic responses to the CRT when the autistic and neurotypical samples were pooled, and the effects of age, gender and diagnostic group membership were controlled. We replicated this analysis in our sample including both total scores on the AQ scale and each subscale score separately. Total scores on the AQ scale showed a very weak, non-significant, negative correlation with analytic responding on the CRT ($r(48) = -0.08$, $p = .563$). Attention to detail showed a moderate positive correlation with analytic responding on the CRT ($r(48) = 0.46$, $p = .001$). Additionally, there was a significant negative correlation between scores on the imagination subscale of the AQ scale and analytic responding on the CRT ($r(48) = -0.32$, $p = .025$). This latter finding indicated that people with lower imagination scores (i.e., who were more autistic-like in this respect) scored *lower* on the CRT.

4. Discussion

The aim of this study was to investigate the development of intuitive and analytic reasoning in autism, using the cognitive reflection test. In particular, we were interested in: (1) whether autistic participants generate the same typical intuitive responses to the CRT items as neurotypical individuals; (2) whether the autistic and neurotypical groups produce a similar number of intuitive and analytic responses; (3) whether developmental changes in performance on the CRT between adolescence and adulthood are similar across the diagnostic groups; and (4) whether analytic responding on the CRT is related to autistic traits. These questions are important from a theoretical perspective as autistic individuals are often described as hyper-rational in their thinking (see, e.g., Rozenkrantz et al. 2021), and at the same time, it has been claimed that they show a reduced tendency to produce the typical intuitive responses to judgment and reasoning problems (e.g., Brosnan and Ashwin 2023).

Although performance on the CRT in autism has been investigated in the past (Brosnan et al. 2016, 2017; Brosnan and Ashwin 2023; Taylor et al. 2022), these studies had some methodological weaknesses in terms of matching the autistic and neurotypical samples on some key characteristics. Only two studies (Brosnan et al. 2017; Taylor et al. 2022) matched the samples on cognitive ability, and these studies reported conflicting findings. To address these issues, we administered the CRT to two samples of autistic and neurotypical participants who were well-matched on age, gender, cognitive ability and educational background. As the two samples were drawn from different age groups (i.e., early adolescents and young adults), we were also able to investigate developmental changes in performance on the CRT in both the autistic and neurotypical samples.

Our results showed that in both diagnostic groups, most participants gave either the typical intuitive or the analytic response to each CRT item (participants from each group did this over 85% of the time). Thus, although the problems were open-ended, autistic participants generated exactly the same types of responses as neurotypical individuals. Moreover, in line with the findings of Taylor et al. (2022), there was no difference between the groups in the proportion of intuitive and analytic responses that they gave. This pattern was exactly the same in both the adolescent and adult samples. Thus, we replicated this result in two independent samples. A Bayesian analysis indicated moderate evidence in the case of both adolescents and young adults that there was no difference between the diagnostic groups in their tendency to give either intuitive or analytic responses to the CRT.

With regard to developmental changes, adults produced significantly more analytic, and significantly fewer intuitive responses than adolescents (see also Primi et al. 2016), and the difference between age groups was exactly the same in both diagnostic groups. This suggests no difference in the development of cognitive reflection between neurotypical and autistic participants. Regarding the large developmental difference between age groups, it should be noted that the adult participants were highly educated (as they were university students), and probably not representative to a general young adult population. Thus, our results may overestimate the typical developmental differences. Nevertheless, the difference between age groups was large, and for this reason they are likely to be indicative of the actual developmental pattern.

Following Brosnan et al. (2017), Brosnan and Ashwin (2023), and Taylor et al. (2022), we also investigated the relationships between autistic traits and performance on the CRT. In line with Taylor et al. (2022), we found no relationship between overall AQ scores and cognitive reflection. Nevertheless, we found a very strong correlation between attention to detail and performance on the CRT in the autistic group. This correlation between attention to detail and cognitive reflection was also present when the diagnostic groups were pooled, and we controlled for the effects of age, gender, and diagnosis. Additionally, we found a moderate *negative* correlation between the imagination subscale of the AQ and analytic thinking on the CRT when the samples were pooled. In other words, participants with *better* imagination skills performed better on the CRT, which is in contrast with the idea that, in general, autistic traits are positively related to cognitive reflection.

In terms of the relation between attention to detail and cognitive reflection, a detail-focused processing style has been linked to systematic processing and literal interpretation of materials, which might also be coupled with a reduced tendency to process the "gist" of experiences or presented materials (cf. McCrory et al. 2007). That is, a detail-focused processing style might make the autistic group less likely to rely on pragmatic or other contextual cues. A recent study has also found a link between attention to detail and analogical reasoning skills in an autistic sample (Morsanyi et al. 2022). It would be interesting to check in future studies if attention to detail is also linked to performance on other types of judgment and reasoning problems in the case of both autistic and neurotypical samples. Investigating the relations between attention to detail and reasoning skills in autism is particularly interesting as it has been proposed that attention to detail might be linked to special skills and talents in autism (e.g., Baron-Cohen et al. 2009; Happé and Vital 2009).

Overall, our findings concur with Taylor et al. (2022) in showing that there is no difference between autistic and neurotypical individuals in cognitive reflection when the two samples are matched on cognitive ability. Moreover, a novel finding of our study is that cognitive reflection develops exactly the same way in autistic individuals as in neurotypical samples. Additionally, also in line with Taylor et al. (2022), we found no evidence for a link between cognitive reflection and autistic traits, with the exception of a link with attention to detail. These findings fail to replicate the results of previous studies by Brosnan and colleagues (Brosnan et al. 2016, 2017; Brosnan and Ashwin 2023). The reason that Brosnan and colleagues found significant group differences may be that they did not match the autistic and neurotypical samples on key cognitive characteristics, including age, cognitive ability, and educational background. This highlights the importance of taking into account the cognitive requirements of reasoning and decision-making tasks when designing studies that compare autistic and neurotypical samples with the aim of understanding potential differences in their reasoning and decision-making skills. Studies in this field remain scarce (although see Morsanyi and Byrne 2020). Nevertheless, methodological rigor is essential to avoid results that turn out to be false positives, and could be used to perpetuate myths and stereotypes about autism.

In terms of the implications of the current findings, it is interesting to consider to what extent performance on the CRT can be considered as an indication of intuitive and analytic reasoning ability in general. The CRT has been an extremely popular measure, which can be attributed to the fact that it correlates moderately or strongly with a range of reasoning and decision-making tasks, as well as with measures of moral judgments and values, religious and paranormal beliefs, and creativity (see, e.g., Pennycook et al. 2015 for an overview). In this respect, it can be considered an ideal measure that captures an essential skill or predisposition that is necessary for analytic thinking (which may be described as the ability to suppress tempting, but incorrect responses in favor of a more effortful, analytic response—e.g., Frederick 2005; Travers et al. 2016). Nevertheless, it has been questioned whether the CRT is a reliable and useful measure of intuitive thinking, or whether it purely measures analytic thinking (cf., Pennycook et al. 2016). Indeed, there is an ongoing debate in the literature whether intuitive and analytic thinking should be best conceptualized as being dependent on separable cognitive processes/systems or whether a unitary system underlies both intuitive and analytic thinking (see, e.g., De Neys 2021; Osman and Stavy 2006). Given that the correlation between CRT performance and performance on other reasoning and judgment tasks tends to be moderate, we can also expect that group differences may be (at least to some extent) task-specific. Thus, finding no group differences in cognitive reflection does not necessarily imply that autistic and neurotypical samples would not differ on other reasoning or decision-making tasks.

Even though autistic and neurotypical participants produced the same proportion of intuitive and analytic responses in our study, we have not measured the time taken to produce those responses. There may be differences between the groups in terms of the time course of producing these responses or they may not produce these responses with the same level of confidence. For example, although analogical reasoning ability has been found to be unimpaired in autism (e.g., Green et al. 2017; Morsanyi and Holyoak 2010), autistic participants took longer and exerted more effort when they had to verify verbal analogies (Morsanyi et al. 2022). Such differences could have an impact on reasoning performance in real-life settings, when responses need to be formulated quickly. Future studies could investigate the metacognitive monitoring and control processes (e.g., Ackerman and Thompson 2017) that accompany reasoning and decision-making in autism.

Mata et al. (2013) found that performance on the CRT was related to metacognitive skills. In particular, analytic thinkers (who scored higher on the CRT) were less overconfident than people who scored lower, and they were better able to discriminate between whether they gave correct vs. incorrect responses, as reflected by their confidence in their responses. Their confidence ratings were also more strongly related to the objective difficulty of the problems. A potential implication of these findings is that metacognitive monitoring skills in relation to reasoning processes may be similar across autistic and neurotypical populations. Nevertheless, this needs to be confirmed empirically, and across a range of tasks.

Existing research into metacognitive skills in autism suggests that there may be differences in metacognitive accuracy between autistic and neurotypical samples, although so far this research has focused on memory performance. According to a meta-analysis (Carpenter and Williams 2023), autistic participants do not differ from matched controls in terms of the accuracy of their judgements of learning (i.e., judgments about whether they will be able to recall newly learnt content). Nevertheless, metacognitive accuracy in autism was found moderately reduced on judgements-of-confidence tasks (where after each response, participants had to rate how confident they were that their answer was correct), although this difference may only be present in the case of autistic children, and not in autistic adults. Overall, research into metacognitive skills in autism remains scarce, and many studies suffer from methodological issues (cf., Carpenter and Williams 2023).

In summary, our findings suggest that there is no general tendency in autism to produce a higher proportion of analytic responses, or to produce fewer intuitive responses than neurotypical individuals. This pattern was replicated in two independent samples (i.e., in adolescents and adults), and the strength of evidence for no group differences was moderate in both samples. Autistic traits in general were also unrelated to both intuitive and analytic reasoning performance, although attention to detail was positively related to analytic reasoning. This raises the possibility that, when producing analytic responses, there may be some differences in the underlying cognitive processes between autistic and neurotypical participants. As reasoning and decision making in autism are under-researched areas, it is important to carry out more research into these topics. These investigations could also involve looking into the metacognitive monitoring and control processes that accompany intuitive and analytic responding in autism. As these skills rely heavily on cognitive resources (which may include general or fluid intelligence, working memory and executive functioning skills), and may also be influenced by relevant knowledge and educational background (see, e.g., Stanovich 2018), it is essential that any studies into reasoning and decision-making skills carefully match the autistic and neurotypical samples on relevant cognitive characteristics to avoid false positives. Future studies could also explore further the role of attention to detail in reasoning performance in autism.

Author Contributions: Conceptualization, K.M. and J.H.; methodology, K.M. and J.H.; formal analysis, K.M. and J.H.; investigation, J.H.; data curation, K.M. and J.H.; writing—original draft preparation, K.M. and J.H.; writing—review and editing, K.M. and J.H.; supervision, K.M.; project administration, K.M. and J.H.; funding acquisition, K.M. and J.H. All authors have read and agreed to the published version of the manuscript.

Funding: The writing of this paper was partially supported by an Economic and Social Research Council Research Centre grant (Centre for Early Mathematics Learning; reference number: ES/W002914/1) to K.M.

Institutional Review Board Statement: The study was conducted in accordance with the Declaration of Helsinki, and approved by the Psychology Research Ethics Committee of Queen's University Belfast (reference number: 81-2013, date of approval: 13 October 2014).

Informed Consent Statement: Informed consent was obtained from all subjects involved in the study.

Data Availability Statement: The data that support the findings of this study are available at https://osf.io/t7yn4/?view_only=d7be657e7a56445897a8ccc336e44c5f.

Conflicts of Interest: The authors declare no conflict of interest.

Appendix A

The CRT-L problems, as used in the current study, with the typical analytic and intuitive responses (Primi et al. 2016).

1. A bat and a ball cost £1.10 in total. The bat costs £1.00 more than the ball. How much does the ball cost? [analytic response = 5 p; intuitive response = 10 p]
2. If it takes 5 min for five machines to make five widgets, how long would it take for 100 machines to make 100 widgets? [analytic response = 5 min; intuitive response = 100 min]
3. In a lake, there is a patch of lily pads. Every day, the patch doubles in size. If it takes 48 days for the patch to cover the entire lake, how long would it take for the patch to cover half of the lake? [analytic response = 47 days; intuitive response = 24 days]
4. If three elves can wrap three toys in 1 h, how many elves are needed to wrap six toys in 2 h? [analytic response = 3 elves; intuitive response = 6 elves]
5. Jerry received both the 15th highest and the 15th lowest mark in the class. How many students are there in the class? [analytic response = 29 students; intuitive response = 30 students]
6. In an athletics team, tall members are three times more likely to win a medal than short members. This year the team has won 60 medals so far. How many of these have been won by short athletes? [analytic response = 15 medals; intuitive response = 20 medals]

References

Ackerman, Rakefet, and Valerie A. Thompson. 2017. Meta-reasoning: Monitoring and control of thinking and reasoning. *Trends in Cognitive Sciences* 21: 607–17. [CrossRef] [PubMed]

American Psychiatric Association. 2013. *Diagnostic and Statistical Manual of Mental Disorders*, 5th ed. Washington, DC: American Psychiatric Association.

Baron-Cohen, Simon, and Matthew K. Belmonte. 2005. Autism: A window onto the development of the social and the analytic brain. *Annual Review of Neuroscience* 28: 109–26. [CrossRef] [PubMed]

Baron-Cohen, Simon, Emma Ashwin, Chris Ashwin, Teresa Tavassoli, and Bhismadev Chakrabarti. 2009. Talent in autism: Hyper-systemizing, hyper-attention to detail and sensory hypersensitivity. *Philosophical Transactions of the Royal Society B: Biological Sciences* 364: 1377–83. [CrossRef]

Baron-Cohen, Simon, Jennifer Richler, Dheraj Bisarya, Nhishanth Gurunathan, and Sally Wheelwright. 2003. The systemizing quotient: An investigation of adults with Asperger syndrome or high–functioning autism, and normal sex differences. *Philosophical Transactions of the Royal Society of London. Series B: Biological Sciences* 358: 361–74. [CrossRef]

Baron-Cohen, Simon, Sally Wheelwright, Amy Burtenshaw, and Esther Hobson. 2007. Mathematical talent is linked to autism. *Human Nature* 18: 125–31. [CrossRef] [PubMed]

Baron-Cohen, Simon, Sally Wheelwright, Richard Skinner, Joanne Martin, and Emma Clubley. 2001. The autism-spectrum quotient (AQ): Evidence from Asperger syndrome/high-functioning autism, males and females, scientists and mathematicians. *Journal of Autism and Developmental Disorders* 31: 5–17. [CrossRef] [PubMed]

Brosnan, Mark, and Chris Ashwin. 2023. Thinking, fast and slow on the autism spectrum. *Autism*, 13623613221132437. [CrossRef]

Brosnan, Mark, Chris Ashwin, and Marcus Lewton. 2017. Brief report: Intuitive and reflective reasoning in autism spectrum disorder. *Journal of Autism and Developmental Disorders* 47: 2595–601. [CrossRef]

Brosnan, Mark, Emma Chapman, and Chris Ashwin. 2014. Adolescents with autism spectrum disorder show a circumspect reasoning bias rather than 'jumping-to-conclusions'. *Journal of Autism and Developmental Disorders* 44: 513–20. [CrossRef]

Brosnan, Mark, Marcus Lewton, and Chris Ashwin. 2016. Reasoning on the autism spectrum: A dual process theory account. *Journal of Autism and Developmental Disorders* 46: 2115–25. [CrossRef]

Campitelli, Guillermo, and Paul Gerrans. 2014. Does the cognitive reflection test measure cognitive reflection? A mathematical modeling approach. *Memory & Cognition* 42: 434–47.

Carpenter, Katie L., and David M. Williams. 2023. A meta-analysis and critical review of metacognitive accuracy in autism. *Autism* 27: 512–25. [CrossRef] [PubMed]

Chiesi, Francesca, Caterina Primi, and Kinga Morsanyi. 2011. Developmental changes in probabilistic reasoning: The role of cognitive capacity, instructions, thinking styles, and relevant knowledge. *Thinking & Reasoning* 17: 315–50.

Chiesi, Francesca, Matteo Ciancaleoni, Silvia Galli, and Caterina Primi. 2012. Using the Advanced Progressive Matrices (Set I) to assess fluid ability in a short time frame: An item response theory–based analysis. *Psychological Assessment* 24: 892–900. [CrossRef] [PubMed]

De Martino, Benedetto, Neil A. Harrison, Steven Knafo, Geoff Bird, and Raymond J. Dolan. 2008. Explaining enhanced logical consistency during decision making in autism. *Journal of Neuroscience* 28: 10746–50. [CrossRef]

De Neys, Wim. 2021. On dual-and single-process models of thinking. *Perspectives on Psychological Science* 16: 1412–27. [CrossRef]

Foster, Jeffrey L., Zach Shipstead, Tyler L. Harrison, Kenny L. Hicks, Thomas S. Redick, and Randall W. Engle. 2015. Shortened complex span tasks can reliably measure working memory capacity. *Memory & Cognition* 43: 226–36. [CrossRef]

Frederick, Shane. 2005. Cognitive reflection and decision making. *Journal of Economic Perspectives* 19: 25–42. [CrossRef]

Frith, Uta. 1989. *Autism: Explaining the Enigma*. Oxford: Blackwell.

Green, Adam E., Lauren Kenworthy, Natalie M. Gallagher, Ligia Antezana, Maya G. Mosner, Samantha Krieg, Katherina Dudley, Allison Ratto, and Benjamin E. Yerys. 2017. Social analogical reasoning in school-aged children with autism spectrum disorder and typically developing peers. *Autism* 21: 403–11. [CrossRef]

Happé, Francesca, and Pedro Vital. 2009. What aspects of autism predispose to talent? *Philosophical Transactions of the Royal Society B: Biological Sciences* 364: 1369–75. [CrossRef] [PubMed]

Happé, Francesca G. 1997. Central coherence and theory of mind in autism: Reading homographs in context. *British Journal of Developmental Psychology* 15: 1–12. [CrossRef]

Hull, Laura, K. V. Petrides, and William Mandy. 2020. The female autism phenotype and camouflaging: A narrative review. *Review Journal of Autism and Developmental Disorders* 7: 306–17. [CrossRef]

Jolliffe, Therese, and Simon Baron-Cohen. 1999. A test of central coherence theory: Linguistic processing in high-functioning adults with autism or Asperger syndrome: Is local coherence impaired? *Cognition* 71: 149–85. [CrossRef] [PubMed]

Kalandadze, Tamar, Courtenay Norbury, Terje Nærland, and Kari-Anne B. Næss. 2018. Figurative language comprehension in individuals with autism spectrum disorder: A meta-analytic review. *Autism* 22: 99–117. [CrossRef]

Kanne, Stephen M., Jennifer Wang, and Shawn E. Christ. 2012. The Subthreshold Autism Trait Questionnaire (SATQ): Development of a brief self-report measure of subthreshold autism traits. *Journal of Autism and Developmental Disorders* 42: 769–80. [CrossRef] [PubMed]

Klaczynski, Paul A. 2001. Analytic and heuristic processing influences on adolescent reasoning and decision-making. *Child Development* 72: 844–61. [CrossRef]

Klaczynski, Paul A., and Jennifer M. Cottrell. 2004. A dual-process approach to cognitive development: The case of children's understanding of sunk cost decisions. *Thinking & Reasoning* 10: 147–74.

Lee, Michael D., and Eric-Jan Wagenmakers. 2014. *Bayesian Cognitive Modeling: A Practical Course*. Cambridge: Cambridge University Press.

Levin, Irwin P., Gary J. Gaeth, Megan Foley-Nicpon, Vitaliya Yegorova, Charles Cederberg, and Haoyang Yan. 2015. Extending decision making competence to special populations: A pilot study of persons on the autism spectrum. *Frontiers in Psychology* 6: 539. [CrossRef]

Liberali, Jordana M., Valerie F. Reyna, Sarah Furlan, Lilian M. Stein, and Seth T. Pardo. 2012. Individual differences in numeracy and cognitive reflection, with implications for biases and fallacies in probability judgment. *Journal of Behavioral Decision Making* 25: 361–81. [CrossRef]

Maenner, Matthew J., Zachary Warren, Ashley Robinson Williams, Esther Amoakohene, Amanda V. Bakian, Deborah A. Bilder, Maureen S. Durkin, Robert T. Fitzgerald, Sarah M. Furnier, Michelle M. Hughes, and et al. 2023. Prevalence and characteristics of autism spectrum disorder among children aged 8 years—Autism and Developmental Disabilities Monitoring Network, 11 Sites, United States, 2020. *MMWR Surveill Summ* 72: 1–14. [CrossRef]

Mata, André, Mário B. Ferreira, and Steven J. Sherman. 2013. The metacognitive advantage of deliberative thinkers: A dual-process perspective on overconfidence. *Journal of Personality and Social Psychology* 105: 353–73. [CrossRef]

McCrory, Eamon, Lucy A. Henry, and Francesca Happé. 2007. Eye-witness memory and suggestibility in children with Asperger syndrome. *Journal of Child Psychology and Psychiatry* 48: 482–89. [CrossRef]

Morsanyi, Kinga, and Keith J. Holyoak. 2010. Analogical reasoning ability in autistic and typically developing children. *Developmental Science* 13: 578–87. [CrossRef]

Morsanyi, Kinga, and Ruth M. J. Byrne, eds. 2020. *Thinking, Reasoning, and Decision Making in Autism*. Oxon: Routledge.

Morsanyi, Kinga, and Simon J. Handley. 2008. How smart do you need to be to get it wrong? The role of cognitive capacity in the development of heuristic-based judgment. *Journal of Experimental Child Psychology* 99: 18–36. [CrossRef]

Morsanyi, Kinga, and Simon J. Handley. 2013. Heuristics and biases: Insights from developmental studies. In *The Development of Thinking and Reasoning*. Edited by Pierre Barrouillet and Caroline Gauffroy. London: Psychology Press, pp. 122–49.

Morsanyi, Kinga, Chiara Busdraghi, and Caterina Primi. 2014. Mathematical anxiety is linked to reduced cognitive reflection: A potential road from discomfort in the mathematics classroom to susceptibility to biases. *Behavioral and Brain Functions* 10: 1–13. [CrossRef] [PubMed]

Morsanyi, Kinga, Dušan Stamenković, and Keith J. Holyoak. 2020. Metaphor processing in autism: A systematic review and meta-analysis. *Developmental Review* 57: 100925. [CrossRef]

Morsanyi, Kinga, Jayne Hamilton, Dušan Stamenković, and Keith J. Holyoak. 2022. Linking metaphor comprehension with analogical reasoning: Evidence from typical development and autism spectrum disorder. *British Journal of Psychology* 113: 479–95. [CrossRef]

Morsanyi, Kinga, Simon J. Handley, and Jonathan S. Evans. 2010. Decontextualised minds: Adolescents with autism are less susceptible to the conjunction fallacy than typically developing adolescents. *Journal of Autism and Developmental Disorders* 40: 1378–88. [CrossRef] [PubMed]

Osman, Magda, and Ruth Stavy. 2006. Development of intuitive rules: Evaluating the application of the dual-system framework to understanding children's intuitive reasoning. *Psychonomic Bulletin & Review* 13: 935–53.

Otero, Inmaculada, Jesus F. Salgado, and Silvia Moscoso. 2022. Cognitive reflection, cognitive intelligence, and cognitive abilities: A meta-analysis. *Intelligence* 90: 101614. [CrossRef]

Pacini, Rosemary, and Seymour Epstein. 1999. The relation of rational and experiential information processing styles to personality, basic beliefs, and the ratio-bias phenomenon. *Journal of Personality and Social Psychology* 76: 972–87. [CrossRef]

Pennycook, Gordon, James Allan Cheyne, Derek J. Koehler, and Jonathan A. Fugelsang. 2016. Is the cognitive reflection test a measure of both reflection and intuition? *Behavior Research Methods* 48: 341–48. [CrossRef]

Pennycook, Gordon, Jonathan A. Fugelsang, and Derek J. Koehler. 2015. Everyday consequences of analytic thinking. *Current Directions in Psychological Science* 24: 425–32. [CrossRef]

Primi, Caterina, Kinga Morsanyi, Francesca Chiesi, Maria Anna Donati, and Jayne Hamilton. 2016. The development and testing of a new version of the cognitive reflection test applying item response theory (IRT). *Journal of Behavioral Decision Making* 29: 453–69. [CrossRef]

Primi, Caterina, Maria Anna Donati, Francesca Chiesi, and Kinga Morsanyi. 2018. Are there gender differences in cognitive reflection? Invariance and differences related to mathematics. *Thinking & Reasoning* 24: 258–79.

Raven, John C. 1998. *Raven's Standard Progressive Matrices*. San Antonio: The Psychological Corp.

Raven, John C., and John H. Court. 2003. *Manual for Raven's Progressive Matrices and Vocabulary Scales*. San Antonio: Harcourt Assessment.

Rozenkrantz, Liron, Anila M. D'Mello, and John D. Gabrieli. 2021. Enhanced rationality in autism spectrum disorder. *Trends in Cognitive Sciences* 25: 685–96. [CrossRef] [PubMed]

Sattler, Jerome M. 2008. *Assessment of Children: Cognitive Foundations*. San Diego: J.M. Sattler.

Sattler, Jerome M., and Ron Dumont. 2004. *Assessment of Children: WISC-IV and WPPSI Supplement*. San Diego: J. M. Sattler Publisher, Inc.

Shah, Punit, Caroline Catmur, and Geoffrey Bird. 2016. Emotional decision-making in autism spectrum disorder: The roles of interoception and alexithymia. *Molecular Autism* 7: 43. [CrossRef]

Stanovich, Keith E. 2018. Miserliness in human cognition: The interaction of detection, override and mindware. *Thinking & Reasoning* 24: 423–44.

Taylor, Emily C., George D. Farmer, Lucy A. Livingston, Mitchell J. Callan, and Punit Shah. 2022. Rethinking fast and slow processing in autism. *Journal of Psychopathology and Clinical Science* 131: 392–406. [CrossRef] [PubMed]

Travers, Eoin, Jonathan J. Rolison, and Aidan Feeney. 2016. The time course of conflict on the Cognitive Reflection Test. *Cognition* 150: 109–18. [CrossRef]

Turner, Marilyn L., and Randall W. Engle. 1989. Is working memory capacity task dependent? *Journal of Memory and Language* 28: 127–54. [CrossRef]

Wechsler, David. 2004. *Wechsler Intelligence Scale for Children—Fourth UK Edition*. London: Harcourt Assessment.

Wechsler, David. 2011. *Wechsler Abbreviated Scale of Intelligence–Second Edition (WASI-II)*. San Antonio: NCS Pearson.

Weller, Joshua A., Nathan F. Dieckmann, Martin Tusler, C. K. Mertz, William J. Burns, and Ellen Peters. 2013. Development and testing of an abbreviated numeracy scale: A Rasch analysis approach. *Journal of Behavioral Decision Making* 26: 198–212. [CrossRef] [PubMed]

Journal of
Intelligence

Article

Actively Open-Minded Thinking and Its Measurement

Keith E. Stanovich [1,*] and Maggie E. Toplak [2]

[1] Department of Applied Psychology and Human Development, University of Toronto, 252 Bloor St. West, Toronto, ON M5S 1V6, Canada
[2] Department of Psychology, York University, 126 Behavioural Science Building, 4700 Keele St., Toronto, ON M3J 1P3, Canada
* Correspondence: keith.stanovich@utoronto.ca

Abstract: Actively open-minded thinking (AOT) is measured by items that tap the willingness to consider alternative opinions, sensitivity to evidence contradictory to current beliefs, the willingness to postpone closure, and reflective thought. AOT scales are strong predictors of performance on heuristics and biases tasks and of the avoidance of reasoning traps such as superstitious thinking and belief in conspiracy theories. Nevertheless, AOT is most commonly measured with questionnaires rather than performance indicators. Questionnaire contamination becomes even more of a danger as the AOT concept is expanded into new areas such as the study of fake news, misinformation, ideology, and civic attitudes. We review our 25-year history of studying the AOT concept and developing our own AOT scale. We present a 13-item scale that both is brief and accommodates many previous criticisms and refinements. We include a discussion of why AOT scales are such good predictors of performance on heuristics and biases tasks. We conclude that it is because such scales tap important processes of cognitive decoupling and decontextualization that modernity increasingly requires. We conclude by discussing the paradox that although AOT scales are potent predictors of performance on most rational thinking tasks, they do not predict the avoidance of myside thinking, even though it is virtually the quintessence of the AOT concept.

Keywords: actively open-minded thinking; myside thinking

1. Actively Open-Minded Thinking and Its Measurement

In many previous publications we have articulated the differences between the concept of rationality and the concept of intelligence (Stanovich 2009, 2011; Stanovich et al. 2016). For our actions to be instrumentally rational, they must be the best means toward our goals, and for our beliefs to be epistemically rational, they must correspond to the way the world is—they must be true. Many components of instrumental rationality and epistemic rationality are assessed by heuristics and biases tasks in the psychological literature (Baron 2008; Evans 2014; Kahneman 2011; Kelman 2011; Koehler and Harvey 2004; Manktelow 2012; Stanovich 2004, 2011). Although there are several broader conceptualizations of rationality (Mele and Rawling 2004; Stanovich 2013), the so-called axiomatic approach to rationality (Luce and Raiffa 1957; Savage 1954)—which defines rationality as adherence to certain types of consistency and coherence relationships—is well-covered by the heuristics and biases literature.

Intelligence is an underlying component that facilitates performance on heuristics and biases tasks, but it is not the only one. Individual differences in thinking dispositions also underlie differences in rational responding. Thus, rationality is the more encompassing construct. The distinction between cognitive capacity (intelligence) and thinking dispositions is an old one in psychology. Cognitive capacities are the types of cognitive processes studied by information processing researchers seeking the underlying cognitive basis of performance on IQ tests. Perceptual speed, discrimination accuracy, working memory capacity, and the efficiency of the retrieval of information stored in long-term memory are examples of cognitive capacities that underlie traditional psychometric intelligence and that have been extensively investigated. Thinking dispositions, in contrast, are better viewed as cognitive styles.

Rational thinking dispositions are those that relate specifically to the adequacy of belief formation and decision making—for example, the tendency to collect information before making up one's mind, the tendency to seek various points of view before coming to a conclusion, the disposition to think extensively about a problem before responding, the tendency to calibrate the degree of strength of one's opinion to the degree of evidence available, the tendency to think about future consequences before taking action, the tendency to explicitly weigh pluses and minuses of situations before making a decision, and the tendency to seek nuance and avoid absolutism. In short, individual differences in rational thinking dispositions include variation in people's goal management, epistemic values, and epistemic self-regulation—differences in the operation of the reflective mind in our tripartite model (Stanovich 2011; Stanovich et al. 2016).

On our Comprehensive Assessment of Rational Thinking instrument (CART; Stanovich et al. 2016), thinking disposition scales are supplemental to the subtests of the CART. Thinking dispositions are not primary measures of rationality themselves, because they are not maximizing concepts like the other constructs on the CART[1]. The most important thinking disposition that we measure on the CART is actively open-minded thinking (AOT).

2. Twenty-Five Years Trying to Measure AOT

Our research group has spent over 25 years trying to understand the AOT concept and attempting to measure it. We were originally inspired by the writings of Baron (1985, 1988, 1993), who first named and discussed AOT as an important thinking disposition (see Baron et al. 2023, for a history of the concept). In Stanovich and West (1997), we conceptualized AOT as a thinking disposition encompassing the cultivation of reflectiveness rather than impulsivity; the desire to act for good reasons; tolerance for ambiguity combined with a willingness to postpone closure; and the seeking and processing of information that disconfirms one's beliefs. The items on the initial version (Stanovich and West 1997) of our AOT scale tapped reasoning styles such as the disposition toward reflectivity using items such as "If I think longer about a problem I will be more likely to solve it," and "Intuition is the best guide in making decisions," (the latter reverse-scored). Other items tapped willingness to consider evidence contradictory to beliefs (e.g., "People should always take into consideration evidence that goes against their beliefs") and the willingness to consider alternative opinions and explanations ("A person should always consider new possibilities"). Some items tapped the willingness to postpone closure ("There is nothing wrong with being undecided about many issues"). Philosophically, the original scale focused strongly on issues of epistemic self-regulation raised in philosophical discussions (Goldman 1986; Harman 1995; Nozick 1993; Samuelson and Church 2015). The scale was a marker for the avoidance of epistemological absolutism; willingness to perspective-switch; and the tendency to consider alternative opinions and evidence.

In this paper, we will focus on the links between AOT and performance on heuristics and biases tasks from the reasoning literature, because that is where the bulk of our empirical offers have been directed. Recently, however, there has been a burgeoning literature (e.g., Ackerman and Thompson 2017; De Neys 2023) examining the meta-reasoning processes that are implicated when solving these types of reasoning tasks. Although we have not investigated connections between AOT and these meta-reasoning processes, there would seem to be a good deal of conceptual overlap. As mentioned above, Baron's (1985, 1988) early conceptualization of AOT and our earliest attempts to operationalize the concept focused heavily on epistemic self-regulation. This emphasis provides a possible link to the meta-reasoning literature (Ackerman and Thompson 2017), where the issues of monitoring ongoing thinking, allocating cognitive resources, and attentional switching are paramount. The concept of cognitive decoupling, which we elaborate later in this paper, provides a potential connection between the extant literature on AOT and the growing literature on how meta-reasoning contributes to task outcomes (e.g., Ackerman and Thompson 2017; Raoelison et al. 2020; Thompson et al. 2013).

We have been investigating actively open-minded thinking for over two decades now and have been continually refining the scale since that initial study (Sá et al. 1999; Stanovich and West 2007; Stanovich et al. 2016; Stanovich and Toplak 2019). For example, Sá et al. (1999) introduced nine new items into the scale in order to measure an aspect of AOT that we termed belief identification. These items were inspired by a theoretical paper by Cederblom (1989) in which he argued for a potential thinking style centered around the extent to which people identify their beliefs with their concept of self (e.g., "Certain beliefs are just too important to abandon no matter how good a case can be made against them" [reverse-scored]). Other additions and subtractions of components occurred over the next decade. By 2007 (Stanovich and West 2007), we had a 41-item instrument that was subsequently trimmed down to 30 items in our CART for adults (Stanovich et al. 2016; 16 items in the short form).

In our initial studies (Stanovich and West 1997) we found that our AOT scale was moderately associated with the ability to evaluate arguments. This association held even when the variance due to cognitive ability was partialled out. In several subsequent studies, we found that our AOT scale predicted performance on a variety of heuristics and biases tasks, often even after partialling cognitive ability (Stanovich and West 1997, 1998a, 1998b, 2000). This performance pattern has been found in a variety of studies conducted in many labs and has been obtained across a plethora of heuristics and biases tasks, including noncausal base-rate tasks, hypothesis evaluation tasks, four-card selection tasks, covariation detection, gambler's fallacy, conjunction fallacy, Bayesian reasoning, framing problems, ratio bias, sample size problems, and probability matching (Bruine de Bruin et al. 2007; Erceg et al. 2022; Finucane and Gullion 2010; Kokis et al. 2002; Parker and Fischhoff 2005; Pennycook et al. 2014; Sa and Stanovich 2001; Sá et al. 2005; Stanovich et al. 2016; Toplak et al. 2007; Toplak and Stanovich 2002; Toplak et al. 2011, 2014a, 2014b; Viator et al. 2020; Weller et al. 2018; West et al. 2008).

It is startling that a questionnaire measure tapping a thinking disposition correlates with so many heuristics and biases tasks and that it is often a unique predictor after cognitive ability is partialled. Equally impressive has been the recent expansion of the AOT concept as an explanatory mechanism in new and diverse areas (Baron 2019; Baron et al. 2015; Baron et al. 2023; Baron et al. 2017), including linking the concept to optimal information acquisition (Haran et al. 2013) and cognitive inhibition skills (Campitelli and Gerrans 2014).

One of the earliest linkages that we observed when we began studying the AOT was that it predicted pseudoscientific beliefs and superstitious thinking. It has consistently correlated with the presence of what we termed contaminated mindware—declarative knowledge that is incorrect and that leads to suboptimal action (see Rizeq et al. 2021; Stanovich 2011; Stanovich et al. 2016). For example, in one of our earliest studies (Stanovich and West 1997), several of the subcomponents of our first AOT test displayed significant negative correlations with superstitious thinking (flexible thinking = −.26; absolutism = −.23; dogmatism = −.19; categorical thinking = −.28). A later study using an updated composite AOT scale found a −.38 correlation between it and superstitious thinking (Toplak et al. 2011). Using a more refined AOT scale, we found a −.44 correlation with a university sample and −.59 with an mTurk sample (Stanovich et al. 2016), as well as correlations ranging from −. 39 to −.53 with Prolific samples (Stanovich and Toplak 2019). These findings have been much replicated, as AOT scales have been found to correlate with a variety of different measures of superstitious thinking and belief in the paranormal (Erceg et al. 2022; Jastrzębski and Chuderski 2022; Pennycook et al. 2020; Rizeq et al. 2021; Svedholm and Lindeman 2013; Svedholm-Hakkinen and Lindeman 2018).

Conspiracy belief scales measure another form of contaminated mindware that has been linked to actively open-minded thinking. These beliefs, given the lack of evidence for them, appear to be remarkably prevalent (Oliver and Wood 2014), and they seem to be part of a cluster of thinking styles that interconnect with superstitious behavior and animistic thinking (Oliver and Wood 2014; Rizeq et al. 2021; Stanovich et al. 2016).

Belief in specific conspiracy theories and general conspiratorial ideation have been found to correlate with AOT scales, but at a somewhat lower level than the correlations obtained with superstitious thinking. For example, Swami et al. (2014) found a significant negative correlation between the Stanovich and West (2007) AOT scale and the Belief in Conspiracy Theories Inventory (Swami et al. 2010, 2011), but the magnitude of the correlation was only −.07. Somewhat stronger results were obtained with the AOT scale that was part of our Comprehensive Assessment of Rational Thinking (Stanovich et al. 2016) and our Conspiracy Beliefs subscale. We found correlations of −.34 and −.26 in university samples but a higher −.48 correlation in an mTurk sample (Stanovich et al. 2016). Using a different AOT scale, we found correlations ranging from −.19 to −.29 in Prolific samples (Stanovich and Toplak 2019). These findings have been much replicated, as AOT scales have been found to correlate with a variety of different measures of conspiracy belief (Binnendyk and Pennycook 2022; Erceg et al. 2022; Jastrzębski and Chuderski 2022; Pennycook et al. 2020; Rizeq et al. 2021; Yelbuz et al. 2022).

3. Examining a Wider Range of Correlates Leads to New Questions about AOT Scale Composition

Recent expansion of the use of the AOT scale into areas such as belief in evolution (Deniz et al. 2008; Sinatra et al. 2003); skeptical processing of fake news (Bronstein et al. 2019); accuracy in future forecasting (Mellers et al. 2015); moral reasoning (Baron et al. 2015); religiosity/ideology (Baron 2019; Stanovich and Toplak 2019); and skeptical attitudes toward alternative medicine (Svedholm-Hakkinen and Lindeman 2018) has raised new questions about the specific compositional structure of the AOT scales in use.

In Stanovich and Toplak (2019), we described how we were led to re-examine the way in which our original scale was constructed and revised over the years by reading some recent studies that inserted the AOT concept into discussions of religion and ideology. The extremity of the results of these studies gave us pause and forced us to think more about the logic of some of the items. First, there were the startling results of Piazza and Landy (2013), who reported some extremely high correlations between an AOT scale and various measures of religiosity: −.58 with an attitudes toward religion scale, −.59 with a religious faith questionnaire, −.63 with a Christian orthodoxy scale, −.58 with self-reported religiosity, and a truly astonishing correlation of −.70 with a morality founded on divine authority scale. Baron et al. (2015) observed a correlation similar to those of Piazza and Landy (−.61) between an AOT measure and a four-item religiosity scale. Likewise, Bronstein et al. (2019) reported a similarly high correlation of −.67 between a short-form AOT scale and a religious fundamentalism measure. Other studies, such as that of Yilmaz and Saribay (2017), have found similarly strong correlations (−.47) between AOT scores and social conservativism.

We were startled by these high correlations (in the −.50 to −.70 range), because we have run over a dozen studies employing versions of AOT scales (see Stanovich et al. 2016) in which religiosity and political ideology have been included as demographics questions. We have very consistently found correlations in the much lower range of −.25 to −.40 between religiosity and the AOT (correlations with ideology are almost always even lower). When puzzling over the cause of these discrepancies in the association between religiosity and AOT that our group observes versus those reported in these other studies, one of the first things we noticed was that most of the other research tended to use short-form AOT scales—often short forms of fewer than 10 items. These are much smaller scales than the 41-item AOT measure that we were using over a decade ago (Stanovich and West 2007) and the 30-item revised measure that we used in the CART (Stanovich et al. 2016).

More important than the sheer number of items, of course, is the specific composition of the short forms. Here, a deeper analysis of the items used across various studies revealed a potential source of the discrepancies between the results. That source appears to reside primarily in the items that Sá, West, and Stanovich introduced into the scale in 1999. Sá et al. (1999) termed these items belief identification items, but Stanovich and Toplak (2019) suggested the broader term belief revision items. Nine of these items were introduced into the AOT scale in our 1999 study. One item of a similar type was already in the earlier scale. Thus, the 41-item scale used in the mid-2000s by Stanovich and West (2007) consisted of 10 belief revision items out of a total of 41 (24.4%). The 30-item updated AOT subtest in our CART (Stanovich et al. 2016) had nine belief identification items (30% of the total). It was immediately of concern to us in perusing the short forms used in other studies that the proportion of belief revision items was substantially higher. Piazza and Landy (2013), in the study that obtained extremely high correlations with a host of variables measuring religiosity, used a seven-item AOT scale that contained five belief revision items (Yilmaz and Saribay 2017, used the same seven-item short form). Baron et al. (2015) used an eight-item AOT short form that contained four belief revision items. Bronstein et al. (2019), who reported a substantial correlation of −.67 with a religious fundamentalism measure, used an eight-item AOT short form that contained five belief revision items. Thus, the three studies displaying religiosity correlations of .55 or above used short forms of the AOT that were composed of 71.4%, 50%, and 62.5% belief identification items. This is much higher than the roughly 24–30% composition that we have used in our versions of this instrument. We conjectured that the high proportion of belief revision items in these other studies was the source of the high correlations with religiosity.

What is the feature of the belief revision items that might be augmenting correlations with religiosity? Consider two such items:

Beliefs should always be revised in response to new information or evidence.
One should disregard evidence that conflicts with your established beliefs. (reverse-scored)

The general thrust of this kind of item is that the subject is being asked what should be done when encountering evidence that conflicts with a prior belief or opinion. It is important to note that no *specific* prior opinion is mentioned in any of these items. It is just the generic word "belief" that is used. Of course, adjusting a prior belief based on new contradictory evidence is more or less easy to do depending upon what the belief is. For example, is it my belief that I voted the right way in the presidential election of 2020? Or is it my belief that the deli counter is better at Albertsons than at Safeway? The latter is obviously going to be a belief that is more easily conditioned by evidence than the former. The generic nature of the word "belief" in these items allows the respondent to insert any imaginary opinion as the belief in question. Potential social desirability considerations may lead most people to insert a belief that is easy to change. Thus, these items are almost inviting someone to fall prey to the bias blind spot—that is, thinking that others are characterized by a particular bias but that you yourself are not (Pronin 2007).

All of the above might be true for a secular person, but a person with strongly held religious convictions might well be prone to see the word "belief" as referring to their spiritual beliefs—a class of beliefs that are not going to be easily altered by evidence. In contrast, a secular person might be much less likely to see the word "belief" as denoting an imaginary opinion that is so strongly held. Our conjecture was that what a religious person does when seeing the generic word "belief" is simulate an *actual* stance (their spirituality) that is much more difficult to reconsider based on evidence than a generic belief or an anodyne one.

To see this, one might imagine a secular person who answered one of our belief revision items affirmatively. To such a person who answered by saying, "Well of course I'd change my belief if I got contradictory evidence, that's what an intelligent person does", we might imagine the conversation continuing. "OK", we might reply, "now imagine the belief is your vote against Trump in 2020. Would you be likely to change that based on new information?" It is doubtful that the item would be so enthusiastically endorsed if we substituted in the *specific* belief "my vote against Trump was a good thing".

To the extent that secular people are inserting an anodyne belief such as the preference for Pepsi over Coke as opposed to a belief strongly related to worldview such as belief in God or a particular religion, then they are advantaged on such items. This advantage, along with the corresponding disadvantage to the religious respondent who might slot in "belief in God" for the term "belief", inflates the negative correlation between AOT and religiosity—such items are harder to agree to on the part of the religious-minded. This non-equivalence never occurred to us at the time we were creating the belief revision items, perhaps because of our own secular biases. No doubt, if the correlations had come out in the other direction, we would have been quicker to notice a problem, since those of us constructing the items were all secularists.

Stanovich and Toplak (2019) showed in a post hoc analysis of data from the CART (Stanovich et al. 2016) that belief revision items on the AOT subtest showed higher correlations with both religiosity and ideology than non-belief revision items did. Importantly, the nonbelief revision items show just as strong correlations with other variables such as superstitious thinking and belief in conspiracy theories. In a new experiment, we demonstrated that subjects high in religiosity are differentially affected by the word "belief" in an item and that using that term inflates correlations between AOT and religiosity—and, to a lesser extent, the correlation between AOT and ideology. Pennycook et al. (2020) reported converging findings, in that they found that a version of the AOT that substituted the word opinion for the word belief reduced the correlation between the AOT and religious beliefs from −.42 to −.20. The correlations with ideology and various social opinions were likewise reduced.

Assessing whether opinions and beliefs are flexibly conditioned by evidence is an important component of actively open-minded thinking, but using the word belief misleadingly inflates correlations in studies where the focus shifts to the larger set of issues that we mentioned above (the relation between AOT and religiosity, ideology, voting behavior, etc.). There is no doubt that a correlation between AOT and religiosity exists. It is just that it is in the range of −.20 to −.30 rather than −.65. The difference matters, because of the contexts in which many of the correlations in the range of −.60 to −.70 have been obtained. That context has been, in many cases, studies that have used only short forms of the AOT with modest reliabilities. If these −.65 to −.70 correlations were corrected for attenuation—or if the two variables were measured as latent constructs—it would not be surprising if the relationship between them approached −1.0.

With individual differences in AOT entirely explained by religiosity, psychological research would then be saying to the public that religiosity and failing to think in an open-minded manner were, for all intents and purposes, the same thing—that being highly religious is virtually synonymous with being close-minded. Our findings, of course, support the weaker conclusion that there is a replicable moderate correlation between actively open-minded thinking and religiosity.

It is increasingly the case that the psychological correlates of worldview, voting behavior, and ideological orientation are becoming points of contention in our divided political culture (Baron 2019; Baron et al. 2023; Crawford and Jussim 2018; Ditto et al. 2019; Duarte et al. 2015; Kahan et al. 2017; Stanovich 2017, 2021). If psychological studies of this type are increasingly becoming an adjunct of politics, it is important that psychology maintain its credibility as a neutral arbiter—a credibility that has been vastly eroded in recent years by empirical evidence of the ideological bias in our science (Ceci and Williams 2018; Crawford and Jussim 2018; Duarte et al. 2015; Ellis 2020; Haidt 2022; Jussim 2019, 2022; Stanovich 2021). Greater intellectual diversity in our own lab years ago might have prevented us from continuing to use items in our AOT scale that inflated negative correlations with religiosity.

4. Scale Structure Changing over Time

Discussing the history of the belief revision items opens up the topic of the composition of AOT scales and their evolution over time. At the very beginning of our studies of AOT, we included in our scale items that we constructed ourselves that were designed to tap the AOT concept, but we also included a variety of items that were included on scales tapping related constructs. For example, in our initial 1997 scale (Stanovich and West 1997), we constructed items tapping the disposition toward reflectivity ("If I think longer about a problem I will be more likely to solve it"; "Difficulties can usually be overcome by thinking about the problem, rather than through waiting for good fortune"; "Intuition is the best guide in making decisions"; and "Coming to decisions quickly is a sign of wisdom", the latter two reverse-scored), willingness to consider evidence contradictory to beliefs (e.g., "People should always take into consideration evidence that goes against their beliefs"), willingness to consider alternative opinions and explanations ("A person should always consider new possibilities", "Considering too many different opinions often leads to bad decisions", the latter reverse-scored), and tolerance for ambiguity combined with a willingness to postpone closure ("There is nothing wrong with being undecided about many issues", "Changing your mind is a sign of weakness", and "Basically, I know everything I need to know about the important things in life", the latter two reverse-scored).

However, to these kinds of items, we added items from extant scales of dogmatism, the openness–values and openness–ideas facets from the Revised NEO Personality Inventory (Costa and McCrae 1992), and scales measuring absolutism and categorical thinking (Epstein and Meier 1989). As discussed above, nine belief revision items were added in 1999 (Sá et al. 1999). Our 41-item scale published in Stanovich and West (2007) has been much used, although it was still, like our earlier scales, quite a conceptual amalgam.

The 30-item AOT scale that we published in the CART, however, represented a more conceptually coherent concept of AOT than our earlier measures. Items from many scales where the construct was related to AOT but not central to the AOT concept were eliminated. So, for instance, items tapping dogmatism, absolutism, and categorical thinking were eliminated, because although these concepts are related to (lack of) AOT, they are not AOT's defining features. Newton et al. (2023), for example, found that dogmatism/categorical thinking was separable from AOT. The openness–ideas facet items from the Revised NEO Personality Inventory were removed because they were less conceptually related to AOT than they were to need for cognition (e.g., "I enjoy working on 'mind-twister'-type puzzles").

We removed the openness–values facet items from the Revised NEO Personality Inventory that were in our 1997 scale for a different reason. Some of these items are contaminated by ideological and/or anti-religious bias. Specifically, they require high scorers to have a progressive worldview—high scores are harder to achieve for conservatives or those higher in religiosity (Charney 2015). For example, the purpose of the item "I believe that we should look to our religious authorities for decisions on moral issues" is clearly to probe whether the individual is prone to rely on authorities to determine moral beliefs. But the specific authority that the subject has to ignore in order to score highly is a religious authority. There is no corresponding item testing whether the subject is equally reliant on secular authorities (see Stanovich and Toplak 2019 for the importance of content in an item like this). Is it more close-minded to rely on a theologian for moral guidance than to rely on a university "bio-ethicist?" Secular subjects are guaranteed to score higher on an item like this, which virtually builds in a correlation between openness and religiosity. Another item—"I believe that the different ideas of right and wrong that people in other societies have may be right for them"—seems to require that full-blown cultural relativism be endorsed in order to receive a high openness score. However, the only ideological niche where such strong relativism is highly endorsed is in an extreme form of multiculturalism that exists only on the political left. Thus, we removed all of the openness–values items from the CART version of the AOT to avoid building ideology into this scale[2]. If there is a correlation between AOT and liberalism, we want that to be an actual fact about human psychology and not something that is an artifact of item construction.

In addition to the removal of these classes of items, the 30-item CART version of the scale contained some new items that were taken from the MRM scale introduced by

Stanovich (2008). MRM refers to the master rationality motive proposed by Stanovich (2008). That motive is the desire to act in accordance with good reasons. Items from this scale were added to the AOT scale on the CART, including items such as "I like to think that my actions are motivated by sound reasons" and "If a belief suits me then I am comfortable, it really doesn't matter if the belief is true (reverse scored)".

5. Toward a New 13-Item Recommended AOT Scale

All of these changes made the 30-item CART version of our AOT scale a much more coherent measure than the much-used 41-item version in Stanovich and West (2007), probably similar in coherence to humility scales[3]. This was even more true of the shortened 16-item version of the CART AOT scale. The greater coherence of the CART version was predictable from the factor analysis of the 41-item Stanovich and West (2007) version conducted by Svedholm-Hakkinen and Lindeman (2018). They argued that a four-factor solution was needed to capture the multidimensional nature of the scale. Interestingly, however, three of the four factors were composed of items from categories that we have since eliminated from the 30-item CART version of the scale. Consistent with our argument above, one of these three factors was labeled Liberalism and was composed largely of more openness–values items of the type that we have eliminated ("I believe that the different ideas of right and wrong that people in other societies have may be right for them"). Another factor was labeled dogmatism by Svedholm-Hakkinen and Lindeman (2018) and consisted of openness–values items that we have eliminated, in addition to dogmatism and categorical thinking items that have also been eliminated ("I think there are many wrong ways, but only one right way, to almost everything"). Finally, they identified a factor they called Belief Personification that also consisted of dogmatism items and categorical thinking items that we have eliminated ("I tend to classify people as either for me or against me").

In fact, Svedholm-Hakkinen and Lindeman (2018) identified just one factor from items we are still using. They labeled the factor Fact Resistance, but it was actually composed primarily of items that we have called belief revision items. Importantly, when they correlated the Stanovich and West (2007) scale with other criterion variables such as superstitious thinking and trust in alternative medicine, the belief revision factor was the primary correlate. Associations with the other three factors were quite low. Overall, the analysis of our 41-item scale by Svedholm-Hakkinen and Lindeman (2018) supports the revisions in the AOT scale that we have made in the CART (Stanovich et al. 2016).

It would be nice to conclude here that the AOT scale in the CART should be the default scale for researchers in this area. Unfortunately, that conclusion would be premature. First of all, at 30 items, the scale is overly long for many investigations with time limits. Even the 16-item shortened form is on the long end for some studies. Many investigations in the AOT literature use scales no longer than 10 items.

Even more importantly, our AOT scale in the CART contains a number of belief revision items that implicate the ideological/religious bias that we discussed earlier. It is becoming increasingly important that AOT scales not contain biases of these types as the investigations using the scale proliferate into politically charged areas (e.g., studies of fake news, conspiracy theories, and politically charged issues such as climate change). Increasingly, the scale is being used to adjudicate issues in the literature of political psychology. It is even more important in these types of areas that the scale not be biased ideologically or in terms of worldview. The conclusion that "liberals are more open-minded" will be specious if based on an instrument that has such a correlation built-in.

For all of these reasons, we feel compelled to present a new recommended AOT scale based on our own research and our reading of the recent AOT literature. That scale is presented in Table 1. In the right-hand column of Table 1, we have listed the primary AOT concepts that are tapped by each item. Several of the items tap more than one AOT concept. Subjects responded on a six-point scale with no neutral point: disagree strongly (1), disagree moderately (2), disagree slightly (3), agree slightly (4), agree moderately (5), agree strongly (6).

Table 1. Recommended 13-Item Actively Open-Minded Thinking Scale.

Item	Code
People should always take into consideration evidence that goes against their opinions.	BR
Changing your mind is a sign of weakness. (R)	REF/ALT/BR/OC
I like to think that my actions are motivated by sound reasons.	REASONS
It is important to stick to your opinions even when evidence is brought to bear against them. (R)	BR/OC/BEL_ID
Intuition is the best guide in making decisions. (R)	REF
Considering too many different opinions often leads to muddled thinking. (R)	AMB/ALT/OC
One should disregard evidence that conflicts with your current opinions. (R)	BR/ALT
Coming to decisions quickly is a sign of wisdom. (R)	REF
Allowing oneself to be convinced by a solid opposing argument is a sign of good character.	BR/BEL_ID/OC
If something I think feels right then I am comfortable, whether or not it is true. (R)	REF/TRUTH
A person should always consider new information.	ALT
People should revise their conclusions in response to relevant new information.	BR
Certain opinions are just too important to abandon no matter how good a case can be made against them. (R)	BR

R = reverse-scored. REF = tendency toward reflection. ALT = concern for alternative explanations. BR = belief revision item. AMB = tolerance for ambiguity. OC = tendency toward overconfidence. BEL_ID = belief identification item. REASONS = value placed on thinking motivated by reasons. TRUTH = value placed on truth. Subjects responded on a six-point scale with no neutral point: disagree strongly (1), disagree moderately (2), disagree slightly (3), agree slightly (4), agree moderately (5), agree strongly (6).

There are 13 items in our recommended scale, 12 of which appeared in the AOT scale used in the CART. However, several of these items, particularly the belief revision items, have been slightly rewritten. For example, to reflect the findings of Stanovich and Toplak (2019), the word belief was removed from items—usually being replaced with the word "opinion" or "something I think" (see Pennycook et al. 2020, who used this substitution). The single item that did not appear in the AOT scale of the CART but is included in Table 1 as part of the recommended scale ("Intuition is the best guide in making decisions", reverse-scored) was originally in the Stanovich and West (2007) scale, and thus, we have a lot of data on that item as well. Based on an analysis of the 12 items in Table 1 that were present in the CART AOT, this scale would have a reliability of .84 (Cronbach's alpha) and would factor as having a single dominant factor (only one eigenvalue > 1).

6. The AOT Measures Psychological Tendencies to Decouple and Decontextualize—A Critical Aspect of Modernity

It is important to measure AOT carefully and without bias for two reasons. First, as noted above, the concept is increasingly being used in broader areas of sociocultural concern such as belief in alternative medical practices; belief in pseudoscience and conspiracies; detection of fake news; moral decision-making; and debates about the origins and correlates of political ideologies (Baron 2019; Bronstein et al. 2019; Pennycook et al. 2020; Stenhouse et al. 2018). Secondly, it is becoming increasingly clear that in the domain of rational thinking, AOT is a uniquely potent predictor. It is ubiquitously linked to subtests in our CART test (Stanovich et al. 2016). The CART is a very comprehensive measure of rational thought and is composed of 20 different subtests (and 4 supplemental scales, which include AOT). Our 30-item AOT scale not only was correlated with every one of the 20 subtests but accounted for variance over and above cognitive ability in the vast majority of them

(17 out of 20). Despite the multifariousness of the rationality construct itself (which is why the CART contains 20 subtests), a particular thinking style—actively open-minded thinking—does permeate almost all of the components (from probabilistic reasoning to avoiding overconfidence and many more).

What are the central features of thinking that make the AOT such a good predictor of rational thinking? We would argue that the common psychological dimension is the tendency to engage in cognitive decoupling (Stanovich 2011; Stanovich and Toplak 2012). To a lesser extent, the items may tap a related tendency toward the decontextualization of problems.

Cognitive decoupling is particularly relevant to the heuristics and biases tasks that operationally define rationality in cognitive science (Baron 2008; Kelman 2011; Stanovich 1999, 2012), because these tasks often create hostile problem-solving environments (Stanovich et al. 2016). Heuristics and biases tasks are designed to trap the cognitive miser (Kahneman 2011; Stanovich 2004, 2018). Tasks from this literature often have an intuitively compelling wrong answer that must be overridden, as in the famous Linda conjunction fallacy problem (Tversky and Kahneman 1983), where even sophisticated responders are tempted by the attractiveness of the wrong answer. Stephen J. Gould's (1991) introspection was that "I know the conjunction is least probable, yet a little homunculus in my head continues to jump up and down, shouting at me—'but she can't be a bank teller; read the description'" (p. 469).

In a probabilistic reasoning task from the heuristics and biases literature, the entire point is to see how dominant or nondominant the statistical interpretation is over the narrative interpretation. The fact that many heuristics and biases tasks can be construed by the subject in different ways (a statistical interpretation versus a narrative interpretation, for instance) is often seen as a weakness of such tasks, when in fact it is the design feature that makes the task diagnostic.

As a result, heuristics and biases tasks create a more hostile reasoning environment than typical IQ test problems, in that the latter do not contain enticing lures toward an incorrect response. Neither is the construal of an intelligence test item left up to the subject. Instead, the instructions to an IQ test item attempt to remove ambiguity in a way that is not true of a heuristics and biases problem. The famous Linda conjunction problem would be a prime case in point. The instructions purposefully do not tell the subject how to weight the conflicting cues—the similarity of the description ("deeply concerned with issues of discrimination and social justice, and also participated in anti-nuclear demonstrations") to the classification ("feminist bank teller") and the subset/superset relationship between feminist bank teller and bank teller. Most subjects detect the two conflicting cues in the problem (De Neys 2014, 2023), but the instructions pragmatically obscure the fact that the correct weighting of the cues is 0/100.

Thus, rational thinking paradigms attempt to measure the *propensity* to use a cognitive skill in a way that IQ tests do not. People who can answer an explicit probability question on a test or can accurately define "control-group" when asked may not invoke these principles when their relevance to a problem is partially disguised.

Cognitive decoupling is implicated in such hostile task environments in two ways (Oaksford and Chater 2012; Stanovich 2011; Stanovich and Toplak 2012). It is implicated in the inhibitory override of the intuitive response triggered by many heuristics and biases tasks, but it is also implicated in the sustained simulation of alternative worlds that is necessary to compute the correct response. The first type of cognitive decoupling—inhibition of the prepotent response—is akin to that studied in the executive functioning literature (Kovacs and Conway 2016; Miyake and Friedman 2012; Nigg 2017). However, the ability to suppress miserly processing gets the job only half done. Suppressing one response is not helpful unless there is a better response available to substitute for it. Where do these better responses come from? One answer is that they can come from processes of hypothetical reasoning and cognitive simulation (Evans 2007, 2010; Evans and Stanovich 2013; Stanovich 2011). When we reason hypothetically, we create temporary models of the world and test out actions (or alternative causes) in that simulated world. Decoupling is necessary in order to prevent our representations of the real world from becoming confused with representations of imaginary situations. The tendency to initiate such decoupling for the purposes of simulation is a dispositional variable, separable from cognitive capacity (the ability to sustain the decoupling).

Given this understanding of the importance of cognitive decoupling in heuristics and biases tasks[4], our conjecture is that AOT scales tap the propensity to engage in these types of cognitive operations. For example, some AOT items relate to avoiding miserly processing and overriding the tendency to fix beliefs quickly or to decide quickly: "Coming to decisions quickly is a sign of wisdom" (reverse-scored). Others tap the willingness to consider possibilities beyond the focal model that is in the mind, e.g., "Considering too many different opinions often leads to muddled thinking" (reverse-scored), "Changing your mind is a sign of weakness" (reverse-scored), and "A person should always consider new information". Additionally, many belief revision items require the subject to hold an existing belief in abeyance while simulating the effect of new information on the original belief—classic cognitive decoupling. All of these types of items are included in the recommended AOT scale in Table 1.

AOT scales capture global attitudes that make people more willing to decouple from strong default responses and consider new and/or conflicting evidence—or, for those responding on the other end of the scale, to be more comfortable with natural responses and accumulated knowledge. The tendency to be comfortable with responses that seem intuitive or that have been imbibed by repetition in familiar environments is probably the factor that accounts for the negative .20 to .30 correlations with religiosity that are observed in the literature in scales that are not overly contaminated with the biasing term "belief".

The need to cognitively decouple is increasingly a requirement of modernity itself (see Stanovich 2004 for a more comprehensive discussion of this point). Modernity is the result of long historical trends that have replaced local/particular traditions with science and rationality as the arbiters of truth claims. This shift coincides with an increase in environments for thinking that are hostile rather than benign. IQ tests do not pick up these hostile aspects of the cognitive environment of modernity—but rational thinking tasks do. In fact, critics who charge that heuristics and biases tasks are artificial are missing an important point. The *kind* of "artificiality" that they display represents a strength of such tasks rather than a weakness. It is a design feature, not a bug. Years ago, Einhorn and Hogarth (1981) made the telling point that "in a rapidly changing world it is unclear what the relevant natural ecology will be. Thus, although the laboratory may be an unfamiliar environment, lack of ability to perform well in unfamiliar situations takes on added importance" (p. 82).

What Einhorn and Hogarth are pointing out is that the argument that laboratory tasks are not like "real life" is becoming less and less true. "Life", in fact, *is becoming more like the tests!* For example, market economies contain agents who will exploit automatic responding for profit (better buy that "extended warranty" on a $150 electronic device!). This puts a premium on overriding intuitive responses that will be exploited by others in a market economy. The danger of such miserly tendencies (Stanovich 2018) in the domain of personal finance is suggested by the well-known finding that consumers of financial services often purchase high-cost products that underperform in terms of investment return when compared to the low-cost strategies recommended by true experts (e.g., dollar-cost averaging into no-load index mutual funds; see Bazerman 2001). The reason is, of course, that the high-cost fee-based products and services are the ones with high immediate recognizability in the marketplace.

Many rational thinking tasks require subjects to decontextualize in a particular way—by "ignoring what they know" or by ignoring irrelevant context (belief bias in syllogisms, the famous Linda problem, etc.). That makes these tasks a good proxy for an aspect of scientific thinking, because, in science, we are often required to ignore what we know or believe. Testing a control group when you fully expect it to underperform compared to an experimental group is a form of ignoring what you believe. Science is a way of systematically ignoring what we know, at least temporarily (during the test), so that we can recalibrate our belief after the evidence is in.

Likewise, many aspects of the contemporary legal system put a premium on detaching prior belief and world knowledge from the process of evidence evaluation. Modernity increasingly requires decontextualizing in the form of stripping away what we personally "know" due to its emphasis on such characteristics as fairness, rule-following despite context, even-handedness, sanctioning of nepotism, unbiasedness, universalism, inclusiveness, and legally mandated equal treatment.

7. The Paradox of AOT and Myside Thinking

A consistent observation in our earliest studies of individual differences in rational thinking was that almost every cognitive bias was correlated with intelligence as measured with a variety of cognitive ability indicators (Stanovich 1999). As discussed above, individual differences in most cognitive biases were also predicted by actively open-minded thinking.

Despite these consistent findings involving almost every other cognitive bias, myside bias has failed to correlate with AOT scales in the same manner as other biases. For example, in our study using Perkins' (1985) argument generation paradigm (Toplak and Stanovich 2003), we found substantial myside biases on several issues (people tended to give more arguments in favor of their position than against), but the degree of myside bias was not correlated with several thinking dispositions, including AOT, dogmatism, and need for cognition. Macpherson and Stanovich (2007) examined myside bias in both argument generation and evidence evaluation and also measured two different thinking dispositions: AOT and need for cognition. None of the four resulting correlations were significant.

In our studies of naturalistic myside bias (Stanovich and West 2007) and argument evaluation (Stanovich and West 2008), relationships between myside bias and rational thinking dispositions were also negligible. Guay and Johnston (2022) examined myside thinking in political reasoning and found that the need for certainty and openness did not predict the magnitude of the myside effect.

Kahan and Corbin (2016) found an interaction between myside thinking and AOT scores, but the interaction was in the opposite of the expected direction. Conservatives and liberals who were high in AOT had more diverging opinions on climate change than conservatives and liberals who were low in AOT[5]. Stenhouse et al. (2018) found no significant interaction between AOT and ideological difference in climate-change attitudes. Although not replicating the interaction observed by Kahan and Corbin (2016), the Stenhouse et al. (2018) results (as well those of Clements and Munro 2021) converged with their results and those of Macpherson and Stanovich (2007) and Stanovich and West (2007) in finding no evidence that higher AOT scores attenuate tendencies toward myside thinking.

In a followup study, Eichmeier and Stenhouse (2019) found a significant correlation between party identification and AOT scores (as have most studies; see Stanovich and Toplak 2019). However, using an argument evaluation paradigm, they found no indication that AOT scores were related to the myside bias observed in the argument strength ratings (see also Beatty and Thompson 2012; Clements and Munro 2021). Thus, the findings from the Stenhouse lab (Eichmeier and Stenhouse 2019; Stenhouse et al. 2018) are exactly parallel to those from the Stanovich lab (Macpherson and Stanovich 2007; Stanovich and Toplak 2019; Stanovich and West 2007). Both find that AOT scores correlate in the .20 to .30 range with ideology/partisanship, but neither lab finds an indication that AOT itself actually predicts the avoidance of myside bias. Although conservatives score lower on AOT scales, they do not display larger myside bias effects than liberals (Ditto et al. 2019; Guay and Johnston 2022; Stanovich 2021).

This convergence of findings is disconcerting, because of all the biases one would expect to be correlated with AOT, it would be myside bias (Baron 1993, 2019; Baron et al. 2023). Baron et al. (2015) argue that "AOT is a set of dispositions aimed at avoiding 'myside bias', the tendency to think in ways that strengthen whatever possible conclusions are already strong" (p. 267). In a later treatment of the concept, Baron et al. (2023) argue that the core of AOT encompasses avoiding myside bias and avoiding overconfidence in favored conclusions. The findings indicating that AOT does not correlate with direct measures of myside bias constitute an embarrassment to this view. The findings also do not fit well with our view, articulated above, that the AOT taps processes of decoupling and decontextualization that support the detachment needed to deal with the hostile environment of heuristics and biases tasks. It seems as though such detachment from the intuitive response is facilitated on normal heuristics and biases tasks such as the Linda problem but is not brought to bear when the problem involves detaching from a favored belief.

We would argue that strength of belief is an issue here that may make the findings a little more congenial to our view. *Belief bias* does indeed correlate with AOT as we would expect, but myside bias does not. Belief bias occurs when people have difficulty evaluating conclusions that conflict with what they know about the world. For example: "all living things need water; roses need water; therefore, roses are living things" is an invalid syllogism. Belief bias has been most extensively studied in the syllogistic reasoning and conditional reasoning literatures (Evans 2017), but it is observed in other paradigms as well (Levin et al. 1993; Stanovich and West 1997, 1998b; Thompson and Evans 2012).

Belief bias is not the same as myside bias. Belief bias occurs when real-world knowledge interferes with reasoning performance. Myside bias is a bias toward searching and interpreting evidence in a manner that tends to favor the hypothesis we want to be true (Mercier 2017; Stanovich et al. 2013). What turns a belief bias into a myside bias? Myside bias refers to processing in favor of existing opinions that are currently highly valued. To use a distinction discussed years ago by Abelson (1988), myside bias concerns the beliefs that individuals hold with high conviction. Convictions—unlike more typical beliefs—are accompanied by emotional commitment and ego preoccupation. Convictions also tend to have undergone more cognitive elaboration (see Abelson 1988; and also Fazio 2007, and Howe and Krosnick 2017, for more contemporary discussions). Skitka et al. (2005) found that attitudes rooted in moral mandates tended to become convictions. Convictions that were rooted in such moral judgments were especially potent predictors of outcome variables (social distance, good will, etc.).

To illustrate the difference between a simple belief and a conviction, imagine a thought experiment where you were on another planet (Zircan), otherwise exactly like Earth, and heard from someone that on planet Zircan roses were never red and were always brown. You would have no trouble acquiring that belief. You would feel no urge to argue with anyone that roses can be red. On planet Zircan, they simply are not, and you would have no trouble giving up your belief that roses can be red. On the other hand, if you were to hear that on planet Zircan it was believed that left-handed people were morally inferior to right-handed people, you likely would not accept that belief and in fact would try to argue against it. You would instead defend your belief that the moral worth of human beings does not depend on whether they are left-handed or right-handed. That belief is a conviction for you in a way that the belief that roses can be red is not.

Convictions often derive from worldviews that spawn so-called protected values—those that resist trade-offs with other values (Baron and Spranca 1997). Protected values (sometimes termed sacred values; see Ditto et al. 2012; Tetlock 2003) are viewed as moral obligations that arise from deontological rules concerning action, and the thought of violating them often provokes anger. Experiments have shown that subjects are reluctant to trade or engage in monetary tradeoffs when protected values are at stake (Baron and Leshner 2000; Bartels and Medin 2007). Interestingly, Fisher and Keil (2014) found that the closer beliefs were to convictions, the more poorly calibrated subjects were—almost always believing that they could provide good arguments for their convictions when in fact they could not.

In further writings on the idea that some beliefs can become convictions, Abelson (1986; Abelson and Prentice 1989) makes the distinction between what he calls testable beliefs and distal beliefs. Testable beliefs are closely tied to the real world and the words we use to describe that world (e.g., roses are red). They can be verified by observations—sometimes easily made personal observations but other times requiring reliance on the expertise of others and the more sophisticated methods of science. In contrast, distal beliefs cannot be directly verified by experience, nor can they be easily confirmed by turning to experts or scientific consensus. For example, you may think that pharmaceutical companies make excessive profits, or that your state should spend more on mental health and less on green initiatives. Certainly, economic statistics and public policy facts might condition distal beliefs such as these (either strengthening or weakening our attachment to them) but they cannot verify our distal beliefs in the same manner that they can verify testable ones. Many distal beliefs embody our values. When they do, they are apt to become convictions, because they will lead to emotional commitment and ego preoccupation, as argued by Abelson (1988). Distal beliefs often derive from a person's general worldview or, in politics, from their ideology.

Myside bias centers on distal beliefs, not testable ones. Belief bias, in contrast, concerns testable beliefs. This is why belief bias is more remediable by education and more correlated with cognitive ability than myside bias (Stanovich 2021). The proposition that health care spending is the second largest item in the US federal budget is a testable belief. The proposition that Americans spend too much on health care is a distal belief. Certainly, economic facts might alter our attitude toward the latter proposition, but they cannot verify this distal belief in the same manner that they can verify testable beliefs. The conclusions that interfere with reasoning in the case of belief bias are testable beliefs. Myside bias, in contrast, occurs when people evaluate and generate evidence in a manner favorable toward their prior opinions and attitudes—where the attitudes in question are convictions.

It is possible that these distinctions (testable versus distal; ego involvement versus noninvolvement; sacred values versus non-sacred) help to explain the curious paradox regarding AOT as a bias predictor—namely, that it predicts a plethora of biases except the one closest to its definition. Building on our view of AOT as a measure of the tendency to detach and decontextualize, one hypothesis might be that with myside bias paradigms, we are seeing the limits of individual detachment. Heuristics and biases tasks, as discussed above, often involve a conflict between a classically normative response and a classically nonnormative one[6]. De Neys (2014, 2023) has shown that in many cases the conflict between the two responses is detected at some cognitive level. The detected conflict might broach awareness to a sufficient degree that tendencies toward detachment can be helpful. In contrast, many myside bias paradigms (particularly the more naturalistic ones; see Stanovich and West 2007) may not provide opportunities for any conflict to be detected, thus neutering the possibility of high AOT subjects using their skills. Alternatively, the involvement of convictions may be overwhelming even in cases where awareness of alternative reactions has taken place.

Relevant here may be another property of myside bias—that it displays little domain generality (Stanovich 2021). People who display a high degree of myside bias in one domain do not tend to show a high degree of myside bias in a different domain. However, different beliefs vary reliably in the degree of myside bias that they engender. In short, it might not be people who are characterized by more or less myside bias but beliefs that differ in how strongly they are structured to repel contradictory ideas. These facts about myside bias have profound implications because they invert the way we think about beliefs. Models that focus on the properties of acquired beliefs, such as memetic theory, may provide better frameworks for the study of myside bias (see Stanovich 2021). The focus of memetics on the properties of beliefs rather than the psychological characteristics of the host is consistent with research showing that the degree of myside bias is better predicted by the former than the latter. They also might render the individual difference findings regarding AOT somewhat less paradoxical.

Detaching from a prepotent response in a heuristics and biases task such as the Linda problem may be vastly easier than using AOT tendencies toward detachment and decoupling to overturn a conviction and/or weaken a commitment to a sacred value. The levels of detachment and decontextualization required for the latter are orders of magnitude higher than the parallel cognitive requirements in a typical heuristics and biases task. This would be consistent with the argument previously made by Stanovich (2021) that myside bias is an outlier bias in the rational thinking literature.

The parallel explanation from within Baron's (2017; Baron 2019; Baron et al. 2015; Baron et al. 2023) conception of the AOT stresses that responses on the scale are an indication of the endorsement of norms or standards of good thinking. However, people adhere more or less to those standards, and they may differentially adhere to them based on the issue in question. Hence, Baron (2017) argues that "People may endorse AOT in a self-report questionnaire about beliefs, and they may behave consistently with it in most domains, but they may have gaps when they are strongly committed to a particular view" (p. 2). This argument is very parallel to that made with respect to our own decoupling/detachment view of what the AOT measures. Since most myside paradigms involve issues where the subjects are indeed likely to be "strongly committed to a particular view", the Baron framework will arrive at the same place as us regarding individual differences in myside tasks that deal with conviction-based beliefs—AOT will not be a good indicator of who avoids myside bias in those domains. That means, however, that AOT will be less useful in some of the domains where we need it most—for example, social policy and politics.

For the first several decades of work in the heuristics and biases tradition (Kahneman and Tversky 1973; Tversky and Kahneman 1974), from the 1970s to the 1990s, myside bias was treated as simply another bias on a growing list of biases (anchoring bias, hindsight bias, availability bias, etc.) and its occurrence in the laboratory paradigms that were designed to study it was deemed non-normative, without much discussion in most papers. That initial stance now seems oversimplified. Myside bias does not act like any other bias in the traditional lists of thinking errors in this literature in terms of individual differences. We have also noted how it shows little domain generality.

Finally, Stanovich (2021) has reviewed the extensive literature showing that it is not easy to demonstrate that myside bias is non-normative. It is unclear whether most myside thinking should even be considered a bias that leads to non-normative responses. For example, many utility-based theories that model beliefs in terms of cost and benefits (Loewenstein and Molnar 2018; Sharot et al. 2023) show that the early dismissal of myside bias as an irrational tendency was premature. Even in the domain of pure epistemic rationality, Stanovich (2021) has discussed how so-called knowledge projection models of myside bias show it is rational in many circumstances (Hahn and Harris 2014; Koehler 1993). The detachment and decoupling tendencies of AOT may not work against the epistemic mechanisms rationally acting as governors and damping down belief change. However, AOT does predict normative responding in many other tasks that do not have so many inertial components as myside bias.

8. Final Thoughts

At the end of a thoughtful chapter on AOT, Baron et al. (2023) mentioned that, in their view, "AOT is thus a moral virtue as well as a personal one" (p. 24). This statement provides a provocative way to probe our feelings about our own work. Specifically, do we believe that AOT is a moral virtue? Referring to AOT in the abstract, that is a difficult question for us, although we have some sympathy with the position. However, we have less sympathy with the idea that *extant* AOT scales are measuring something we would want to call a moral virtue. Undoubtedly, AOT scales are measuring something very central to the kind of thinking that is tapped on heuristics and biases tasks. Because such tasks tap aspects of rationality, AOT scales are measuring something very important. Nevertheless, the inability of AOT scales to associate with the avoidance of myside bias in a variety of paradigms is a very troubling finding. Avoiding myside bias is the quintessence of the most important theoretical treatments of actively open-minded thinking (Baron 1993, 2019; Baron et al. 2023). To say that we have a measure that predicts much other rational thinking but not the quintessence of the concept is a disconcerting conclusion.

It is a disconcerting conclusion in just the way that the conclusion to Steve Pinker's (2021) recent book on rationality was disconcerting. In the first nine chapters of his book, we learn that humans have many tools of rationality at their disposal. The brain is full of automatic propensities that have been honed over millennia to optimally regulate our responses to stimuli in environments that are not rapidly changing. Also available to us are all the cultural tools of rational thought that have been discovered throughout history (probabilistic reasoning, signal detection theory, expected utility theory, logic, game theory, scientific inference). By the process of cultural ratcheting, we have accomplished any number of supreme achievements such as curing illness, decoding the genome, and uncovering the most minute constituents of matter. All of this exists alongside surveys showing that 41% of the population believes in extrasensory perception, 32% in ghosts and spirits, and 25% in astrology—just a few of the pseudoscientific beliefs that Pinker (2021) mentions. These facts highlight what Pinker calls the rationality paradox: "How, then, can we understand this thing called rationality which would appear to be the birthright of our species yet is so frequently and flagrantly flouted?" (p. 6).

Pinker (2021) admits that the solution to this "pandemic of poppycock" (p. 286) is not to be found in correcting the many thinking biases that he covers in his book. We cannot remediate this kind of rational thinking through providing information. People captured by this poppycock have too much mindware—not too little (see Stanovich et al. 2016). Pinker notes that "nothing from the cognitive psychology lab could have predicted QAnon, nor are its adherents likely to be disabused by a tutorial in logic or probability" (p. 287). This admission uncomfortably calls to mind a quip by Scott Alexander (2021) that "of the fifty-odd biases discovered by Kahneman, Tversky, and their successors, forty-nine are cute quirks, and one is destroying civilization. This last one is confirmation bias—our tendency to interpret evidence as confirming our pre-existing beliefs instead of changing our minds." This quip is not literally correct, because the "other 49" are not "cute quirks" with no implications in the real world. In his final chapter, Pinker describes and cites research showing that these biases have been linked to real-world outcomes in the financial, occupational, health, and legal domains (as we did in the last chapter of our book on the CART, Stanovich et al. 2016). They are not just cute quirks. Nevertheless, the joke hits home. That is why Stanovich (2021) wrote a whole book on the one bias that *is* "destroying civilization"—myside bias.

Relationships involving the AOT parallel the Alexander quip—it predicts most of the "other 49" and fails to predict "the one that is destroying civilization". It is clear why it is a bias that is difficult to avoid. It involves convictions—our most sacred and emotionally salient beliefs. To avoid it, people would need to be more skeptical of the strongest beliefs that they have acquired. They would have to learn to treat their beliefs less like possessions and more like contingent hypotheses (see Stanovich 2021). People would also need to be particularly skeptical of the beliefs that were acquired in their early lives—those that were passed on by parents, relatives, and their peers. It is likely that these beliefs have not been subjected to selective tests, because they were acquired during a developmental period when their host lacked reflective capacities.

All of this is heavy lifting at the individual level, however, and it is sobering to admit that AOT scales do not provide accurate measures of the tendency to avoid myside bias. If we want to get at people's attitudes toward scientific evidence on a contested issue, we actually have to take a domain-specific belief that a person has on the matter, present them with contradictory evidence, and see how they assimilate that contradictory evidence (as some studies have done; see Ditto et al. 2019). You cannot just ask people on a questionnaire whether it is good to pay attention to contradictory evidence.

Ultimately, we all need to rely on the "institutions of rationality" (Rauch 2021) that provide the epistemic tools to deal with what Pinker (2021), aptly channeling the work of

(Dan Kahan 2016; Kahan et al. 2017), calls the "Tragedy of the Belief Commons". Cultural institutions can enforce rules whereby people benefit from rational tools without having to learn the tools themselves. Pinker (2021) describes some institutional reforms within the media and the internet but laments the state of our universities and their "suffocating left-wing monoculture, with its punishment of students and professors who question dogmas on gender, race, culture, genetics, colonialism, and sexual identity and orientation" (p. 313). He describes how "on several occasions correspondents have asked me why they should trust the scientific consensus on climate change, since it comes out of institutions that brook no dissent" (p. 314). In short, the public is coming to know that universities have approved positions on certain topics, and thus is quite rationally reducing its confidence in research that comes out of them. It is thus consistent with the earlier evidence we have reviewed on how myside bias is independent of intelligence and educational status that university faculties—full of cognitive ability and educational attainment—cannot free themselves from myside bias.

Author Contributions: Conceptualization, K.E.S. and M.E.T.; writing—original draft preparation, K.E.S.; writing—review and editing, K.E.S. and M.E.T. All authors have read and agreed to the published version of the manuscript.

Funding: Preparation of this manuscript was supported by the Social Sciences and Humanities Research Council (SSHRC) grant number 435-2018-0874.

Institutional Review Board Statement: Not applicable.

Informed Consent Statement: Not applicable.

Data Availability Statement: Not applicable.

Conflicts of Interest: The authors declare no conflict of interest.

Notes

1 Optimal functioning does not result from maximizing cognitive styles. Instead, rationality, plotted against most thinking dispositions, is an inverted U-shaped function. One does not maximize rationality by maximizing the reflectivity/impulsivity dimension, for example, because a person doing so might get lost in interminable pondering and never make a decision. One does not maximize the dimension of belief flexibility either, because such a person might end up with a pathologically unstable personality. Reflectivity and belief flexibility are "good" cognitive styles only in the sense that most people are too low in both dimensions (Baron 2008). Most people would be more rational if they increased their degrees of reflectivity and belief flexibility, but this does not mean that either of these thinking dispositions should always be maximized.

2 Most of the other facets of openness, as measured in scales tapping the Big Five personality dimensions, are not related to AOT, as conceptualized in the rational thinking literature—for example, facets such as openness to fantasy, aesthetics, feelings, and actions.

3 For example, the comprehensive intellectual humility scale developed by Krumrei-Mancuso and Rouse (2016) includes five items measuring belief identification (e.g., "I feel small when others disagree with me on topics that are close to my heart", reverse-scored), five items measuring belief revision (e.g., "I am open to revising my important beliefs in the face of new information"), six items measuring respect for the opinion of others (e.g., "I can have great respect for someone, even when we don't see eye-to-eye on important topics"), and six items measuring the avoidance of overconfidence (e.g., "My ideas are usually better than other people's ideas"). Likewise, the short humility scale studied by Leary et al. (2017) mixes belief revision items ("I reconsider my opinions when presented with new evidence") with items tapping respect for alternative opinions ("I recognize the value in opinions that are different from my own").

4 Our characterization of heuristics and biases tasks as implicating decoupling holds only for those instances where the task truly does create a hostile environment (see Stanovich 2018) for a particular subject because the normative response has not been practiced to automaticity. In cases where the normative response has been practiced to an automaticity so great that it dominates the intuitive response, there is no response conflict and cognitive decoupling is not required (see De Neys 2023; Thompson et al. 2018).

5 Most of the criticisms of this study by Baron (2017) would not reverse the direction of the association between polarization and AOT. They would, instead, drive the association toward zero, thus making the Kahan and Corbin (2016) results more consistent with the conclusion drawn in this section.

6 Stanovich (2018) discusses all the complexities involved in pinpointing the cognitive source of the two responses—particularly the classically normative one, which can originate from either System 1 or System 2.

References

Abelson, Robert P. 1986. Beliefs are like possessions. *Journal of the Theory of Social Behaviour* 16: 223–50. [CrossRef]

Abelson, Robert P. 1988. Conviction. *American Psychologist* 43: 267–75. [CrossRef]

Abelson, Robert P., and Deborah A. Prentice. 1989. Beliefs as possessions: A functional perspective. In *Attitudes, Structure, and Function*. Edited by Anthony R. Pratkanis, Steven J. Breckler and Anthony G. Greenwald. Hillsdale: Erlbaum, pp. 361–81.

Ackerman, Rakefet, and Valerie Thompson. 2017. Meta-reasoning: Monitoring and control of thinking and reasoning. *Trends in Cognitive Sciences* 21: 607–17. [CrossRef]

Alexander, Scott. 2021. Book Review: The Scout Mindset. *Astral Codex Ten*, September 29. Available online: https://astralcodexten.substack.com/p/book-review-the-scout-mindset(accessed on 13 January 2023).

Baron, Jonathan. 1985. *Rationality and Intelligence*. Cambridge: Cambridge University Press.

Baron, Jonathan. 1988. *Thinking and Deciding*, 1st ed. Cambridge: Cambridge University Press.

Baron, Jonathan. 1993. Why teach thinking?—An essay. *Applied Psychology: An International Review* 42: 191–214. [CrossRef]

Baron, Jonathan. 2008. *Thinking and Deciding*, 4th ed. Cambridge: Cambridge University Press.

Baron, Jonathan. 2017. Comment on Kahan and Corbin: Can polarization increase with actively open-minded thinking? *Research & Politics* 4: 2053168016688122. [CrossRef]

Baron, Jonathan. 2019. Actively open-minded thinking in politics. *Cognition* 188: 8–18. [CrossRef] [PubMed]

Baron, Jonathan, and Mark Spranca. 1997. Protected values. *Organizational Behavior and Human Decision Processes* 70: 1–16. [CrossRef]

Baron, Jonathan, and Sarah Leshner. 2000. How serious are expressions of protected values? *Journal of Experimental Psychology: Applied* 6: 183–94. [CrossRef]

Baron, Jonathan, Burcu Gürçay, and S. Emlen Metz. 2017. Reflective thought and actively open-minded thinking. In *Individual Differences in Judgment and Decision Making*. Edited by Maggie E. Toplak and Joshua Weller. London: Psychology Press, pp. 107–26.

Baron, Jonathan, Ozan Isler, and Onurcan Yilmaz. 2023. Actively open-minded thinking and the political effects of its absence. In *Divided: Open-Mindedness and Dogmatism in a Polarized World*. Edited by Victor Ottati and Chadly Stern. Oxford: Oxford University Press.

Baron, Jonathan, Sydney Scott, Katrina Fincher, and S. Emlen Metz. 2015. Why does the Cognitive Reflection Test (sometimes) predict utilitarian moral judgment (and other things)? *Journal of Applied Research in Memory and Cognition* 4: 265–84. [CrossRef]

Bartels, Daniel M., and Douglas L. Medin. 2007. Are morally-motivated decision makers insensitive to the consequences of their choices? *Psychological Science* 18: 24–28. [CrossRef] [PubMed]

Bazerman, Max H. 2001. Consumer research for consumers. *Journal of Consumer Research* 27: 499–504. [CrossRef]

Beatty, Erin L., and Valerie A. Thompson. 2012. Effects of perspective and belief on analytic reasoning in a scientific reasoning task. *Thinking & Reasoning* 18: 441–60.

Binnendyk, Jabin, and Gordon Pennycook. 2022. Intuition, reason, and conspiracy beliefs. *Current Opinion in Psychology* 47: 101387. [CrossRef] [PubMed]

Bronstein, Michael V., Gordon Pennycook, Adam Bear, David G. Rand, and Tyrone D. Cannon. 2019. Belief in fake news is associated with delusionality, dogmatism, religious fundamentalism, and reduced analytic thinking. *Journal of Applied Research in Memory and Cognition* 8: 108–17. [CrossRef]

Bruine de Bruin, Wändi, Andrew M. Parker, and Baruch Fischhoff. 2007. Individual differences in adult decision-making competence. *Journal of Personality and Social Psychology* 92: 938–56. [CrossRef]

Campitelli, Guillermo, and Paul Gerrans. 2014. Does the cognitive reflection test measure cognitive reflection? A mathematical modeling approach. *Memory & Cognition* 42: 434–47.

Ceci, Steven, and Wendy Williams. 2018. Who decides what is acceptable speech on campus? Why restricting free speech is not the answer. *Perspectives on Psychological Science* 13: 299–323. [CrossRef]

Cederblom, Jerry. 1989. Willingness to reason and the identification of the self. In *Thinking, Reasoning, and Writing*. Edited by E. Maimon, B. Nodine and F. O'Conner. New York: Longman, pp. 147–59.

Charney, Evan. 2015. Liberal bias and the five-factor model. *Behavioral and Brain Sciences* 38: e139. [CrossRef]

Clements, Zakary A., and Geoffrey D. Munro. 2021. Biases and their impact on opinions of transgender bathroom usage. *Journal of Applied Social Psychology* 51: 370–83. [CrossRef]

Costa, Paul T., and Robert R. McCrae. 1992. *Revised NEO Personality Inventory*. Odessa: Psychological Assessment Resources.

Crawford, Jarret, and Lee Jussim, eds. 2018. *The Politics of Social Psychology*. New York: Routledge.

De Neys, Wim. 2014. Conflict detection, dual processes, and logical intuitions: Some clarifications. *Thinking & Reasoning* 20: 169–87.

De Neys, Wim. 2023. Advancing theorizing about fast-and-slow thinking. *Behavioral and Brain Sciences*, 1–68. [CrossRef] [PubMed]

Deniz, Hasan, Lisa Donnelly, and Irfan Yilmaz. 2008. Exploring the factors related to acceptance of evolutionary theory among Turkish preservice biology teachers. *Journal of Research in Science Teaching* 45: 420–43. [CrossRef]

Ditto, Peter H., Brittany S. Liu, and Sean P. Wojcik. 2012. Is anything sacred anymore? *Psychological Inquiry* 23: 155–61. [CrossRef]

Ditto, Peter H., Brittany S. Liu, Cory J. Clark, Sean P. Wojcik, Eric E. Chen, Rebecca H. Grady, Jared B. Celniker, and Joanne F. Zinger. 2019. At least bias Is bipartisan: A meta-analytic comparison of partisan bias in liberals and conservatives. *Perspectives on Psychological Science* 14: 273–91. [CrossRef]

Duarte, José L., Jarret T. Crawford, Charlotta Stern, Jonathan Haidt, Lee Jussim, and Philip E. Tetlock. 2015. Political diversity will improve social psychological science. *Behavioral and Brain Sciences* 38: e130. [CrossRef]

Eichmeier, April, and Neil Stenhouse. 2019. Differences that don't make much difference: Party asymmetry in open-minded cognitive styles has little relationship to information processing behavior. *Research & Politics* 6: 2053168019872045.

Einhorn, Hillel J., and Robin M. Hogarth. 1981. Behavioral decision theory: Processes of judgment and choice. *Annual Review of Psychology* 32: 53–88. [CrossRef]

Ellis, J. M. 2020. *The Breakdown of Higher Education*. New York: Encounter.

Epstein, Seymour, and Petra Meier. 1989. Constructive thinking: A broad coping variable with specific components. *Journal of Personality and Social Psychology* 57: 332–50. [CrossRef] [PubMed]

Erceg, Nikola, Zvonimir Galić, and Andreja Bubić. 2022. Normative responding on cognitive bias tasks: Some evidence for a weak rationality factor that is mostly explained by numeracy and actively open-minded thinking. *Intelligence* 90: 101619. [CrossRef]

Evans, Jonathan St B. T. 2007. *Hypothetical Thinking: Dual Processes in Reasoning and Judgment*. New York: Psychology Press.

Evans, Jonathan St B. T. 2010. *Thinking Twice: Two Minds in One Brain*. Oxford: Oxford University Press.

Evans, Jonathan St B. T. 2014. *Reasoning, Rationality and Dual Processes*. London: Psychology Press.

Evans, Jonathan St B. T. 2017. Belief bias in deductive reasoning. In *Cognitive Illusions*, 2nd ed. Edited by Rüdiger Pohl. London: Routledge, pp. 165–81.

Evans, Jonathan St B. T., and Keith E. Stanovich. 2013. Dual-process theories of higher cognition: Advancing the debate. *Perspectives on Psychological Science* 8: 223–41. [CrossRef]

Fazio, Russell H. 2007. Attitudes as object–evaluation associations of varying strength. *Social Cognition* 25: 603–37. [CrossRef] [PubMed]

Finucane, Melissa L., and Christina M. Gullion. 2010. Developing a tool for measuring the decision-making competence of older adults. *Psychology and Aging* 25: 271–88. [CrossRef] [PubMed]

Fisher, Matthew, and Frank C. Keil. 2014. The illusion of argument justification. *Journal of Experimental Psychology: General* 143: 425–33. [CrossRef] [PubMed]

Goldman, Alvin I. 1986. *Epistemology and Cognition*. Cambridge: Harvard University Press.

Gould, Stephen Jay. 1991. *Bully for the Brontosaurus*. New York: Norton.

Guay, Brian, and Christopher D. Johnston. 2022. Ideological asymmetries and the determinants of politically motivated reasoning. *American Journal of Political Science* 66: 285–301. [CrossRef]

Hahn, Ulrike, and Adam J. L. Harris. 2014. What does it mean to be biased: Motivated reasoning and rationality. In *Psychology of Learning and Motivation*. Edited by Gordon H. Ross. New York: Academic Press, vol. 61, pp. 41–102.

Haidt, Jonathan. 2022. The Two Fiduciary Duties of Professors. *Heterodox Blog*, September 20. Available online: https://heterodoxacademy.org/blog/the-two-fiduciary-duties-of-professors/ (accessed on 13 January 2023).

Haran, Uriel, Ilana Ritov, and Barbara A. Mellers. 2013. The role of actively open-minded thinking in information acquisition, accuracy, and calibration. *Judgment and Decision Making* 8: 188–201. [CrossRef]

Harman, Gilbert. 1995. Rationality. In *Thinking*. Edited by Edward Smith and Daniel Osherson. Cambridge: The MIT Press, vol. 3, pp. 175–211.

Howe, Lauren C., and Jon A. Krosnick. 2017. Attitude strength. *Annual Review of Psychology* 68: 327–51. [CrossRef]

Jastrzębski, Jan, and Adam Chuderski. 2022. Analytic thinking outruns fluid reasoning in explaining rejection of pseudoscience, paranormal, and conspiracist beliefs. *Intelligence* 95: 101705. [CrossRef]

Jussim, Lee. 2019. The Threat to Academic Freedom ... from Academics. *Medium*, December 27. Available online: https://medium.com/@leej12255/the-threat-to-academic-freedom-from-academics-4685b1705794 (accessed on 13 January 2023).

Jussim, Lee. 2022. Diversity Is Diverse: Social Justice Reparations and Science. *Substack*, December 30. Available online: https://unsafescience.substack.com/p/diversity-is-diverse (accessed on 13 January 2023).

Kahan, Dan M. 2016. The politically motivated reasoning paradigm, part 1: What politically motivated reasoning is and how to measure it. In *Emerging Trends in the Social and Behavioral Sciences: An Interdisciplinary, Searchable, and Linkable Resource*. Edited by Robert A. Scott and Stephen Michael Kosslyn. New York: Wiley Online library. [CrossRef]

Kahan, Dan M., and Jonathan C. Corbin. 2016. A note on the perverse effects of actively open-minded thinking on climate-change polarization. *Research & Politics* 3: 1–5. [CrossRef]

Kahan, Dan M., Ellen Peters, Erica Cantrell Dawson, and Paul Slovic. 2017. Motivated numeracy and enlightened self-government. *Behavioural Public Policy* 1: 54–86. [CrossRef]

Kahneman, Daniel. 2011. *Thinking, Fast and Slow*. New York: Farrar, Straus & Giroux.

Kahneman, Daniel, and Amos Tversky. 1973. On the psychology of prediction. *Psychological Review* 80: 237–51. [CrossRef]

Kelman, Mark. 2011. *The Heuristics Debate*. New York: Oxford University Press.

Koehler, Derek J., and Nigel Harvey, eds. 2004. *Blackwell Handbook of Judgment and Decision Making*. Oxford: Blackwell.

Koehler, Jonathan J. 1993. The influence of prior beliefs on scientific judgments of evidence quality. *Organizational Behavior and Human Decision Processes* 56: 28–55. [CrossRef]

Kokis, Judite V., Robyn Macpherson, Maggie E. Toplak, Richard F. West, and Keith E. Stanovich. 2002. Heuristic and analytic processing: Age trends and associations with cognitive ability and cognitive styles. *Journal of Experimental Child Psychology* 83: 26–52. [CrossRef] [PubMed]

Kovacs, Kristof, and Andrew R. A. Conway. 2016. Process Overlap Theory: A Unified Account of the General Factor of Intelligence. *Psychological Inquiry* 27: 151–77. [CrossRef]

Krumrei-Mancuso, Elizabeth J., and Steven V. Rouse. 2016. The Development and Validation of the Comprehensive Intellectual Humility Scale. *Journal of Personality Assessment* 98: 209–21. [CrossRef] [PubMed]

Leary, Mark, Kate Diebels, Erin Davisson, Katrina Jongman-Sereno, Jennifer Isherwood, Kaitlin Raimi, Samantha Deffler, and Rick Hoyle. 2017. Cognitive and interpersonal features of intellectual humility. *Personality and Social Psychology Bulletin* 43: 793–813. [CrossRef]

Levin, Irwin P., Edward A. Wasserman, and Shu-fang Kao. 1993. Multiple methods of examining biased information use in contingency judgments. *Organizational Behavior and Human Decision Processes* 55: 228–50. [CrossRef]

Loewenstein, George, and Andras Molnar. 2018. The renaissance of belief-based utility in economics. *Nature Human Behaviour* 2: 166–67. [CrossRef]

Luce, R. Duncan, and Howard Raiffa. 1957. *Games and Decisions*. New York: Wiley.

Macpherson, Robyn, and Keith E. Stanovich. 2007. Cognitive ability, thinking dispositions, and instructional set as predictors of critical thinking. *Learning and Individual Differences* 17: 115–27. [CrossRef]

Manktelow, Ken I. 2012. *Thinking and Reasoning*. Hove: Psychology Press.

Mele, Alfred, and Piers Rawling, eds. 2004. *The Oxford Handbook of Rationality*. Oxford: Oxford University Press.

Mellers, Barbara, Eric Stone, Pavel Atanasov, Nick Rohrbaugh, S. Emlen Metz, Lyle Ungar, Michael M. Bishop, Michael Horowitz, Ed Merkle, and Philip Tetlock. 2015. The psychology of intelligence analysis: Drivers of prediction accuracy in world politics. *Journal of Experimental Psychology: Applied* 21: 1–14. [CrossRef] [PubMed]

Mercier, Hugo. 2017. Confirmation bias—Myside bias. In *Cognitive Illusions*, 2nd ed. Edited by Rüdiger Pohl. New York: Routledge, pp. 99–114.

Miyake, Akira, and Naomi P. Friedman. 2012. The nature and organization of individual differences in executive functions four general conclusions. *Current Directions in Psychological Science* 21: 8–14. [CrossRef]

Newton, Christie, Justin Feeney, and Gordon Pennycook. 2023. On the Disposition to Think Analytically: Four Distinct Intuitive-Analytic Thinking Styles. Manuscript. Available online: https://osf.io/7kf5b/ (accessed on 13 January 2023).

Nigg, Joel. 2017. On the relations among self-regulation, self-control, executive functioning, effortful control, cognitive control, impulsivity, risk-taking, and inhibition for developmental psychopathology. *Journal of Child Psychology and Psychiatry* 58: 361–83. [CrossRef]

Nozick, Robert. 1993. *The Nature of Rationality*. Princeton: Princeton University Press.

Oaksford, Mike, and Nick Chater. 2012. Dual processes, probabilities, and cognitive architecture. *Mind & Society* 11: 15–26.

Oliver, J. Eric, and Thomas J. Wood. 2014. Conspiracy theories and the paranoid style(s) of mass opinion. *American Journal of Political Science* 58: 952–66. [CrossRef]

Parker, Andrew M., and Baruch Fischhoff. 2005. Decision-making competence: External validation through an individual differences approach. *Journal of Behavioral Decision Making* 18: 1–27. [CrossRef]

Pennycook, Gordon, James Allan Cheyne, Derek J. Koehler, and Jonathan A. Fugelsang. 2020. On the belief that beliefs should change according to evidence: Implications for conspiratorial, moral, paranormal, political, religious, and science beliefs. *Judgement and Decision Making* 15: 476–98. [CrossRef]

Pennycook, Gordon, James Allan Cheyne, Nathaniel Barr, Derek J. Koehler, and Jonathan A. Fugelsang. 2014. Cognitive style and religiosity: The role of conflict detection. *Memory & Cognition* 42: 1–10.

Perkins, David N. 1985. Postprimary education has little impact on informal reasoning. *Journal of Educational Psychology* 77: 562–71. [CrossRef]

Piazza, Jared, and Justin F. Landy. 2013. "Lean not on your own understanding": Belief that morality is founded on divine authority and non-utilitarian moral thinking. *Judgment and Decision Making* 8: 639–61. [CrossRef]

Pinker, Steven. 2021. *Rationality*. New York: Penguin.

Pronin, Emily. 2007. Perception and misperception of bias in human judgment. *Trends in Cognitive Sciences* 11: 37–43. [CrossRef] [PubMed]

Raoelison, Matthieu, Valerie A. Thompson, and Wim De Neys. 2020. The smart intuitor: Cognitive capacity predicts intuitive rather than deliberate thinking. *Cognition* 204: 104381. [CrossRef]

Rauch, Jonathan. 2021. *The Constitution of Knowledge: A Defense of Truth*. Washington, DC: Brookings Institution Press.

Rizeq, Jala, David B. Flora, and Maggie E. Toplak. 2021. An examination of the underlying dimensional structure of three domains of contaminated mindware: Paranormal beliefs, conspiracy beliefs, and anti-science attitudes. *Thinking & Reasoning* 27: 187–211.

Sa, Walter C., and Keith E. Stanovich. 2001. The domain specificity and generality of mental contamination: Accuracy and projection in judgments of mental content. *British Journal of Psychology* 92: 281–302. [CrossRef]

Sá, Walter C., Carol N. Kelley, Caroline Ho, and Keith E. Stanovich. 2005. Thinking about personal theories: Individual differences in the coordination of theory and evidence. *Personality and Individual Differences* 38: 1149–61. [CrossRef]

Sá, Walter C., Richard F. West, and Keith E. Stanovich. 1999. The domain specificity and generality of belief bias: Searching for a generalizable critical thinking skill. *Journal of Educational Psychology* 91: 497–510. [CrossRef]

Samuelson, Peter L., and Ian M. Church. 2015. When cognition turns vicious: Heuristics and biases in light of virtue epistemology. *Philosophical Psychology* 28: 1095–113. [CrossRef]

Savage, Leonard J. 1954. *The Foundations of Statistics*. New York: Wiley.
Sharot, Tali, Max Rollwage, Cass R. Sunstein, and Stephen M. Fleming. 2023. Why and When Beliefs Change. *Perspectives on Psychological Science* 18: 142–51. [CrossRef]
Sinatra, Gale M., Sherry A. Southerland, Frances McConaughy, and James W. Demastes. 2003. Intentions and beliefs in students' understanding and acceptance of biological evolution. *Journal of Research in Science Teaching* 40: 510–28. [CrossRef]
Skitka, Linda J., Christopher W. Bauman, and Edward G. Sargis. 2005. Moral conviction: Another contributor to attitude strength or something more? *Journal of Personality and Social Psychology* 88: 895–917. [CrossRef]
Stanovich, Keith E. 1999. *Who Is Rational? Studies of Individual Differences in Reasoning*. Mahwah: Erlbaum.
Stanovich, Keith E. 2004. *The Robot's Rebellion: Finding Meaning in the Age of Darwin*. Chicago: University of Chicago Press.
Stanovich, Keith E. 2008. Higher-order preferences and the Master Rationality Motive. *Thinking & Reasoning* 14: 111–27.
Stanovich, Keith E. 2009. *What Intelligence Tests Miss: The Psychology of Rational Thought*. New Haven: Yale University Press.
Stanovich, Keith E. 2011. *Rationality and the Reflective Mind*. New York: Oxford University Press.
Stanovich, Keith E. 2012. On the distinction between rationality and intelligence: Implications for understanding individual differences in reasoning. In *The Oxford Handbook of Thinking and Reasoning*. Edited by Keith J. Holyoak and Robert G. Morrison. New York: Oxford University Press, pp. 343–65.
Stanovich, Keith E. 2013. Why humans are (sometimes) less rational than other animals: Cognitive complexity and the axioms of rational choice. *Thinking & Reasoning* 19: 1–26.
Stanovich, Keith E. 2017. Were the Trump Voters Irrational? *Quillette*, September 28. Available online: https://quillette.com/2017/09/28/trump-voters-irrational/ (accessed on 13 January 2023).
Stanovich, Keith E. 2018. Miserliness in human cognition: The interaction of detection, override and mindware. *Thinking & Reasoning* 24: 423–44.
Stanovich, Keith E. 2021. *The Bias That Divides Us: The Science and Politics of Myside Thinking*. Cambridge: MIT Press.
Stanovich, Keith E., and Maggie E. Toplak. 2012. Defining features versus incidental correlates of Type 1 and Type 2 processing. *Mind & Society* 11: 3–13.
Stanovich, Keith E., and Maggie E. Toplak. 2019. The need for intellectual diversity in psychological science: Our own studies of actively open-minded thinking as a case study. *Cognition* 187: 156–66. [CrossRef] [PubMed]
Stanovich, Keith E., and Richard F. West. 1997. Reasoning independently of prior belief and individual differences in actively open-minded thinking. *Journal of Educational Psychology* 89: 342–57. [CrossRef]
Stanovich, Keith E., and Richard F. West. 1998a. Individual differences in rational thought. *Journal of Experimental Psychology: General* 127: 161–88. [CrossRef]
Stanovich, Keith E., and Richard F. West. 1998b. Who uses base rates and P(D/~H)? An analysis of individual differences. *Memory & Cognition* 26: 161–79.
Stanovich, Keith E., and Richard F. West. 2000. Individual differences in reasoning: Implications for the rationality debate? *Behavioral and Brain Sciences* 23: 645–726. [CrossRef]
Stanovich, Keith E., and Richard F. West. 2007. Natural myside bias is independent of cognitive ability. *Thinking & Reasoning* 13: 225–47.
Stanovich, Keith E., and Richard F. West. 2008. On the failure of intelligence to predict myside bias and one-sided bias. *Thinking & Reasoning* 14: 129–67.
Stanovich, Keith E., Richard F. West, and Maggie E. Toplak. 2013. Myside bias, rational thinking, and intelligence. *Current Directions in Psychological Science* 22: 259–64. [CrossRef]
Stanovich, Keith E., Richard F. West, and Maggie E. Toplak. 2016. *The Rationality Quotient: Toward a Test of Rational Thinking*. Cambridge: MIT Press.
Stenhouse, Neil, Teresa A. Myers, Emily K. Vraga, John E. Kotcher, Lindsey Beall, and Edward W. Maibach. 2018. The potential role of actively open-minded thinking in preventing motivated reasoning about controversial science. *Journal of Environmental Psychology* 57: 17–24. [CrossRef]
Svedholm, Annika M., and Marjaana Lindeman. 2013. The separate roles of the reflective mind and involuntary inhibitory control in gatekeeping paranormal beliefs and the underlying intuitive confusions. *British Journal of Psychology* 104: 303–19. [CrossRef] [PubMed]
Svedholm-Hakkinen, Annika M., and Marjaana Lindeman. 2018. Actively open-minded thinking: Development of a shortened scale and disentangling attitudes towards knowledge and people. *Thinking & Reasoning* 24: 21–40.
Swami, Viren, Martin Voracek, Stefan Stieger, Ulrich S. Tran, and Adrian Furnham. 2014. Analytic thinking reduces belief in conspiracy theories. *Cognition* 133: 572–85. [CrossRef] [PubMed]
Swami, Viren, Rebecca Coles, Stefan Stieger, Jakob Pietschnig, Adrian Furnham, Sherry Rehim, and Martin Voracek. 2011. Conspiracist ideation in Britain and Austria: Evidence of a monological belief system and associations between individual psychological differences and real-world and fictitious conspiracy theories. *British Journal of Psychology* 102: 443–63. [CrossRef] [PubMed]
Swami, Viren, Tomas Chamorro-Premuzic, and Adrian Furnham. 2010. Unanswered questions: A preliminary investigation of personality and individual difference predictors of 9/11 conspiracist beliefs. *Applied Cognitive Psychology* 24: 749–61. [CrossRef]
Tetlock, Philip E. 2003. Thinking the unthinkable: Sacred values and taboo cognitions. *Trends in Cognitive Sciences* 7: 320–24. [CrossRef]
Thompson, Valerie A., and Jonathan St B. T. Evans. 2012. Belief bias in informal reasoning. *Thinking & Reasoning* 18: 278–310.

Thompson, Valerie A., Gordon Pennycook, Dries Trippas, and Jonathan St B. T. Evans. 2018. Do smart people have better intuitions? *Journal of Experimental Psychology: General* 147: 945–61. [CrossRef]

Thompson, Valerie A., Jamie A. Prowse Turner, Gordon Pennycook, Linden J. Ball, Hannah Brack, Yael Ophir, and Rakefet Ackerman. 2013. The role of answer fluency and perceptual fluency as metacognitive cues for initiating analytic thinking. *Cognition* 128: 237–51. [CrossRef]

Toplak, Maggie E., and Keith E. Stanovich. 2002. The domain specificity and generality of disjunctive reasoning: Searching for a generalizable critical thinking skill. *Journal of Educational Psychology* 94: 197–209. [CrossRef]

Toplak, Maggie E., and Keith E. Stanovich. 2003. Associations between myside bias on an informal reasoning task and amount of post-secondary education. *Applied Cognitive Psychology* 17: 851–60. [CrossRef]

Toplak, Maggie E., Eleanor Liu, Robyn MacPherson, Tony Toneatto, and Keith E. Stanovich. 2007. The reasoning skills and thinking dispositions of problem gamblers: A dual-process taxonomy. *Journal of Behavioral Decision Making* 20: 103–24. [CrossRef]

Toplak, Maggie E., Richard F. West, and Keith E. Stanovich. 2011. The Cognitive Reflection Test as a predictor of performance on heuristics and biases tasks. *Memory & Cognition* 39: 1275–89.

Toplak, Maggie E., Richard F. West, and Keith E. Stanovich. 2014a. Assessing miserly processing: An expansion of the Cognitive Reflection Test. *Thinking & Reasoning* 20: 147–68.

Toplak, Maggie E., Richard F. West, and Keith E. Stanovich. 2014b. Rational thinking and cognitive sophistication: Development, cognitive abilities, and thinking dispositions. *Developmental Psychology* 50: 1037–48. [CrossRef]

Tversky, Amos, and Daniel Kahneman. 1974. Judgment under uncertainty: Heuristics and biases. *Science* 185: 1124–31. [CrossRef]

Tversky, Amos, and Daniel Kahneman. 1983. Extensional versus intuitive reasoning: The conjunction fallacy in probability judgment. *Psychological Review* 90: 293–315. [CrossRef]

Viator, Ralph E., Nancy L. Harp, Shannon B. Rinaldo, and Blair B. Marquardt. 2020. The mediating effect of reflective-analytic cognitive style on rational thought. *Thinking & Reasoning* 26: 381–413.

Weller, Joshua, Andrea Ceschi, Lauren Hirsch, Riccardo Sartori, and Arianna Costantini. 2018. Accounting for individual differences in decision-making competence: Personality and gender differences. *Frontiers in Psychology* 9: 2258–58. [CrossRef]

West, Richard F., Maggie E. Toplak, and Keith E. Stanovich. 2008. Heuristics and biases as measures of critical thinking: Associations with cognitive ability and thinking dispositions. *Journal of Educational Psychology* 100: 930–41. [CrossRef]

Yelbuz, Büşra Elif, Ecesu Madan, and Sinan Alper. 2022. Reflective thinking predicts lower conspiracy beliefs: A meta-analysis. *Judgment and Decision Making* 1: 720–44. [CrossRef]

Yilmaz, Onurcan, and S. Adil Saribay. 2017. The relationship between cognitive style and political orientation depends on the measures used. *Judgment & Decision Making* 12: 140–47.

Journal of
Intelligence

MDPI

Article

Alpha Suppression Is Associated with the Tip-of-the-Tongue (TOT) State Whereas Alpha Expression Is Associated with Knowing That One Does Not Know

Edmund Qian-Long Shen, David Friedman, Paul Alexander Bloom and Janet Metcalfe *

Department of Psychology, Columbia University, New York, NY 10027, USA
* Correspondence: jm348@columbia.edu

Abstract: The tip-of-the-tongue (TOT) state is a spontaneously occurring metacognitive state that indicates that the answer to a query is almost, but not quite, at hand, i.e., that resolution is imminent. Since the time of William James, a distinctive feeling of nagging frustration has been observed to be associated with TOT states. On a more positive note, TOT states are also associated with intense goal-directed curiosity and with a strong desire to know that translates into successful mental action. The present study showed that prior to the presentation of resolving feedback to verbal queries—if the individual was in a TOT state—alpha suppression was in evidence in the EEG. This alpha suppression appears to be a marker of a spontaneously occurring, conscious, and highly motivating goal-directed internal metacognitive state. At the same time, alpha expression in the same time period was associated with the feeling of not knowing, indicating a more discursive state. Both alpha and alpha suppression were observed broadly across centro-parietal scalp electrodes and disappeared immediately upon presentation of the resolving feedback. Analyses indicated that the occurrence of alpha suppression was associated with participants' verbal affirmations of being in a TOT state, which is also related to subsequent expression of a late positivity when feedback is provided, and to enhanced memory.

Keywords: tip-of-the-tongue (TOT); curiosity; alpha; alpha suppression; spectral analysis; ERP; consciousness; metacognition

1. Introduction

In their seminal article on meta-reasoning, Ackerman and Thompson (2017) proposed that when faced with a query or problem, the first metacognition that comes into play consists of a higher order evaluation or feeling which they call a 'judgment of solvability'. This metacognition, they note, may trigger the impulse to actively engage in the effort needed to achieve the goal/answer. The experience, as well as the neural correlates of this initial metacognitive state, judgment, feeling, intuition, or appraisal—this second order 'hunch' that we may be able to come up with the answer—are the focus of the present article.

As noted by Polanyi (1974): "This question goes back to antiquity. Plato set it in the Meno. He said that if we know the solution of a problem, there is no problem—and if we don't know the solution, we do not know what we are looking for and cannot expect to find anything ... The problem is ineluctable and can be answered only by recognizing a kind of intuition". Polanyi goes on, somewhat poetically: "I have spoken of our powers to perceive a coherence bearing on reality, with its yet hidden future manifestations, the kind of foreknowledge we call a [solvable] problem. And we know that the scientist produces problems, has hunches, and, elated by these anticipations, pursues the quest that should fulfil these anticipations ... Think of Einstein, when as a boy he came across the speculative dilemma of a light source pursuing its own ray... Through years of sometimes despairing enquiry, he kept up his conviction that the discovery he was seeking was within his ultimate reach and that it would prove worth the torment of its pursuit; and again Einstein proved right... The scientist is faced with similar decisions. An intuition which would merely point him out problems could not tell him which problem to choose. He must be able to estimate the gap separating him from discovery and he must also be able roughly to assess whether the importance of a possible discovery would warrant the investment of the powers and resources needed for its pursuit. Without this strategic intuition, he would waste his opportunities on wild goose chases and be soon out of a job" (p. 249).

We propose that this potentially highly motivating metacognition that is experienced in at least some cases before serious attempts at solution begin, is, essentially, a judgment that the gap is not too great. Indeed, the experience that the answer that is sought is knowable by the problem solver, that it is within reach, or, indeed, that it is 'almost known' fuels its pursuit. Motivation is increased by the metacognitive feeling that the solver 'almost knows' the answer (Berlyne 1954), or that is within their Region of Proximal Learning (Metcalfe and Kornell 2005; Metcalfe et al. 2020; Metcalfe 2021). Metcalfe and Jacobs (2021) have called this kind of goal-directed motivation to know 'Curiosity1.' (Note that they also posited that there is a conceptually different kind of curiosity that they called 'Curiosity2'. This latter is discursive and exploratory, and, according to Jacobs and Metcalfe (forthcoming) may also come into play, but in a different way, during problem solving.) Although highly salient Curiosity1-type feelings of imminence (Smith 1994) may not occur in all cases in which a person is provided with a solvable problem to solve, of course, when such metacognitive feelings do occur, they can be highly motivating (Schwartz and Metcalfe 2011). The strongest and most well studied case of 'almost knowing' is the tip-of-the-tongue (TOT) state, a distinctive metacognitive experience that intrudes upon people's consciousness spontaneously.

It *is* something, in the Nagelian sense that it is something to be a bat (Nagel 1974, i.e., it is conscious), to be in a TOT state (Schwartz et al. 2000): it is a conscious state. As Carruthers (1989) has argued, much—indeed, perhaps most—of the time people are in states that do not involve consciousness. Furthermore, even metacognition can happen unconsciously. It has been argued (Schwartz and Metcalfe 2011) that conscious metacognition does not run continuously: metacognitive states, including feelings of knowing, judgments of learning, or confidence evaluations, do not impinge themselves upon the subject most of the time. Indeed, while many metacognitive judgments—judgments about one's confidence in some answer, say, about whether one has understood, or about whether one is making progress—can be voluntarily accessed on demand, they do not routinely surface into consciousness spontaneously (Schwartz and Metcalfe 2011) without being solicited. TOT experiences are the exception. They occur with an immediacy in awareness rivaling visual, auditory, or pain qualia, and they are compelling to mental action.

There are many reports of the phenomenological 'feel' of TOT states, notably the famous quotation of William James (1890, p. 251) who said: "The state of our consciousness is peculiar. There is a gap therein; but no mere gap. It is a gap that is intensely active. A sort of wraith of the name is in it, beckoning us in a given direction, making us at moments tingle with the sense of our closeness and then letting us sink back without the longed-for term. If wrong names are proposed to us, this singularly definite gap acts immediately so as to negate them. They do not fit into its mold. And the gap of one word does not feel like the gap of another, all empty of content as both might seem necessarily to be when described as gaps . . . " In another example, Brown and McNeill (1966) observed that: "The signs of it [a TOT] were unmistakable; [the subject] would appear to be in mild torment, something like the brink of a sneeze, and if he found the word, his relief was considerable" (p. 326). The TOT state, then, appears to be experientially distinctive and not particularly pleasant.

Schwartz (1999) has argued that this state of consciousness is universally experienced, as evidenced by the existence of a linguistic expression denoting the state in all 51 languages that he investigated. An exception is Q'eqchi', a language that functionally has no written communications (although there is a written guide, there are no books, newspapers or other widespread written communication in Q'eqchi') and has no distinct expression for the TOT state. Nevertheless, Brennen et al. (2007) showed that speakers of Q'eqchi' immediately understand the phenomenology of TOTs. They both claimed to experience such a state, and, when given trivia questions such as, "Who is the President of the United States?", "Who is Argentina's most famous ever footballer?", or "Who is the indigenous Guatemalan from Quiché who won the Nobel Peace Prize in 1992?[1]" they indicated that although they could not say the name now, they knew that they would remember it later, and expressed the kinds of typical frustration associated with being in a TOT state. They appeared to have the subjective, phenomenological experience. American Sign Language, too, has an expression for this state, according to Schwartz (1999), although it is tip-of-the-fingers-related rather than being tongue-related.

Furthermore, the TOT state has been documented to have functional significance: it acts as a trigger to mental action. It is highly motivationally relevant (Ryan et al. 1982) and is associated with goal-directed Curiosity1 (Bloom et al. 2018; Brooks et al. 2021; Fastrich et al. 2018; Loewenstein 1994; Metcalfe et al. 2017; Metcalfe and Jacobs 2021). Indeed, the TOT state has been taken as the quintessential exemplar of the feeling state underpinning people's compelling goal-directed need to know. Metcalfe et al. (2017) showed that people were approximately twice as curious to know answers to questions when they were in a TOT state as compared to when they were not. Furthermore, when curious, information is better learned (Gruber et al. 2014; Kang et al. 2009; Marvin et al. 2020; Murayama et al. 2019; Xu and Metcalfe 2016) and serves as a reward (FitzGibbon et al. 2021). Cleary et al. (2021) have shown that when they were in TOT states, participants were more willing to accept the risk of taking a potentially penalizing test, and Lau et al. (2020) have shown that, when curious, people are willing to endure adverse circumstances including even the possibility of electric shock, in order to obtain the answer.

The TOT state, then, is one of the most widely studied of metacognitive states (e.g., Bock 1987; Brown 2012; Brown and McNeill 1966; Funnell et al. 1996; Gardiner et al. 1973; Heine et al. 1999; Koriat and Lieblich 1974, 1977; Perfect and Hanley 1992; Schwartz 1999; Schwartz and Cleary 2016; Schwartz and Metcalfe 2014; Schwartz and Pournaghdali 2020). It has implications for our understanding of many cognitive processes including reasoning, problem solving, lexical access, and psycholinguistic processes (e.g., Bowers et al. 1990; Burke et al. 1991; Gollan and Brown 2006; Miozzo and Caramazza 1997; Nordmann et al. 2013); it presages changes in cognition with normal aging (MacKay and Burke 1990; Schwartz and Frazier 2005); it is implicated in decision-making and perceptual biasing (Cleary 2019; Cleary and Claxton 2015), and it is well known in investigations of memory retrieval (Gollan and Brown 2006; Koriat 2000; Schwartz 2008; Schwartz and Smith 1997).

Some neural correlates of the TOT state have been investigated: fMRI studies have pointed to activation of several brain regions including the anterior cingulate cortex, dorsolateral prefrontal cortex, and inferior frontal cortex (Kikyo et al. 2001; Maril et al. 2001, 2005). In addition, people show an increased late positivity in the ERP record starting at around 250 ms after receiving informational feedback when they had been in a TOT state as compared to when they had not been in a TOT state (Bloom et al. 2018; and see Diaz et al. 2007; Galdo-Alvarez et al. 2009).

As has often been noted, the TOT state (and curiosity, along with it, see Metcalfe et al. 2022, Experiment 2) is resolved immediately upon definitively attaining or being given the answer. Once one knows for certain that one has the answer, the nagging phenomenological state disappears (Loewenstein 2007). Thus, there should be rapid changes in mental and brain processes with different neural involvement implicated pre- and post-feedback. The neural processes involved upon resolution, presumably, should be different from those prior to resolution. Observations of such rapid state changes are more detectable with EEG than they are with fMRI.

Ryals et al. (2021) have shown that a finely modulated pupillary dilation is observable while the person is in the TOT state. This finding suggests that a particular state of conscious awareness—that we hypothesized might be obvious in EEG spectral analysis insofar as brain oscillations are often thought to reflect feeling states—may accompany affirmations of being in a TOT state. We have been able to find only one previous EEG spectral study of the TOT state (Resnik et al. 2014). The 10 participants in that study were of mean age 70.2 years. While it is well known that the TOT state is much more frequent in older than younger adults, it is not known whether the mental processes might also differ. Our study used college students. Our materials were also different: the previous study used a pictorial face-naming task whereas our study used general information queries.

Our primary interest in the present paper, then, was in investigating potential EEG neural correlates of the TOT state itself, prior to its resolution, in order to obtain further evidence regarding this state of conscious awareness. While previous work has identified spectral EEG correlates of various metacognitive processes (Jin et al. 2019; Klimesch 1997, 2012), only a few studies have investigated oscillatory brain signals specifically associated with the TOT state (Resnik et al. 2014). Because relatively little work has been conducted in this area, we did not have specific a priori hypotheses concerning particular frequency bands or electrodes. Thus, we took an exploratory approach towards identifying such correlates by conducting a broad search for spectral EEG measures associated with the experience of TOT states. As these analyses identified differences in the alpha frequency band (8–12 Hz), we provide an overview of the alpha suppression literature and offer post hoc explanations for these findings in the Discussion.

We were primarily interested in investigating neural correlates of the TOT state *prior to* its resolution; that is, the markers of the actual state, rather than the consequences of having been in a TOT state previously. However, we also sought to compare these contemporaneous markers to those that ensued upon resolution of the TOT state; that is, once feedback was provided. Accordingly, we analyzed data that had been collected in our lab (Bloom et al. 2018) but had not previously been examined regarding the neural correlates of the TOT state itself. Here, we will present some post-feedback results, but will primarily examine the pre-feedback spectral correlates of TOTs—in short, we will focus on the TOT state itself.

2. Methods

2.1. Participants

Participants were 30 healthy adults studying at Columbia University with ages ranging from 18 to 32 years old (16 male, 13 female, 1 did not report gender), with a mean age of 21.27 years and SD of 4.26 years. Participants were recruited through flyers in buildings on the Columbia University campus and emails to lists of previous study participants who had consented to being contacted. All participants were fluent in English, and reported being right-handed and free of medications known to impact the central nervous system. All participants provided written informed consent, and were compensated at a rate of USD 15/hour. All protocols were approved by the Columbia University and New York State Psychiatric Institute institutional review boards.

One participant did not report experience of any TOT states, and was thus excluded from all analyses. To ensure an adequate numbers of trials, the EEG recordings were examined to determine whether each participant had at least 5 usable trials (see EEG preprocessing) for both the tip-of-the-tongue (TOT) and no-TOT conditions. Three subjects failed this criterion by having less than 5 usable TOT trials, and were thus excluded during preprocessing. A final total of 26 participants were included in all EEG analyses.

2.2. Design

The study was a within-participants quasi-experimental design. All participants were read a set of general information questions during the initial question period. After being asked each question, they were given 3 s to try to answer, in the pre-feedback period. All responses in which people provided the correct answer were excluded. If they had not given the answer within the 3 s pre-feedback interval, they then made a TOT judgment, which was followed by a 500 ms fixation period after which correct answer feedback was given. They were then given 1 s to process the feedback. Once all questions had been presented, participants were given a surprise retest on all questions. Question order was randomized for each participant. EEG was recorded during the pre-feedback and the post-feedback phases (see EEG procedures below). In the analyses, below, the answer to the TOT probe was treated as if it were an independent variable with two levels (TOT vs. no-TOT).

2.3. Stimuli

The stimuli were 150 general information questions taken from Nelson and Narens (1980), and updated and corrected by Bloom et al. (2018). For example, one question was "What is the capital of Jamaica?" (Answer: Kingston). All correct answers were a single word. The questions were displayed with PsychoPy2 in a large white font with a gray background on a computer monitor (Peirce 2007) for 3000 ms. Answers were displayed for 1000 ms. All stimuli and experiment PsychoPy code are available at https://osf.io/nkh9u/ (accessed on 8 May 2021).

2.4. Procedure

First, EEG caps were applied. Next, participants were told that they would be given a series of questions to try to answer verbally and, if they could not answer within the given time, to report whether or not they were in a TOT state. Participants were also instructed that they would get to view the answers to each question for which they could not generate an answer within the allotted time. Before starting the study, an experimenter explained the TOT state as a phenomenon in which "you feel sure you know the answer and think you can get it—it is imminent—but you cannot think of it at the moment." Participants completed one practice trial prior to beginning the experiment to ensure that they understood the instructions and were responding appropriately.

As is shown in Figure 1, during each trial, the question text appeared onscreen and was read aloud by an experimenter sitting next to the participant in the EEG booth. Upon finishing reading the question, the experimenter pressed a key on a keyboard to start the 3 s pre-feedback response period. If a verbal answer was given by the participant during this period (whether correct or not), the experimenter pressed a key to move directly to the next trial (i.e., the next question) without providing the answer to the question. If the participant did not give a response they were prompted to indicate whether they were experiencing a TOT state or not (yes/no). The experimenter recorded their responses with a keyboard press, which triggered a 500 ms period with a fixation cross on the screen, followed by a 1000 ms period during which the feedback (correct answer) displayed. Following feedback, a blank screen was shown for 750–1250 ms (jittered), before the next trial began. All participants completed 150 trials.

Following the TOT/feedback phase, experimenters removed the EEG cap, and participants were given a 2 min break. Next, participants were told that there would be a test phase, and were given instructions on how to complete the test. The participant left the EEG booth for the final test phase during which he or she saw the same 150 questions in a different randomized order, and were allowed to type in their answer. They were told to give their best guess in response to each question even if they were not sure of the answer. No feedback was provided during this phase. After the test phase, participants filled out a demographics questionnaire, and were debriefed, compensated, and were allowed to view the answers to all of the general information questions if they desired.

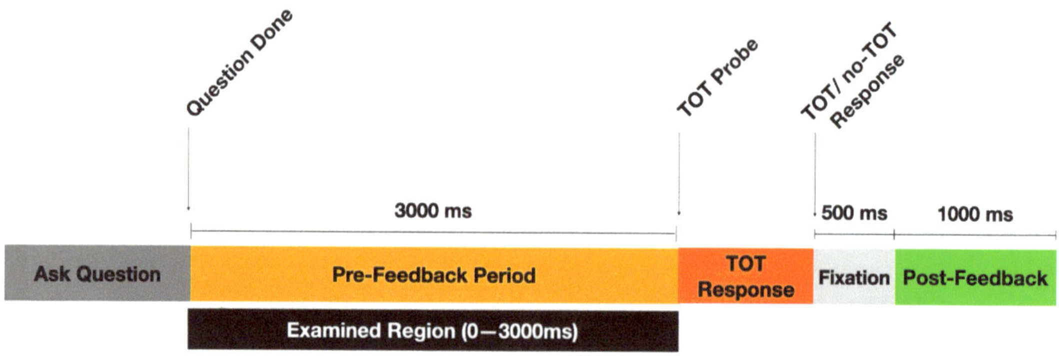

Figure 1. Timeline of individual trials. Note: 'Ask Question' represents a window of variable length when the experimenter was reading aloud the general information question to the subject. 'Examined Region' represents the 3000 ms window—corresponding to the pre-feedback/subject response period—that was selected for spectral and time-frequency analysis.

2.5. Response Scoring

Participants' final recall responses were evaluated against the correct answers using an approximate pattern matching algorithm (AGREP, generalized Levenshtein edit distance = .1), and a naive experimenter checked all of the algorithmically-based scoring. The final recall results were reported in Bloom et al. (2018).

2.6. Electroencephalographic (EEG) Recording and Processing

Brain electrical activity was recorded during the TOT and feedback phases from 62 scalp sites (sintered Ag/AgCl) mounted in an Electrocap (Neuromedical Supplies). The EEG data were digitized at 500 Hz (DC; high-frequency cutoff of 100 Hz; right-forehead ground). Electrodes were placed on the outer canthus of each eye to record horizontal eye movements, and directly above and below the left eye for vertical movements. Activity was originally referenced to the nose and re-referenced offline to the average of the left and right mastoids. Impedances were maintained below 10 kΩ throughout the experiment. Prior to analyses, all recordings were filtered using a 0.1–25 Hz IIR-Butterworth bandpass filter to remove DC drift and muscle movements. Bad channels and data periods were systematically removed using the EEGLAB artifact subspace reconstruction (ASR) algorithm, with rejection criteria (flatline rejection threshold = 5 s, high-frequency noise threshold = 4 SD, minimum acceptable correlation = 0.7) as recommended by Delorme's group-analysis preprocessing pipeline (Delorme and Makeig 2004; Kothe et al. 2019). Interpolation for removed channels was done across all trials. Independent component analysis (ICA) decomposition was conducted with the Picard algorithm, and artifactual components were labeled and rejected via ICLabel (rejection probability threshold = 0.90) (Ablin et al. 2018). As no feedback was presented for trials in which participants verbally recalled the answers during the 3 s response period after question presentation, these trials were not included in the EEG analyses. Only trials in which no answer was given and participants were either in a TOT or no-TOT state were included in analyses.

To address the pre-feedback time period, the EEG data were analyzed during the 3000 ms answering period after the experimenter had finished asking the general information question and initiated (via button press) the onset of the EEG epoch. This time period included mental activity that was presumably associated with attempts at determining the answer to the question and, importantly, the TOT or no-TOT mindset. The examination region was defined as the 3000 ms between the 'question-done' key-press and the TOT probe because this was the timespan with the cleanest resolution, reflecting primarily internal cognitive phenomena rather than responses to external stimuli. Since the experimenter was talking (asking the question aloud) during the 1 s before the question-done trigger, that period was filled with a high amount of literal noise. Those auditory stimuli would translate into substantial EEG noise, rendering it difficult if not impossible to parse whether ERP/ event-related spectral perturbations (ERSP) phenomena observed during that window were due to sound of the experimenter's voice as compared to the participant's cognitive state. The 3000 ms of silence between the question-done trigger and the subsequent TOT probe was thus the cleanest window for epoching. To address the post-feedback time period, the EEG data were analyzed during the 1000 ms after feedback was displayed.

Trials in which participants provided the wrong answer were excluded because of the problem of variable epoch length. An incorrect answer could have been given several hundred milliseconds after the question-done trigger in some trials, and nearly 3 s after the trigger in other trials. Such a wide range of response times is incompatible with ERP and ERSP analyses, which require consistent epoch durations in order to generate averaged curves across participants. Comparing TOT trials with 'don't know' trials is thus the cleanest design for an EEG-based study of this particular phenomenon, insofar as the two trial types have consistent, stable epochs which can be meaningfully compared with one another.

2.7. Spectral Analyses

Overall spectral power of the 3000 ms pre-feedback and 1000 ms post-feedback periods was calculated using the EEGLAB *std spec* algorithm, applying fast-Fourier transforms (FFTs) to each trial. Log power at each frequency was calculated and expressed in microvolts squared (μV^2). After performing FFTs for all individual electrodes, the spectral decompositions of the channels were then aggregated to generate the averaged scalp topography over a specified frequency range—given our focus on alpha power differences, we computed averaged topographies specifically for spectral power in the 8 Hz to 12 Hz band.

ERSPs were computed for the 3000 ms pre-feedback period using the EEGLAB *newtimef* algorithm. In accordance with typical *newtimef* parameters, the number of wavelet cycles was set to increase linearly with frequency by a factor of 0.8, beginning at the lower-bound frequency of 3 Hz. Time-frequency images were time-locked to the question-done cue, as is shown in Figure 1, and power in decibels (dB) over the 3000 ms window was computed with the subtraction of a 500 ms pre-question-done baseline. Time-frequency images averaged over electrodes identified to be significant in the scalp topographies were then computed for the TOT and no-TOT conditions.

3. Results

3.1. Full-Period Spectral Decompositions

3.1.1. Individual-Electrode Analysis

As is shown in Figure 2, Column A, examination of the main midline electrodes (Fz, Cz, Pz, Oz) revealed that the TOT state was associated with lower alpha spectral power during the 3000 ms pre-feedback window. All four main midline electrodes included regions of alpha power—particularly in the 10 Hz to 11 Hz range—where differences between the TOT and no-TOT conditions were significant ($p < 0.01$) after application of the stringent Bonferroni multiple-comparisons correction. Alpha spectral differences between the two conditions were also more pronounced towards the back of the head, with broader bands of significance in the parietal (9.5–11 Hz) and occipital (9–11 Hz) electrodes than the frontal and central electrodes (10.3–10.6 Hz).

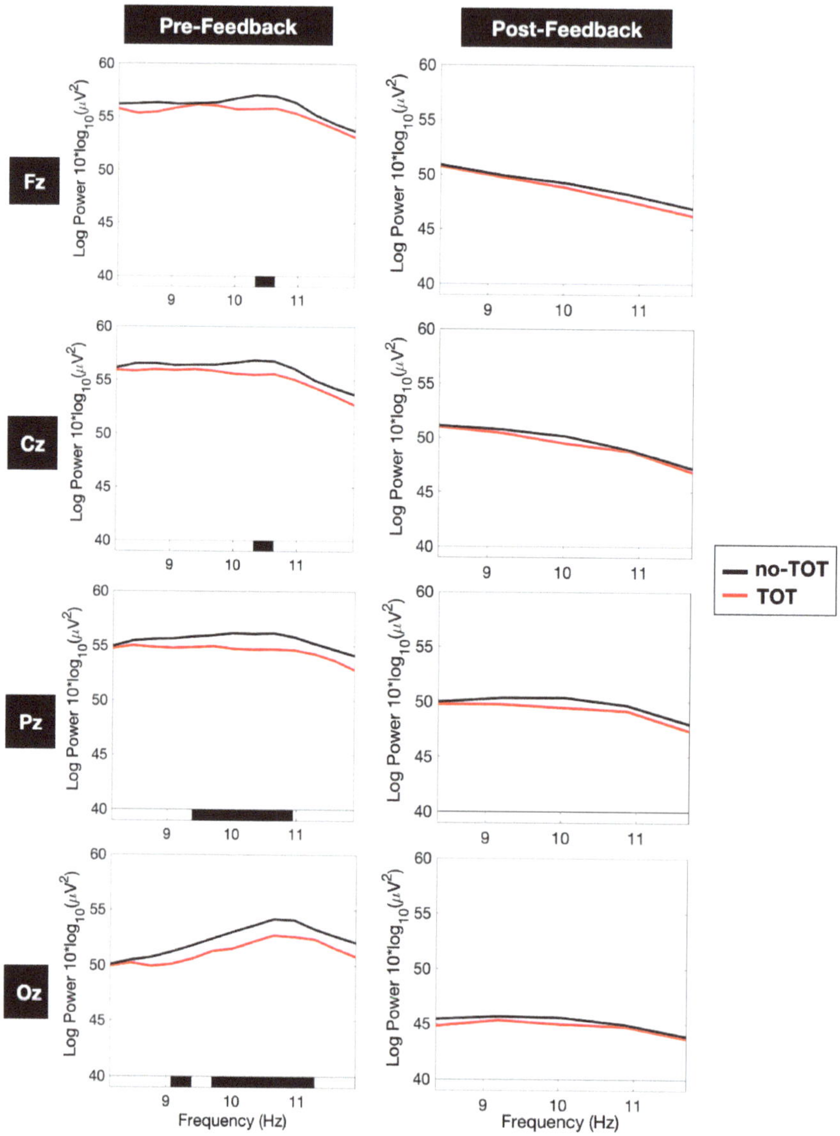

Figure 2. Frequency-power plots of four main midline electrodes during the pre-feedback and post-feedback periods. Note: 'pre-feedback' and 'post-feedback' refer to the 3000 ms and 1000 ms windows depicted in Figure 1. Fz, Cz, Pz, and Oz denominate the frontal, central parietal, and occipital electrodes placed along the midline of the participant's scalp. Black bars under each plot represent frequency banks where the difference between the TOT and no-TOT conditions was statistically significant at the 1% level after applying the Bonferroni multiple-comparisons correction.

No statistically significant spectral differences were observed during the 1000 ms post-feedback window, as shown in Figure 2, Column B. Alpha power was not significantly different when participants were receiving the correct answer, regardless of whether they had been in a TOT state or not.

3.1.2. Averaged Scalp Topographies

Scalp topographies over all 62 electrode sites of aggregate log power in the alpha band (8–12 Hz) supported the results of the individual-electrode analyses. As is shown in Figure 3, in the pre-feedback window, alpha power was significantly lower for TOT trials than no-TOT trials after Bonferroni correction. TOT-associated alpha suppression was stronger over the posterior regions of the scalp, and skewed slightly toward the right hemisphere, as is shown in Figure 3, Columns C and D. Thirty-one electrodes on the scalp—primarily those on the central, parietal, and occipital regions—were significant at the 1% level for lower alpha power in the TOT state, as Figure 3, Column D shows. No differences were observed for topographies of delta (0.5–4 Hz), theta (4–7 Hz), or beta (13–30 Hz) waves over the same 3000 ms window.

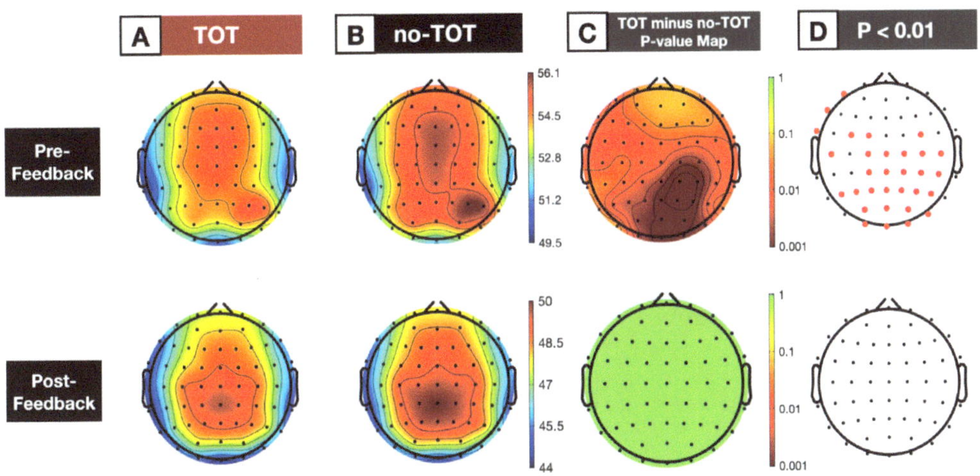

Figure 3. Whole-scalp averaged spectral topography for TOT versus no-TOT trials. Note: Columns (**A**,**B**) depict the topography of log power over the alpha band (8–12 Hz) for TOT and no-TOT conditions, respectively. Pre-feedback power was based on a spectral analysis of the 3000 ms examined period after the question-done cue. Post-feedback power was based on a spectral analysis of the 1000 ms period after the 500 ms fixation cross that indicated the onset of correct answer feedback. Column (**C**) represents the Bonferroni-corrected p values for differences between TOT and no-TOT trials at each point across the scalp. Column (**D**) highlights in red the electrodes where difference in TOT versus no-TOT was significant at the 1% level after Bonferroni multiple-comparisons correction.

As is shown in the lower portion of Figure 3, Column C, no significant spectral differences were observed for TOT versus no-TOT trials during the post-feedback period. Bonferroni-corrected p-value maps indicated that there were no significant differences in alpha power over the scalp during the time period in which participants were receiving the correct answer; that is, in the 1000 ms window after making their TOT/no-TOT response. No individual electrodes were significant at the 1% level for differences in aggregate spectral power on the 8–12 Hz band (see, Figure 3, lower portion of Column D). In addition, no differences were observed for delta, theta, or beta wave topographies over the same window.

As was the case with the frequency-power analyses of the main midline electrodes, whole-scalp topographies identified TOT-associated alpha suppression specifically in the 3000 ms pre-feedback window, when the question was unresolved and the participant was cogitating about a potential response.

3.1.3. ERSP Analysis

An averaged time-frequency image of the significant electrodes that are shown in Figure 3 (AF7, F7, FC1, FC3, FC4, FT7, Cz, C1, C2, C4, C5, C6, CPz, CP1, CP2, CP4, CP6, Pz, P1, P2, P3, P4, P5, P6, POz, PO3, PO4, PO8, Oz, O1, O2) was computed to examine how alpha power evolved over time in the pre-feedback window. (A similar analysis was conducted over the 9 central-parietal electrodes that were previously-reported to be significant in post-feedback TOT-associated activation in Bloom et al. (2018). This alternative analysis yielded highly similar results to those presented here.)

As is shown in Figure 4, alpha power was low in the first 1000 ms (−3.0 dB) of the pre-feedback window relative to the 500 ms baseline period of comparison, for both the TOT and no-TOT conditions. Past 1000 ms, however, consistent alpha power was observed in the no-TOT ERSP image. While both the TOT and no-TOT images demonstrated some degree of alpha power in the 1000 ms to 3000 ms window, the no-TOT image reveals continuous perturbations of +0.6 to +0.8 dB in the alpha band relative to baseline. By contrast, the TOT image demonstrates less consistent alpha power, with perturbations primarily between +0.1 dB to +0.5 dB and only several plotted pixels reaching +0.7 dB. The given ERSP images thus provide a more granular portrait of the alpha differences noted in Figures 2 and 3. Differences in alpha power in the TOT and no-TOT states appear primarily to be the result of a divergence that occurred at approximately 1000 ms after the question-done marker, with more consistent and higher-magnitude alpha-range power in the no-TOT state.

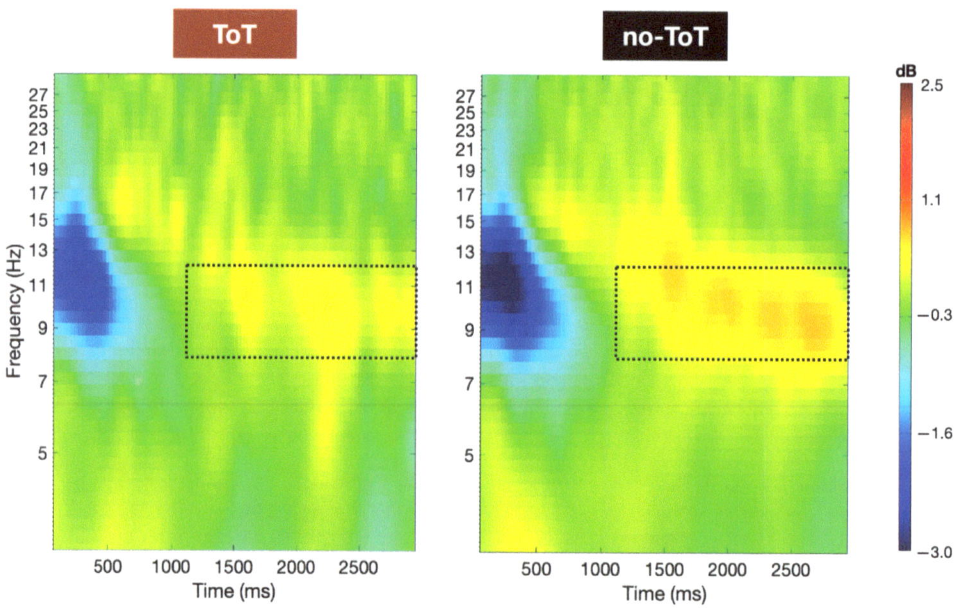

Figure 4. Time-frequency decomposition of spectral power during the pre-feedback period for the significant electrodes. Note: Plots represent the averaged time-frequency decompositions of the 31 electrodes which were found to have Bonferroni-corrected $p < 0.01$ significant differences in alpha power over the pre-feedback window: AF7, F7, FC1, FC3, FC4, FT7, Cz, C1, C2, C4, C5, C6, CPz, CP1, CP2, CP4, CP6, Pz, P1, P2, P3, P4, P5, P6, POz, PO3, PO4, PO8, Oz, O1, O2. The dotted box represents a window of particular interest (1000–3000 ms) in the alpha band (8–12 Hz), where higher consistent levels of alpha power were present in the no-TOT state relative to the TOT state.

3.2. Event-Related Potential Analyses

We also conducted an analysis of the ERPs, time locked to the offset of the question (the question-done signal), and a reanalysis[2] of the ERPs following the correct answer feedback, as shown in Figure 5. This analysis revealed no differences between the TOT and the no-TOT conditions in the pre-feedback interval. Indeed, the analysis showed virtually no deviation from baseline. The right side of Figure 5 also shows our post-feedback reanalysis of the data that were previously presented in Bloom et al. (2018), using the updated preprocessing pipeline described above. Despite the more stringent rejection criteria of the current processing, we still observed greater post-feedback positivity over the central and parietal electrodes for TOT trials compared to no-TOT trials.

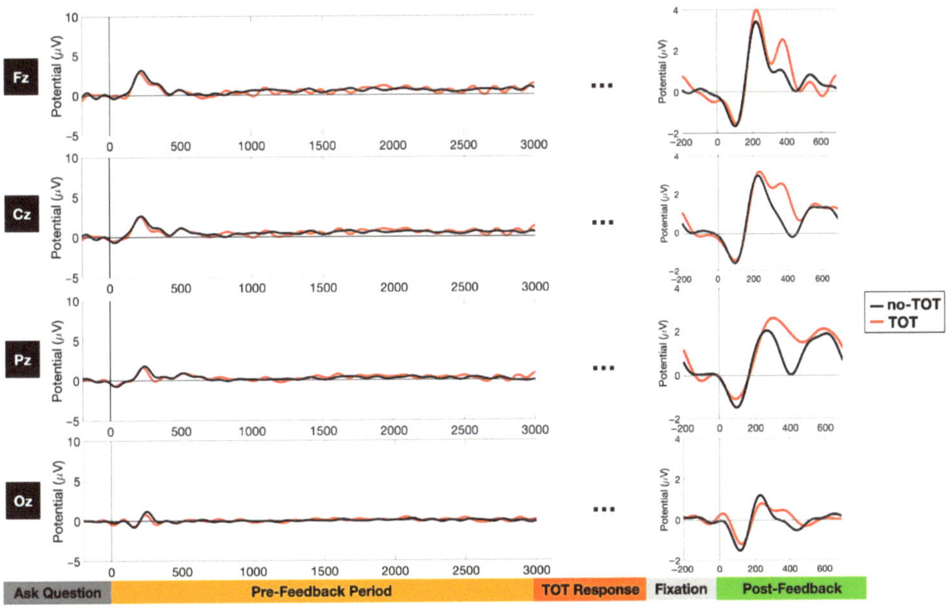

Figure 5. ERP plots of four main midline electrodes during the pre-feedback and feedback periods. Note: Fz, Cz, Pz, and Oz designate the frontal, central parietal, and occipital electrodes placed along the midline of the participant's scalp. The pre-feedback ERPs were time-locked to the question-done cue, as illustrated in Figure 1. The post-feedback ERPs were time-locked to the start of the feedback window following a 500 ms fixation cross, also as illustrated in Figure 1. The post-feedback results are re-analyses, with a more stringent preprocessing algorithm, of results presented in Bloom et al. (2018). Ellipses (...) represent windows in the TOT response and fixation periods for which ERPs were not computed.

The current combination of ERSP and ERP data thus indicate that TOT in the pre-feedback window is associated with substantial spectral differences (alpha suppression) but no ERP differences, while having been in a TOT state prior to feedback results in changes in the post-feedback window that are characterized by ERP differences but no spectral differences.

4. Discussion

The most important result presented here is the finding that just after people had been presented with a question, a neural correlate of what appeared to be a metacognitive evaluation of solvability—in the form of alpha or alpha suppression—was in evidence during the time period after which they had processed the question but had not yet discovered the answer. When people knew that they did not know, they showed a strong propensity toward alpha enhancement. When they were in a TOT state, however, alpha *suppression* was in evidence. This suppression occurred during the time period after which they had processed the question but had not yet discovered the answer. As Figures 2–4 indicate, after about one second of consideration of the question, if people had not yet given an answer to the question, they either lapsed into alpha (when they knew that they did not know), or they exhibited alpha suppression (when they experienced a TOT).

The spectral analysis also indicated that there were no other frequency bandwidths that were different when the person was in the TOT state as compared to the 'don't know' state. Neither theta nor delta was in evidence, in either condition. Insofar as both of these are associated with sleep and also with deeply relaxed mental states, this is not surprising. It is more surprising that no gamma or beta were in evidence in either condition, insofar as these frequencies are associated with concentration, which might have been expected if people were actively trying to retrieve the answer to the question in the TOT state. In short, the only band of activation that was in evidence when people were trying to answer the question in the TOT and no-TOT states was between 8 and 12 HZ, and activation in this particular bandwidth was more prominent when people were in the don't know state; it was suppressed when people indicated that they were in the TOT state. Interestingly, the only other study that used a spectral analysis to investigate TOTs (Resnik et al. 2014) also found alpha suppression.

There was no significant difference between TOT and no-TOT conditions in the pre-feedback ERPs in our study. This is in contrast to a difference in pattern that was shown in an ERP study by Buján et al. (2012). Rather than presenting verbal questions, though, these authors provided pictures of faces for 1 s, then left a blank period of time for one second before asking subjects to indicate whether they knew the name, did not know the name, or thought they knew it but could not retrieve it—which, later, they classified as a TOT item if the person was also able to state the target's profession, and choose the correct name in a 3-alternative forced choice test. They found that starting at around 800 ms after the onset of the picture of the face, the ERPs for items that were labelled known and not known converged and exhibited a higher positive amplitude than did those of the TOT items. The ERPs for TOT items diverged and tended toward the baseline. The divergence in the ERP tracings was quite distinctive by about 1350 ms. It is puzzling to us that what, purportedly, was increased processing in the TOT condition was marked by a decrease in amplitude—a return to baseline. In any event, the authors suggested that this ERP pattern indicted that "in TOT, the processing resources may still be involved in the search for lexical–phonological information about the person, reflecting a continued but unsuccessful effort in this category in order to resolve the conflict, whereas in know and in DK (don't know) the processing resources may be completely released" (p. 989).

The ERP differences between no-TOT and TOT items observed by Buján et al. (2012) prior to feedback were not found in our study. The most likely explanation for this difference between studies is that the famous face pictorial stimuli used in the Buján et al. (2012) experiment afforded a better alignment of the onset of processing than did the verbal queries that we had used. Our queries were of variable length and participants may have finished reading some of them before the experimenter had finished reading the question aloud and hit the button (the signal to which we aligned) indicating that presentation was complete. The explanation given by Buján et al. (2012) for their results, though, fits well with our spectral findings. They suggest that a short time after the question was presented, if the participant knew that they did not know the answer, their efforts to find the answer flagged. In our study, heightened alpha was exhibited at this point. For TOT items, though, people's efforts toward solution likely persisted. This was exemplified, in our study, by alpha suppression.

ERP differences, reflecting differences in processing after the feedback was given, were in evidence in our study. There was an enhanced late positivity when people had been in a TOT state as compared to when they had not, and this late positivity, as well as their subjective claim about having been in a TOT state, was associated with enhanced recall (Bloom et al. 2018). We show these ERP results in Figure 5, using a slightly different preprocessing methodology. The spectral analysis of the post-feedback period, however, as is shown in the bottom panels of Figure 3, did not show any significant differences between TOT and no-TOT items.

5. Conclusions

The primary result of interest in this study was that while increased alpha was related to the participant's judgment that he or she did not know the answer to a query, alpha suppression was associated with their being in a TOT state. The TOT state is a well-known, tantalizing, spontaneously occurring, conscious, phenomenological state in which the participant feels like they 'almost know'—a metacognitive marker that the answer is attainable. Our spectral analysis results are consistent with much research about the mental correlates of both alpha and alpha suppression as well as with the characteristics of TOT states.

Spectral analysis has been extensively employed in studies investigating state differences in relation to attention. It has been widely acknowledged that both alpha expression and alpha suppression are functional correlates of brain activation, and the attendant mental processes (Klimesch 1999). Event-related synchronization of alpha band power (i.e., increases in alpha power) has often been interpreted as indicating reduced information processing in the underlying neuronal networks. It is associated with what some investigators call "cortical idling" and the corresponding mental idling (Pfurtscheller 1999). Suppression of alpha band activity is related to the increased excitability of neurons involved in implicated cortical areas, and has traditionally been related to enhanced attention and information processing (Neuper and Pfurtscheller 2001; Pfurtscheller and Lopes da Silva 1999).

Consistent with this interpretation, mind wandering has been shown to be associated with high alpha power (e.g., Baldwin et al. 2017). Similarly, Arnau et al. (2020), using a mind-wandering probe task in synchrony with EEG, showed both that the primary task response accuracy was reduced when people were mind wandering, and that alpha power was increased when they were mind wandering. Like our own results, their effect was spatially widely distributed over the scalp. Thus, the increase in alpha power seen in trials in which people indicated that they have no idea of what the answer was, could be taken as a neural indicator that they had simply gone off task (though see below for an alternative interpretation).

The finding of alpha suppression when the person was experiencing a TOT state suggests that rather than cognitively idling or mind wandering, participants were continuing to engage their full attention to the task at hand—trying to come up with the answer to the query. The tip-of-the-tongue state, of course, is well known to be highly cognitively engaging and attention focusing as the person attempts to mentally zero in on the specific required word that is the answer.

Alpha suppression has been shown to be associated with convergent thinking, in general. It is notable that in TOT studies, the solution is singular—not any answer will do. It is also characterized as being the prototype of Curiosity 1, where Curiosity 1 is defined in terms of goal seeking (convergent) rather than goal avoidant (divergent), or of what computer scientists (see Wilson et al. 2014) call exploitation rather than exploration. Although perhaps the most parsimonious interpretation of our finding of high alpha when people do not know the answer is simply that they go off task, a second, and more charitable interpretation of the increase in alpha is also possible. It could be a reasonable strategy, in a circumstance in which one does not know the answer, to engage in divergent thinking. Indeed, Schooler (forthcoming) has argued that mind wandering is related to exploratory, divergent, creative thinking. If one fails to come up with the solution directly, then seeking the solution in a less focused manner—letting the mind wander elsewhere—could be strategic. In support of this possibility, a review by Fink and Benedek (2014), concluded that alpha power is related to creative thinking. For example, Jauk et al. (2012) manipulated the kind of thinking by instructing participants to solve an association task either by finding the most common solution (in the convergent condition) or the most uncommon solution (in the divergent condition). Raters assessed the answers given in the divergent condition as being more creative. The study found that alpha synchronization was associated with divergent thinking, whereas alpha suppression accompanied convergent thinking.

A number of visual attention studies have shown that alpha suppression is associated with selective attention to a particular region in external visual space associated with an expectation or anticipation (triggered by a cue) that a target is about appear in that location. Bacigalupo and Luck (2019), for example, showed that a cortically circumscribed location, contralateral in the brain to the location of the expected appearance of the target, exhibited alpha suppression. In the TOT state, though, rather than attention being focused *outwardly*, in anticipation of an externally presented target, it, presumably is focused *inwardly* on the individual's own mental events (see Cooper et al. 2003). In the case of TOT, alpha suppression seems to indicate an inwardly focused attentional expectation—it is a marker that the person, him/herself, is about to come up with the answer. Unfortunately, this expectation-of-an-internal-event-about-to-happen interpretation of our data is clouded by the fact that we actually presented, onscreen, the answer to every question as feedback 3 s after each question was posed. Accordingly, our result might have reflected external-solution anticipation rather than self-solution anticipation. Two considerations challenge this interpretation. First, if the alpha suppression that we observed occurred merely as an anticipation that the solution was about to be presented externally, it should have occurred regardless of whether the person was in the TOT state or not. However, it did not occur with no-TOT items. Second, the only other study that has looked at the spectral components of TOT (Resnik et al. 2014) did not present the solution externally as feedback (at least not until the end of the entire experiment, which was after a considerable temporal delay). Their study, like ours, showed alpha suppression when people were in the TOT state. That alpha suppression is a marker of an internally focused attentional state is consistent with both the decreased dual task performance (Ryan et al. 1982), and the increased pupillary responses (Ryals et al. 2021) that have been observed in conjunction with TOT states.

The alpha suppression observed in our study was widely distributed over the scalp, as is shown in Figure 3. Interestingly, in the MEG TOT study on older adults (Resnik et al. 2014), alpha suppression was tightly localized to the left ventral temporal region. The reason for this difference in distribution is currently unknown, and requires further research. It may, of course, be attributable to differences in the populations (young versus older adults), or methods (picture naming versus general information). It is also possible, however, that it is due to the diversity of the content of what people were seeking when they were in the TOT states in the two experiments. The stimuli that were presented in the Resnik et al. (2014) study were much more uniform than were ours: they were always the names of famous people associated with the face pictures. Our answers, by contrast, traversed a large range of possible topics. It is possible that our distributed pattern occurred because of the diversity of the answers sought. We cannot, from the present data, distinguish among these possibilities.

Other correlates of alpha suppression also line up with what is known about the TOT state. For example, the TOT state is associated with negative affect (Schwartz 1999). Interestingly, although the state itself may be experienced as frustrating (Schwartz et al. 2000), Cleary (2019) has found that items that are emotionally positively valanced are more likely to evoke TOTs. (It may be frustrating to be unable to remember the pleasurable target, whereas one may not care to remember a negatively valanced target and so the frustration of trying without immediate success is avoided.) Alpha suppression is associated with somewhat negative, rather than positive, affect. A highly focused but not especially pleasant state may be conducive to cognitive success. Kuhbandner et al. (2009, 2016), for instance, showed that people's affective state influenced how easy it was for them to consciously perceive a masked target stimulus: a slightly negative affective state was more effective in facilitating perception. Furthermore, these authors, like others, also found that alpha was associated with the positive and alpha suppression with the negative affective state.

Finally, to return to the relation of the TOT state to precognition and Meno's paradox, as noted by Polanyi at the outset: it seems, at first blush, puzzling and perhaps even spooky that people might perceive a hidden reality and be motivated by such an intuition. In fact, though, this idea may not be so mystifying as it seems. People often have partial information about the answer. Indeed, many studies of TOTs show that people are aware of such partial information. They often know the first letter, the number of syllables, the gender, items that are semantically similar, and they often have access to knowledge that items that come to mind spuriously are incorrect (see Schwartz and Metcalfe 2011; Wade and Kidd 2019). Interestingly, an EEG study by Magazzini et al. (2016) linked partial awareness of words to alpha suppression in which the visual perception of the word is masked and sometimes participants can detect it and sometimes not. The authors observed alpha suppression in the partial-seeing case.

High alpha power is associated with off-task thought, with mind wandering, and with divergent, creative thinking (e.g., Arnau et al. 2020; Bozhilova et al. 2020). It seems antithetical to the kind of focused determination that is called for if the individual knew that the answer was near and that he or she was hot on the trail. Yet this kind of divergent, playful thinking could be helpful if the individual had no sense that the answer was attainable and beckoning or even that the answer was blocked and that the trail had gone cold. Casting about, mind wandering, or even just relaxing and taking a deep breath, could be strategic when one knows that one does not know. This state does not typically exhibit anything like the intrusive phenomenological feel of the TOT state, which forces itself upon the consciousness of the subject. Indeed, alpha states are often characterized by a relaxed lack of self-awareness.

By way of contrast, alpha suppression is associated with focused goal-directed attention: people are expectant that a solution will be forthcoming; they may have slightly negative disquieting feelings; they intuit partial information about the answer; they feel that it is knowable and even imminent. This latter metacognitive state that is associated with the TOT state and with alpha suppression (and that can be observed in the lab in a small way), may, when played out at scale in the real world, be the kind of 'judgment of solvability' that underlies the intuition that a scientific dilemma may be within the scientist's ultimate reach and that it could prove worth the torment of its pursuit.

Author Contributions: All authors contributed to all aspects of this project. All authors have read and agreed to the published version of the manuscript.

Funding: This project was supported by National Science Foundation Grant 1824193 awarded to Janet Metcalfe. The authors are solely responsible for the content of this article.

Institutional Review Board Statement: The study was conducted in accordance with the Declaration of Helsinki, and this research was approved by the Institutional Review Board of Columbia University under protocol AAAR9602.

Informed Consent Statement: Informed consent was obtained from all subjects involved in the study.

Data Availability Statement: All stimuli and experiment PsychoPy code are available at https://osf.io/nkh9u/. EEG data are available by request from pab2163@columbia.edu.

Acknowledgments: We thank Matti Vuorre and Judy Xu for their help with the original study on which this subsequent investigation is based.

Conflicts of Interest: The authors declare no conflict of interest.

Notes

1 Maradona (or maybe Messi). Rigoberta Menchú, for her work for the rights of indigenous peoples and reconciliation between ethnic groups.

2 This post-feedback analysis was already reported, in detail, in Bloom et al. (2018). We reanalyzed their post-feedback data here because we used different preprocessing software for our other analyses and wanted consistency throughout.

References

Ablin, Pierre, Jean-François Cardoso, and Alexandre Gramfort. 2018. Faster independent component analysis by preconditioning with Hessian approximations. *IEEE Transactions on Signal Processing* 66: 4040–49. Available online: https://arxiv.org/abs/1706.08171 (accessed on 8 May 2021). [CrossRef]

Ackerman, Rakefet, and Valerie A. Thompson. 2017. Meta-reasoning: Monitoring and control of thinking and reasoning. *Trends in Cognitive Sciences* 21: 607–17. [CrossRef] [PubMed]

Arnau, Stefan, Christoph Löffler, Jan Rummel, Dirk Hagemann, Edmund Wascher, and Anna-Lena Schubert. 2020. Intertrial alpha power indicates mind wandering. *Psychophysiology* 57: e13581. [CrossRef] [PubMed]

Bacigalupo, Felix, and Steven J. Luck. 2019. Lateralized suppression of alpha-band EEG activity as a mechanism of target processing. *The Journal of Neuroscience* 39: 900–17. [CrossRef] [PubMed]

Baldwin, Carryl L., Daniel M. Roberts, Daniela Barragan, John D. Lee, Neil Lerner, and James S. Higgins. 2017. Detecting and quantifying mind wandering during simulated driving. *Frontiers in Human Neuroscience* 11: 406. [CrossRef] [PubMed]

Berlyne, Daniel E. 1954. A theory of human curiosity. *British Journal of Psychology* 45: 180–91. [CrossRef]

Bloom, Paul A., David Friedman, Judy Xu, Matti Vuorre, and Janet Metcalfe. 2018. Tip-of-the-tongue states predict enhanced feedback processing and subsequent memory. *Consciousness and Cognition* 101: 221–33. [CrossRef]

Bock, J. Kathryn. 1987. Coordinating words and syntax in speech plans. In *Progress in the Psychology of Language*. Edited by Albert Ellis. London: Erlbaum, pp. 337–90.

Bowers, Kenneth S., Glenn Regehr, Claude Balthazard, and Kevin Parker. 1990. Intuition in the context of discovery. *Cognitive Psychology* 22: 72–110. [CrossRef]

Bozhilova, Natali, Ruth Cooper, Jonna Kuntsi, Philip Asherson, and Giorgia Michelini. 2020. Electrophysiological correlates of spontaneous mind wandering in attention-deficit/hyperactivity disorder. *Behavioural Brain Research* 391: 112632. [CrossRef]

Brennen, Tim, Anne Vikan, and Ragnhild Dybdahl. 2007. Are tip-of-the-tongue states universal? *Evidence from an Unwritten Language Memory* 15: 167–76.

Brooks, Gregory, Haopei Yang, and Stefan Köhler. 2021. Feeling-of-knowing experiences breed curiosity. *Memory* 29: 153–67. [CrossRef]

Brown, Alan S. 2012. *Tip of the Tongue State*. New York: Psychology Press.

Brown, Roger, and David McNeill. 1966. The "tip of the tongue" phenomenon. *Journal of Verbal Learning & Behavior* 5: 325–37.

Buján, Ana, Santiago Galdo-Álvarez, Mónica Lindín, and Fernando Diaz. 2012. An event-related potentials study of face naming: Evidence of phonological retrieval deficit in the tip-of-the-tongue state. *Psychophysiology* 49: 980–90. [CrossRef]

Burke, Deborah M., Donald G. MacKay, Joanna S. Worthley, and Elizabeth Wade. 1991. On the tip of the tongue: What causes word finding failures in young and older adults? *Journal of Memory and Language* 30: 542–79. [CrossRef]

Carruthers, Peter. 1989. Brute experience. *The Journal of Philosophy* 86: 258–69. [CrossRef]

Cleary, Anne M. 2019. The biasing nature of the tip-of-the-tongue experience: When decisions bask in the glow of the tip-of-the-tongue state. *Journal of Experimental Psychology: General* 148: 1178–91. [CrossRef] [PubMed]

Cleary, Anne M., and Alexander B. Claxton. 2015. The tip-of-the-tongue heuristic: How tip-of-the-tongue states confer perceptibility on inaccessible words. *Journal of Experimental Psychology: Learning, Memory, and Cognition* 41: 1533–39. [CrossRef] [PubMed]

Cleary, Anne M., Katherine L. McNeely-White, Shaylyn A. Russell, Andrew M. Huebert, and Hannah Hausman. 2021. The tip-of-the-tongue state as a form of access to information: Use of tip-of-the-tongue states for strategic adaptive test-taking. *Journal of Applied Research in Memory and Cognition* 10: 131–42. [CrossRef]

Cooper, Nicholas R., Rodney J. Croft, Samuel J. Dominey, Adrian P. Burgess, and John H. Gruzelier. 2003. Paradox lost? Exploring the role of alpha oscillations during externally vs. internally directed attention and the implications for idling and inhibition hypotheses. *International Journal of Psychophysiology: Official Journal of the International Organization of Psychophysiology* 47: 65–74. [CrossRef]

Delorme, Arnaud, and Scott Makeig. 2004. EEGLAB: An open source toolbox for analysis of single-trial EEG dynamics including independent component analysis. *Journal of Neuroscience Methods* 134: 9–21. [CrossRef]

Diaz, Fernando, Mónica Lindín, Santiago Galdo-Alvarez, David Facal, and Onésimo Juncos-Rabadán. 2007. An event-related potentials study of face identification and naming: The tip-of-the-tongue state. *Psychophysiology* 44: 50–68. [CrossRef] [PubMed]

Fastrich, Greta M., Tyson Kerr, Alan D. Castel, and Kou Murayama. 2018. The role of interest in memory for trivia questions: An investigation with a large-scale database. *Motivation Science* 4: 227–50. [CrossRef] [PubMed]

Fink, Andreas, and Mathias Benedek. 2014. EEG alpha power and creative ideation. *Neuroscience and Biobehavioral Reviews* 44: 111–23. [CrossRef]

FitzGibbon, Lily, Johnny King L. Lau, and Kou Murayama. 2021. The seductive lure of curiosity: Information as a motivationally salient reward. *Current Opinion in Behavioral Sciences* 35: 21–27. [CrossRef]

Funnell, Margaret, Janet Metcalfe, and Kyrana Tsapkini. 1996. In the mind but not in the tongue: Feeling of knowing in anomic patient H.W. In *Implicit Memory and Metacognition*. Edited by Lynne M. Reder. Hillsdale: Erlbaum, pp. 171–94.

Galdo-Alvarez, Santiago, Mónica Lindín, and Fernando Díaz. 2009. The effect of age on event-related potentials (ERP) associated with face naming and the tip-of-the-tongue (TOT) state. *Biological Psychology* 81: 14–23. [CrossRef] [PubMed]

Gardiner, Fohn M., Fergus I. M. Craik, and Fraser A. Bleasdale. 1973. Retrieval difficulty and subsequent recall. *Memory & Cognition* 1: 213–16.

Gollan, Tamar H., and Alan S. Brown. 2006. From tip-of-the-tongue (TOT) data to theoretical implications in two steps: When more TOTs means better retrieval. *Journal of Experimental Psychology: General* 135: 462–83. [CrossRef]

Gruber, Matthias J., Bernard D. Gelman, and Charan Ranganath. 2014. States of curiosity modulate hippocampus-dependent learning via the dopaminergic circuit. *Neuron* 84: 486–96. [CrossRef]

Heine, Marilyn K., Beth A. Ober, and Gregory K. Shenaut. 1999. Naturally occurring and experimentally induced tip-of-the-tongue experiences in three adult age groups. *Psychology and Aging* 14: 445–57. [CrossRef]

Jacobs, William J., and Janet Metcalfe. Forthcoming. The role of Curiosity1 and Curiosity2 in the emergence of insight. In *The Emergence of Insight*. Edited by Stephen M. Smith, Carola Salvi Salvi and Jennifer Wiley. Cambridge: Cambridge University Press.

James, William. 1890. *The Principles of Psychology*. New York: Holt.

Jauk, Emanuel, Mathias Benedek, and Aljoscha C. Neubauer. 2012. Tackling creativity at its roots: Evidence for different patterns of EEG alpha activity related to convergent and divergent modes of task processing. *International Journal of Psychophysiology* 84: 219–25. [CrossRef]

Jin, Christina Yi, Jelmer P. Borst, and Marieke K. Van Vugt. 2019. Predicting task-general mind-wandering with EEG. *Cognitive, Affective, and Behavioral Neuroscience* 19: 1059–73. [CrossRef] [PubMed]

Kang, Min Jeong, Ming Hsu, Ian M. Krajbich, George Loewenstein, Samuel M. McClure, Joseph Tao-yi Wang, and Colin F. Camerer. 2009. The wick in the candle of learning: Epistemic curiosity activates reward circuitry and enhances memory. *Psychological Science* 20: 963–73. [CrossRef] [PubMed]

Kikyo, Hideyuki, Kenichi Ohki, and Kensuke Sekihara. 2001. Temporal characterization of memory retrieval processes: An fMRI study of the "tip of the tongue" phenomenon. *European Journal of Neuroscience* 14: 887–92. [CrossRef]

Klimesch, Wolfgang. 1997. EEG-alpha rhythms and memory processes. *International Journal of Psychophysiology* 26: 319–40. [CrossRef] [PubMed]

Klimesch, Wolfgang. 1999. EEG alpha and theta oscillations reflect cognitive and memory performance: A review and analysis. *Brain Research Reviews* 29: 169–95. [CrossRef] [PubMed]

Klimesch, Wolfgang. 2012. Alpha-band oscillations, attention, and controlled access to stored information. *Trends in Cognitive Sciences* 16: 607–17. [CrossRef]

Koriat, Asher. 2000. The feeling of knowing: Some metatheoretical implications for consciousness and control. *Consciousness and Cognition* 9: 149–71. [CrossRef]

Koriat, Asher, and Israel Lieblich. 1974. What does a person in a "TOT" state know that a person in a "don't know" state doesn't know. *Memory & Cognition* 2: 647–55.

Koriat, Asher, and Israel Lieblich. 1977. A study of memory pointers. *Acta Psychologica* 41: 151–64. [CrossRef]

Kothe, Christian, Makoto Miyakoshi, and Antoine Delorme. 2019. clean_rawdata (Version 2.7) [Computer Software]. Available online: https://github.com/sccn/clean_rawdata/wiki (accessed on 8 May 2021).

Kuhbandner, Christof, Philipp Spachtholz, and Bernhard Pastötter. 2016. Bad things come easier to the mind but harder to the body: Evidence from brain oscillations. *Cognitive Affective Behavioral Neuroscience* 16: 768–78. [CrossRef]

Kuhbandner, Christof, Simon Hanslmayr, Markus A. Maier, Reinhard Pekrun, Bernhard Spitzer, Bernhard Pastötter, and Karl-Heinz Bäuml. 2009. Effects of mood on the speed of conscious perception: Behavioural and electrophysiological evidence. *Social Cognitive and Affective Neuroscience* 4: 286–93. [CrossRef] [PubMed]

Lau, Johnny King L., Hiroki Ozono, Kei Kuratomi, Asuka Komiya, and Kou Murayama. 2020. Shared striatal activity in decisions to satisfy curiosity and hunger at the risk of electric shocks. *Nature Human Behaviour* 4: 531–43. [CrossRef] [PubMed]

Loewenstein, George. 1994. The psychology of curiosity: A review and reinterpretation. *Psychological Bulletin* 116: 75. [CrossRef]

Loewenstein, George. 2007. *Exotic Preferences*. New York: Oxford University Press.

MacKay, Donald G., and Deborah M. Burke. 1990. Cognition and Aging: A Theory of New Learning and the Use of Old Connections. In *Advances in Psychology*. Edited by F. Gordon A. Stone and Robert West. Amsterdam: North-Holland, vol. 71, pp. 213–63. [CrossRef]

Magazzini, Lorenzo, Philipp Ruhnau, and Nathan Weisz. 2016. Alpha suppression and connectivity modulations in left temporal and parietal cortices index partial awareness of words. *NeuroImage* 133: 279–87. [CrossRef]

Maril, Anat, Anthony D. Wagner, and Daniel L. Schacter. 2001. On the tip of the tongue: An event-related fMRI study of semantic retrieval failure and cognitive conflict. *Neuron* 31: 653–60. [CrossRef]

Maril, Anat, Jon S. Simons, Josh J. Weaver, and Daniel L. Schacter. 2005. Graded recall success: An event-related fMRI comparison of tip of the tongue and feeling of knowing. *NeuroImage* 24: 1130–38. [CrossRef]

Marvin, Caroline B., Ellen Tedeschi, and Daphna Shohamy. 2020. Curiosity as the impulse to know: Common behavioral and neural mechanisms underlying curiosity and impulsivity. *Current Opinion in Behavioral Sciences* 35: 92–98. [CrossRef]

Metcalfe, Janet. 2021. The Region of Proximal Learning and Curiosity. In *In Their Own Words: What Scholars Want You to Know about Why and How to Apply the Science of Learning in Your Academic Setting*. Edited by Catherine E. Overson, Christopher M. Hakala, Lauren L. Kordonowy and Victor A. Benassi. Washington, DC: American Psychological Association's Division 2 [Society for the Teaching of Psychology].

Metcalfe, Janet, and Nate Kornell. 2005. A Regional of Proximal Learning model of metacognitively guided study-time allocation. *Journal of Memory and Language* 52: 463–77. [CrossRef]

Metcalfe, Janet, and William J. Jacobs. 2021. The two faces of curiosity in creative cognition: Curiosity1, Curiosity2 (and their interaction). In *International Handbook of Creative Cognition*. Edited by Linden J. Ball and Frederic Vallee-Tourangeau. London: Routledge.

Metcalfe, Janet, Bennett L. Schwartz, and Paul A. Bloom. 2017. The Tip-of-the-tongue (TOT) State and curiosity. *Cognitive Research: Principles and Implications* 2: 31.

Metcalfe, Janet, Bennett L. Schwartz, and Teal S. Eich. 2020. Epistemic curiosity in the Region of Proximal Learning. *Current Opinion in Behavioral Sciences* 35: 40–47. [CrossRef]

Metcalfe, Janet, Matti Vuorre, Emily Towner, and Teal S. Eich. 2022. Curiosity: The effects of feedback and confidence on the desire to know. *Journal of Experimental Psychology: General. Advance online publication.* [CrossRef] [PubMed]

Miozzo, Michele, and Alfonso Caramazza. 1997. Retrieval of lexical–syntactic features in tip-of-the tongue states. *Journal of Experimental Psychology: Learning, Memory, and Cognition* 23: 1410–23. [CrossRef] [PubMed]

Murayama, Kou, Lily FitzGibbon, and Michiko Sakaki. 2019. Process account of curiosity and interest: A reward learning model of knowledge acquisition. *Educational Psychology Review* 31: 875–95. [CrossRef]

Nagel, Thomas. 1974. What is it like to be a bat? *The Philosophical Review* 83: 435–50. [CrossRef]

Nelson, Thomas O., and Louis Narens. 1980. Norms of 300 general-information questions: Accuracy of recall, latency of recall, and feeling-of-knowing ratings. *Journal of Verbal Learning and Verbal Behavior* 19: 338–68. [CrossRef]

Neuper, Christa, and Gert Pfurtscheller. 2001. Event-related dynamics of cortical rhythms: Frequency-specific features and functional correlates. *International Journal of Psychophysiology: Official Journal of the International Organization of Psychophysiology* 43: 41–58. [CrossRef]

Nordmann, Emily, Alexandra A. Cleland, and Rebecca Bull. 2013. Cat Got Your Tongue? Using the Tip-of-the- Tongue state to investigate fixed expressions. *Cognitive Science* 37: 1553–64. [CrossRef]

Peirce, Jonathan W. 2007. PsychoPy—Psychophysics software in Python. *Journal of Neuroscience Methods* 162: 8–13. [CrossRef]

Perfect, Timothy J., and J. Richard Hanley. 1992. The tip-of-the-tongue phenomenon: Do experimenter-presented interlopers have any effect? *Cognition* 45: 55–75. [CrossRef]

Pfurtscheller, Gert. 1999. EEG Event-Related Desynchronization (ERD) and Event-Related Synchronization (ERS). In *Electroencephalography: Basic Principles, Clinical Applications and Related Fields*, 4th ed. Edited by Ernst Niedermeyer and F. H. Lopes da Silva. Baltimore: Williams and Wilkins, pp. 958–67.

Pfurtscheller, Gert, and F. H. Lopes da Silva. 1999. Event-related EEG/MEG synchronization and desynchronization: Basic principles. *Clinical Neurophysiology: Official Journal of the International Federation of Clinical Neurophysiology* 110: 1842–57. [CrossRef]

Polanyi, Michael. 1974. Genius in Science. In *Methodological and Historical Essays in the Natural and Social Sciences. Boston Studies in the Philosophy of Science*. Edited by Robert S. Cohen and Marx W. Wartofsky. Dordrecht: Springer, vol. 14. [CrossRef]

Resnik, Karmen, David Bradbury, Gareth R. Barnes, and Alex P. Leff. 2014. Between thought and expression, a magnetoencephalography study of the "tip-of-the-tongue" phenomenon. *Journal of Cognitive Neuroscience* 26: 2210–23. [CrossRef] [PubMed]

Ryals, Anthony J., Megan E. Kelly, and Anne M. Cleary. 2021. Increased pupil dilation during tip-of-the-tongue states. *Consciousness and Cognition* 92: 103152. [CrossRef] [PubMed]

Ryan, Michael P., C. Raymond Petty, and Richard M. 1982. Motivated remembering efforts during tip-of-the-tongue states. *Acta Psychologica* 51: 137–47. [CrossRef]

Schooler, Jonathan. Forthcoming. The case for curious mind wandering (mind wondering). In *The Emergence of Insight*. Edited by Stephen M. Smith, Carola Salvi Salvi and Jennifer Wiley. Cambridge: Cambridge Press.

Schwartz, Bennett L. 1999. Sparkling at the end of the tongue: The etiology of tip-of-the-tongue phenomenology. *Psychonomic Bulletin and Review* 6: 379–93. [CrossRef]

Schwartz, Bennett L. 2008. Working memory load differentially affects tip-of-the-tongue states and feeling-of-knowing judgment. *Memory & Cognition* 36: 9–19.

Schwartz, Bennett L., and Ali Pournaghdali. 2020. Tip of the tongue states: Past and future. *In Memory Quirks*. New York: Routledge, chp. 13. [CrossRef]

Schwartz, Bennett L., and Anne M. Cleary. 2016. Tip-of-the-tongue states, déjà vu and other metacognitive oddities. In *Oxford Handbook of Metamemory*. Edited by John Dunlosky and Sean Tauber. Oxford: Oxford University Press, pp. 95–108.

Schwartz, Bennett L., and Janet Metcalfe. 2011. Tip-of-the-tongue (TOT) states: Retrieval, behavior, and experience. *Memory & Cognition* 39: 737–49.

Schwartz, Bennett L., and Janet Metcalfe. 2014. Tip-of-the-tongue (TOT) states: Mechanisms and metacognitive control. In *Tip-of-the-Tongue States and Related Phenomena*. Edited by Bennett L. Schwartz and Alan S. Brown. Cambridge: Cambridge University Press.

Schwartz, Bennett L., and Leslie D. Frazier. 2005. Tip-of-the-tongue states and aging: Contrasting psycholinguistic and metacognitive perspectives. *Journal of General Psychology* 132: 377–91. [CrossRef]

Schwartz, Bennett L., and Steven M. Smith. 1997. The Retrieval of Related Information Influences Tip-of- the-Tongue States. *Journal of Memory and Language* 36: 68–86. [CrossRef]

Schwartz, Bennett L., Donald M. Travis, Anthony M. Castro, and Steven M. Smith. 2000. The phenomenology of real and illusory tip-of-the-tongue states. *Memory & Cognition* 28: 18–27.
Smith, Steven M. 1994. Frustrated feelings of imminent recall: On the tip-of-the tongue. In *Metacognition: Knowing about Knowing*. Edited by Janet Metcalfe and Arthur P. Shimamura. Cambridge: MIT Press, pp. 27–46.
Wade, Shirlene, and Celeste Kidd. 2019. The role of prior knowledge and curiosity in learning. *Psychonomic Bulletin & Review* 26: 1377–87.
Wilson, Robert C., Andra Geana, John M. White, Elliot A. Ludvig, and Jonathan D. Cohen. 2014. Humans use directed and random exploration to solve the explore–exploit dilemma. *Journal of Experimental Psychology: General* 143: 2074–81. [CrossRef] [PubMed]
Xu, Judy, and Janet Metcalfe. 2016. Studying in the region of proximal learning reduces mind wandering. *Memory & Cognition* 44: 681–95.

Journal of
Intelligence

MDPI

Article

Tracing Cognitive Processes in Insight Problem Solving: Using GAMs and Change Point Analysis to Uncover Restructuring

Mario Graf [1,*], Amory H. Danek [2], Nemanja Vaci [3] and Merim Bilalić [4]

1 Institute of Psychology, University of Klagenfurt, 9020 Klagenfurt, Austria
2 Department of Psychology, Heidelberg University, 69117 Heidelberg, Germany
3 Department of Psychology, Sheffield University, Sheffield S10 2BP, UK
4 Department of Psychology, University of Northumbria at Newcastle, Newcastle upon Tyne NE1 8ST, UK
* Correspondence: mario.graf@aau.at

Abstract: Insight problems are likely to trigger an initial, incorrect mental representation, which needs to be restructured in order to find the solution. Despite the widespread theoretical assumption that this restructuring process happens suddenly, leading to the typical "Aha!" experience, the evidence is inconclusive. Among the reasons for this lack of clarity is that many measures of insight rely solely on the solvers' subjective experience of the solution process. In our previous paper, we used matchstick arithmetic problems to demonstrate that it is possible to objectively trace problem-solving processes by combining eye movements with new analytical and statistical approaches. Specifically, we divided the problem-solving process into ten (relative) temporal phases to better capture possible small changes in problem representation. Here, we go a step further to demonstrate that classical statistical procedures, such as ANOVA, cannot capture sudden representational change processes, which are typical for insight problems. Only nonlinear statistical models, such as generalized additive (mixed) models (GAMs) and change points analysis, correctly identified the abrupt representational change. Additionally, we demonstrate that explicit hints reorient participants' focus in a qualitatively different manner, changing the dynamics of restructuring in insight problem solving. While insight problems may indeed require a sudden restructuring of the initial mental representation, more sophisticated analytical and statistical approaches are necessary to uncover their true nature.

Keywords: insight problems; representational change; problem-solving; generalized additive models; change points analysis

1. Introduction

In cognitive science, the temporal dynamics of problem-solving processes have always been an important topic of investigation. Most problems are assumed to be solved gradually, by piecing together information in order to arrive at a solution (Newell and Simon 1972). To investigate these problems, several tools have been developed, which allow for the observation of each step of the problem-solving process (e.g., Tower of Hanoi, Hobbits and Orcs problem). In the case of "insight problems", the solution often comes seemingly out of nowhere (Duncker 1945), despite the problem appearing unsolvable just a moment earlier. To be solved, insight problems are thought to require a fundamental, sudden change in the way the problem is perceived, a process referred to as restructuring or representational change (Ohlsson 1992; Wertheimer 1925). The restructuring from the initial, incorrect mental representation to the correct one is the key component in modern theories such as representational change theory (RCT) (Knoblich et al. 1999; Ohlsson 1984, 1992, 2011).

Although the sudden nature of the underlying restructuring process is a main theoretical assumption about insight, the evidence for this claim is inconclusive. Ohlsson (1992) even hypothesized that "the sudden appearance of the complete solution in consciousness is an illusion caused by our lack of introspective access to our cognitive processes (...)"

(p. 17). To truly understand the temporal nature of insight, the cognitive component of insight (restructuring) must be examined with appropriate tools. Observing changes in solvers' mental problem representation is a methodological and statistical challenge, which is addressed in the present work. Among the reasons for this lack of clarity is that many measures of insight rely solely on the solvers' subjective experience of the solution process. Using matchstick arithmetic problems, we demonstrate that it is possible to objectively trace problem-solving processes.

We first review the research on representational change, focusing on the experimental designs. After that, we describe a novel analytical approach that improves upon previous attempts. Finally, and arguably most importantly, we show that this analytical approach needs to be combined with appropriate statistical tools in order to work properly. We demonstrate the feasibility of this approach by re-analyzing eye-tracking data from an already published study (Bilalić et al. 2019). The paper is accompanied by an online supplement, with technical details, such as data and code for the analysis, which is freely available at https://osf.io/pwuhs/?view_only=7c52bda4e6fa481e826e5d7570b6ef3 (accessed on 25 April 2023).

1.1. Temporal Dynamics of the Restructuring Process

In 1994, Durso and colleagues conducted an early study on the temporal dynamics of insight problem solving. They asked participants to rate the relatedness of word pairs in a word puzzle and found that, on average, solution-relevant pairs were rated as increasingly similar as participants approached a solution. The authors concluded that "[l]ike dynamite, the insightful solution explodes on the solver's cognitive landscape with breathtaking suddenness, but if one looks closely, a long fuse warns of the impending reorganization" (Durso et al. 1994, p. 98). Novick and Sherman (2003; Experiment 2) provided similar evidence. They asked participants to indicate within a short time window (250 ms after stimulus offset) whether presented anagrams were solvable. They found that, although participants could not find the solution within the allotted time, they were increasingly better at differentiating between solvable and unsolvable anagrams as the presentation time of the anagrams increased. The authors concluded that solvers gradually accumulate information relevant for solving the anagrams.

Several studies have focused on the concept of restructuring in insight problem solving, but have typically not measured the dynamics of the solving process (e.g., Ash et al. 2012; Ash and Wiley 2006, 2008; Fleck and Weisberg 2013; MacGregor and Cunningham 2009). However, a number of studies have attempted to measure the temporal dynamics of restructuring, using different methods to acquire trace data. Some used repeated ratings of problem elements, either regarding their similarity (Durso et al. 1994) or with regard to their relevance for the solution (Cushen and Wiley 2012; Danek et al. 2020). Others recorded eye movements (Ellis et al. 2011; Knoblich et al. 2001; Bilalić et al. 2019; Tseng et al. 2014) or employed solvability judgments (Novick and Sherman 2003). In some of these studies, both incremental and sudden solution patterns were found (Cushen and Wiley 2012; Danek et al. 2020; Novick and Sherman 2003), whereas other studies found only incremental patterns (Durso et al. 1994).

1.2. Eye Movements and Matchstick Arithmetic Problems

Here, we will take a closer look at using eye movement recordings to measure the temporal dynamics of restructuring in insight problems (for a comprehensive overview on eye movements, please see Holmqvist et al. 2011). In general, eye movements provide an objective measure of cognitive processes, as they are closely linked to attention (e.g., Just and Carpenter 1976; Rayner 1995; Reingold et al. 2001). Specifically, eye fixations reveal when people pay attention to certain features of a problem and for how long. More importantly, eye tracking is particularly useful when participants might not remember or even concurrently report that they are paying attention to these elements (Bilalić and McLeod 2014; Bilalić et al. 2008, 2010; Kuhn and Land 2006; Kuhn et al. 2009). This is particularly relevant in the case of insight problems, where it is possible that people are not aware of the dynamics of their solution process.

We use the matchstick arithmetic problems introduced by Knoblich et al. (1999). Matchstick arithmetic problems are suitable for investigation with eye tracking, as was powerfully demonstrated by the seminal study of Knoblich et al. (2001). A matchstick arithmetic problem consists of a false arithmetic statement written using Roman numerals, arithmetic operators, and equal signs, all formed using matchsticks (Knoblich et al. 1999, 2001; see also Figure 1 below). The task is to transform the false arithmetic statement into a true statement by moving only a single stick. Four types of matchstick arithmetic problems have been defined with varying levels of difficulty, depending on the constraints that need to be relaxed and the tightness of the chunks that need to be decomposed. These problem types were theoretically derived from the representational change theory (Ohlsson 1992) and have been empirically confirmed (Knoblich et al. 1999; Öllinger et al. 2006, 2008). The use of matchstick arithmetic problems enables us to build on a well-researched task domain. It is known which problem type should elicit the restructuring process (Knoblich et al. 1999; Öllinger et al. 2006, 2008), and it is possible to contrast it with a type which requires no restructuring. Additionally, based on Knoblich's study (2001), predictions about eye movement patterns can be made. Furthermore, the matchstick arithmetic domain is well suited for eye tracking because each problem consists of individual matchsticks that do not overlap, allowing for precise differentiation of fixations. In other words, we can determine at any point in time which aspect of the problem is attended to.

Figure 1. Matchstick arithmetic problem. Participants are required to transform the false arithmetic statement to a true statement by moving a single matchstick. This problem requires restructuring, because the initial assumption that only the matchsticks from values can be manipulated needs to be changed. In this case, the operator "+" can be decomposed and its vertical matchstick moved to make another "=" sign (VI = VI = VI). The "+" sign is the critical element that needs to be changed for solution.

Knoblich et al. (2001) investigated constraint relaxation type problems, which are considered to require restructuring; see Figure 1 for an example. They found that for this problem (constraint relaxation type), both solvers and non-solvers examined the values in the beginning and spent most of their time doing so. This can be seen as an indication that participants were using an initial incorrect problem representation, triggered by previous knowledge, where only values can be changed. Only in the final third of the problem-solving period did later solvers change their mental representation, as demonstrated by their eye movements. Solvers started to pay attention more to the operators and less to the values. In contrast, non-solvers remained stuck in their initial representation, as they continued to attend to values rather than to operators. Similar results for the same problem were found by another eye-tracking study (Tseng et al. 2014).

The Knoblich et al. (2001) study provides strong evidence for the claim that in problems that require constraint relaxation, a restructuring of the problem representation took place. However, it did not answer the question of whether this change was a sudden or a gradual one. In the final third of the allotted time, solvers paid attention to the important but previously ignored features, which could be interpreted as a result of sudden restructuring. It is nevertheless not that clear, since the final period may have lasted minutes, given that they took around five minutes to solve the problem. Thus, the restructuring might have been a continuous process over time. On the other hand, an eye-tracking study on anagrams by Ellis et al. (2011; see also Ellis and Reingold 2014) found that participants started disregarding the irrelevant problem elements several seconds before they came up with the solution. The viewing times on that problem elements were decreasing gradually. Most intriguingly, both participant groups, those who experienced pop-out insight-like solutions and those who did not, displayed the same gradual accumulation of solution knowledge.

1.3. Metacognitive Processes and Insight Problems

There is evidence that the problem-solving process benefits from hints (e.g., Bowden 1997; Bilalić et al. 2019; Ammalainen and Moroshkina 2021; Becker et al. 2021; Korovkin and Savinova 2021; Spiridonov et al. 2021). This is the case even when hints were unreportable; that is, hints even work when presented briefly below the threshold of consciousness. Ammalainen and Moroshkina (2021) found evidence that hints can influence the problem-solving ability, which can be both, positive and negative. In a positive way, hints which are helpful to find the solution increase solution rates. On the other hand, misleading hints can negatively affect solution rates by distracting problem solvers and leading to a decrease in their success rate. In our paper (Bilalić et al. 2019), we also provided hints when participants were unable to find the correct solution after a certain time.

These hints serve two purposes: a practical and a theoretical one. On a practical level, they provide an additional check on the main assumption behind the restructuring process. On a theoretical level, they serve as explicit clues that tap into metacognitive processes (Takeuchi et al. 2019; Metcalfe and Shimamura 1994). Hints make participants aware of important aspects in the problem, drawing their attention towards elements that may have been neglected. They also change participants' knowledge about the problem, potentially affecting the way they solve insight problems (Bowden 1997; Bilalić et al. 2019; Korovkin and Savinova 2021).

2. Methods

The present work is a re-analysis of our paper (Bilalić et al. 2019). In our paper, we also combined solving of insight and non-insight problems with eye tracking. We presented first a non-insight matchstick problem and then the matchstick insight problem depicted here (see Figure 1) to 61 participants (5 male; M_{age} = 22.8; SD_{age} = 6.5). The study was designed to take into account the methodological issue discussed in the previous section. It built upon previous attempts that utilized more time periods and sometimes presented the last 5 or 10 s separately (see also Bilalić et al. 2008, 2010, 2014). In the 2019 study, we provided a more fine-grained temporal analysis of the solution process by using ten time periods of equal length for our eye movement analysis[1] (for more information, please refer to Bilalić et al. 2019). We demonstrated that the restructuring is a gradual process on the insight problem as the solvers started paying attention to the important aspects of the problem long before they found the solution. Here, we provide another set of data where the jump is sudden; that is, the solvers started paying attention to the important aspects immediately before they found the solution (as reported by Knoblich et al. 2001). This is done to illustrate (1) how classical ways of analyzing data, such as ANOVA, are inappropriate for discovering the sudden changes, and (2) how other non-linear approaches are required.

We expected that all participants would initially focus on the values. Solvers would shift their attention towards the critical element (the "+" operator), while non-solvers would remain fixated on the values. The first question of interest is whether the representational shift in eventual solvers will be sudden or rather incremental. The second question of interest is whether the explicit cue, that is, the hint, will produce a sudden rearrangement of attention towards the critical elements (here "+", but also "=" because "=" is also an operator). In our design, we included hints for participants who had not solved the problem within five minutes. The hint provided at this point was 'You can change the operators, too.' We were interested in whether the hints change the dynamics of problem solving, specifically whether the solution process remains sudden even after receiving an explicit cue.

3. Results

The problem proved difficult as only 34% found the solution. After the hint was provided, an additional 11% of participants were able to find the solution. We present the eye data analysis below, with a particular focus on the critical element of the problem, the plus sign (+). Additionally, when analyzing the impact of hints, we also focused on the equal sign (=) as the hints should also affect the attention drawn to this operator through metacognitive control. For analysis of other problem elements, please refer to the supplementary materials.

3.1. Is Insight Sudden or Incremental? (Solvers vs. Non-Solvers: First 5 Min Analysis)

In Figure 2, raw data and means for each bin of the critical element for the first five minutes are presented.[2] The solving pattern follows the typical sudden pattern, where there is not much difference between eventual solvers and non-solvers with regard to the time spent on the critical element (+) until the end of the first five minutes. Solvers suddenly increase their dwell time just before announcing the solution, while non-solvers continue to observe the critical element sporadically until the end of the solving period.

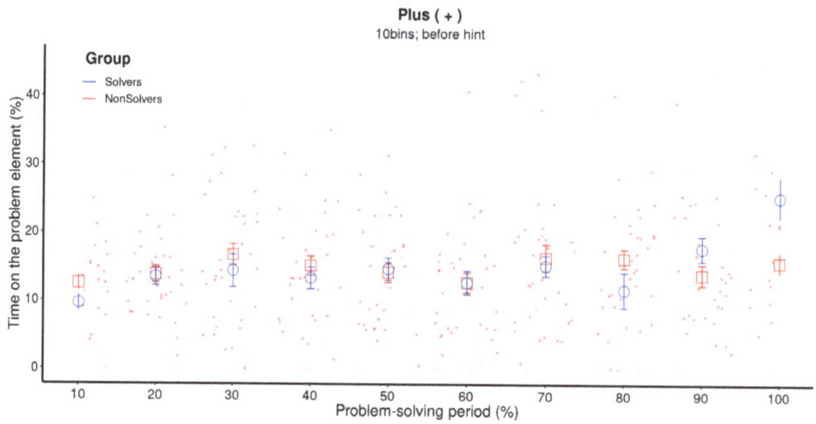

Figure 2. Raw data and means for each bin of the critical element (+). The raw data represents every data point of each participant over the entire problem-solving period. The problem-solving period was divided in 10 proportional bins, each representing 10% of the total problem-solving time. The error bars represent the 68% confidence interval. This figure illustrates a nonlinear increase in the amount of time that solvers spend on the critical element. In the case of solvers, the 100% bin means the participant provided a solution.

The crucial question is how to analyze the temporal changes presented in Figure 2. The traditional method, which we had chosen in our previous paper (Bilalić et al. 2019), is to use an analysis of variance (ANOVA) where the bins and groups are factors that predict the amount of time spent on the critical element. However, ANOVA not only requires a completely balanced dataset, but it also ignores the clustered nature of data (van Rij et al. 2020). Furthermore, it is based on linear regression, which is not suitable for capturing sudden attentional shifts, which are nonlinear in nature. In order to capture the sudden shift as depicted in Figure 2 (the 100% bin for the solvers), ANOVA would need to adjust the linear trend throughout the whole problem-solving period. In other words, a sudden trend may appear as an incremental one as ANOVA adjusts by increasing previous periods (see Figure 3, left panel).

ANOVA can be expressed as linear regression, where an additional quadratic polynomial term is included next to the linear one, in an attempt to capture the shift. However, even in this case, the predicted shift by the ANOVA model would begin earlier, namely at the 80% bin, than it does in the raw data (see Figure 3, right panel). The general limitation of linear regression, with or without polynomial terms, is that it heavily relies on previous trends. If the change is sudden, the previous time periods will also be adjusted accordingly.

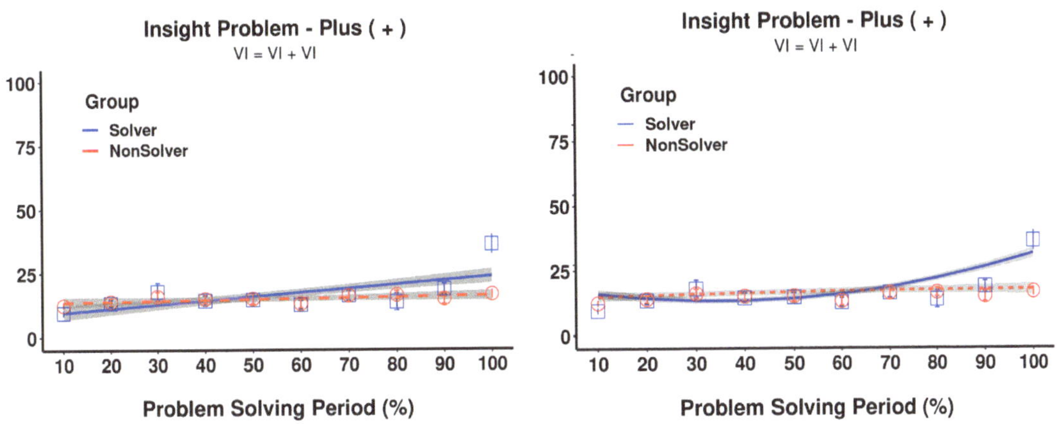

(a) (b)

Figure 3. Estimated model means based on (**a**) ANOVA with linear term; (**b**) ANOVA with both linear and quadratic terms. Y-Axis: Time on the problem element (%). Please refer to supplementary material for the detailed analysis.

One way around this problem is generalized additive (mixed) modeling (GAM). These models are specifically designed to handle nonlinear relationships, as they are data-driven and use non-linear mixed-effects regression (van Rij et al. 2020). A key benefit of GAMs is that they do not require the user to specify the shape of the nonlinear regression line, as the model determines this based on the data. However, while GAMs have a high level of flexibility in modeling nonlinear changes in time series data, they only allow for the exploration of changes in the function and do not provide parametric estimates such as standard error of estimate or its impact on predictive accuracy of the model. More specifically, GAMs do not provide parametric estimates, which means that they do not give us a set of parameters that describe the shape of the nonlinear function. However, the present work intends to demonstrate the advantages and downsides of the available analysis tools in question, which is why GAMs are included here.

Arguably the most reliable way of checking the assumption of suddenness is the use of change point analysis, which looks for significant deviance from previous trends (Raftery and Akman 1986). Unlike the standard regression analysis (ANOVA) and nonlinear GAMs, change point regression estimates the moment of the function inflection. In other words, it includes the possibility to estimate additional parameters, such as intercept and slope of regression, time point when the function changes, and how the intercept and/or slope of regression changes (see the figures of the MCP analysis for illustrations). This makes the technique particularly valuable in detecting increasing patterns as one would expect several points of change in the attentional pattern on the way towards the solution. In this instance, we use the one implemented in the Multiple Change Points package (MCP; Lindeløv 2020).

Below, we address the three main questions using both GAM and MCP analysis. In the supplemental material, we provide the model-estimated values for each case, which include the results and, in the case of the MCPs, how well the model fits the data and which model was used. We begin with the GAM analysis of solvers and non-solvers for the first five minutes to determine whether the insight is sudden or incremental. Figure 4 provides the estimated trend lines for both solvers and non-solvers, as well as the time periods (shaded in orange) where the difference between the two is statistically significant. The model estimates closely follow the raw data (see Figure 2), and the difference between solvers and non-solvers is indeed significant at the beginning of the solving phase, as well as at the 90% bin and the 100% bin.

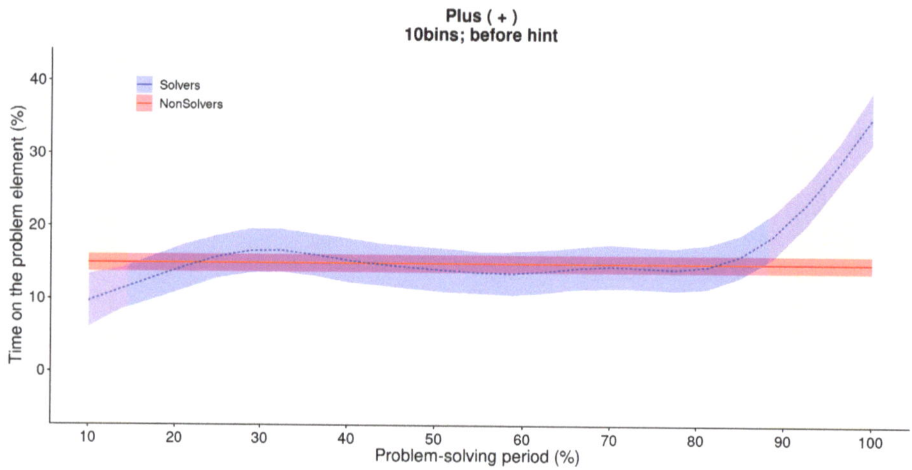

Figure 4. GAM: the difference between two estimated trend lines for solvers and non-solvers of the critical element (+). This figure illustrates that the GAM also found a nonlinear increase in the amount of time that non-solvers spend on the critical element. The orange area determines where the differences between solvers and non-solvers were significant.

Figure 5 shows the results of the MCP analysis for the same data as the GAM above. Similarly to the GAM, the MCP analysis identified a change point around the 90% bin for the solvers, which captures an attentional shift they made. While some non-solvers also shifted their attention towards the "+" sign at the end, it was not as clear as in the case of the solvers.

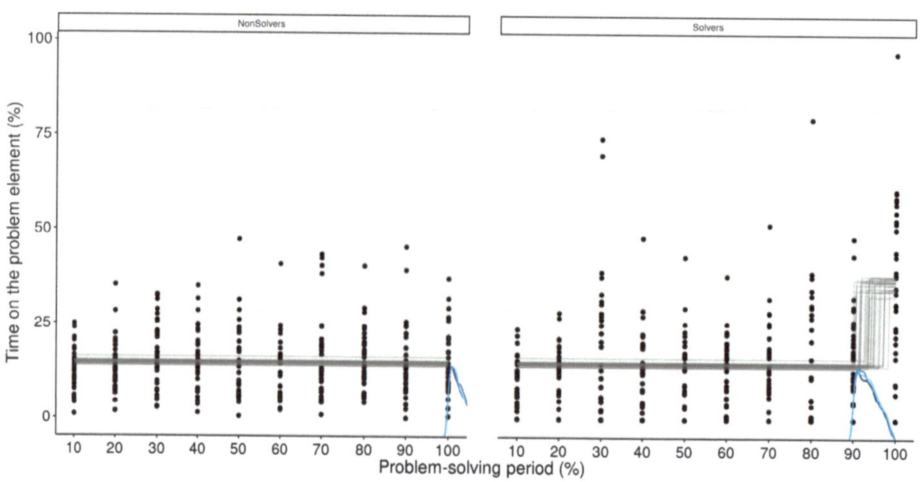

Figure 5. MCP analysis of the critical element (+) for non-solvers and solvers. This figure illustrates every data point of each participant over the problem-solving period. Lines at the bottom of the figure illustrate the posterior density (estimated likelihood) of the change point for each MCMC chain. There is a nonlinear increase in the amount of time that solvers spend on the critical element.

3.2. Do Explicit Cues Rearrange Attentional Distribution? (An Immediate Change after the Hint)

Figure 6 illustrates the impact of providing an explicit hint to the non-solvers from the first five minutes (presented here as a single group; solvers from the first five minutes are not included in this graph). The attentional shift from values towards operators, "+" and "=", is substantial immediately after the hint. The operator "=" is attended to twice as much immediately after the hint than before. The change for "+" is slightly less dramatic at first (only 4%), but by the 20% bin, the dwell time has doubled in comparison to before the hint was provided. Note that only non-solvers are shown here, since solvers did not receive any hints.

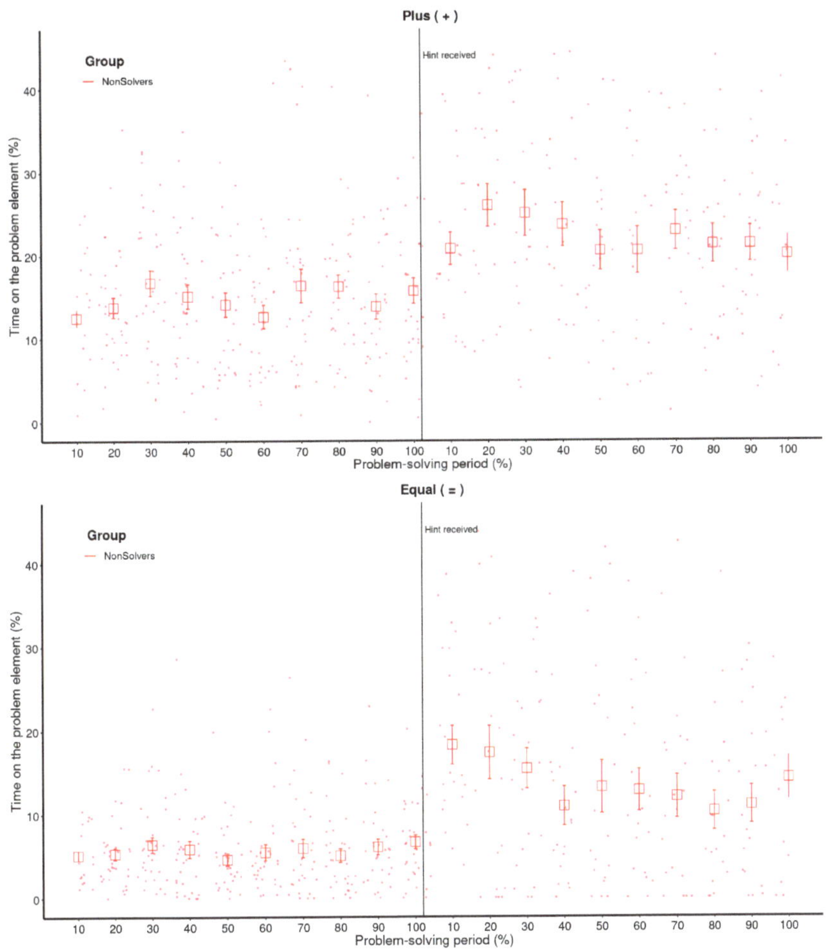

Figure 6. Raw data and means for each bin of the operators ("+" upper panel; "=" lower panel) in the period before and after the hint. The raw data represents every data point of each participant (non-solvers only) over the problem-solving periods. Each of both problem-solving periods (before and after the hint was provided) were divided in 10 proportional bins, each representing 10% of the total problem-solving time. It is necessary to view the problem-solving periods as distinct periods; therefore, each period is labeled from beginning to end (10% to 100%) to differentiate them. The error bars represent the 68% confidence interval. This figure illustrates the attentional shifts from values (mostly attended to before the hint) towards operators (attended to after the hint was given).

The GAM analysis effectively captures the attentional shift, as depicted in Figure 7. However, it predicts that the change occurs prior to the hint being provided, starting already at the 90% bin, which is not a correct reflection of the actual data. While GAM is considerably more flexible than regressions with polynomial terms, the same problem of interdependence of neighboring phases remains. The shift caused by the explicit cue is so drastic that the GAM needs to adjust the increase to begin earlier in order to account for it.

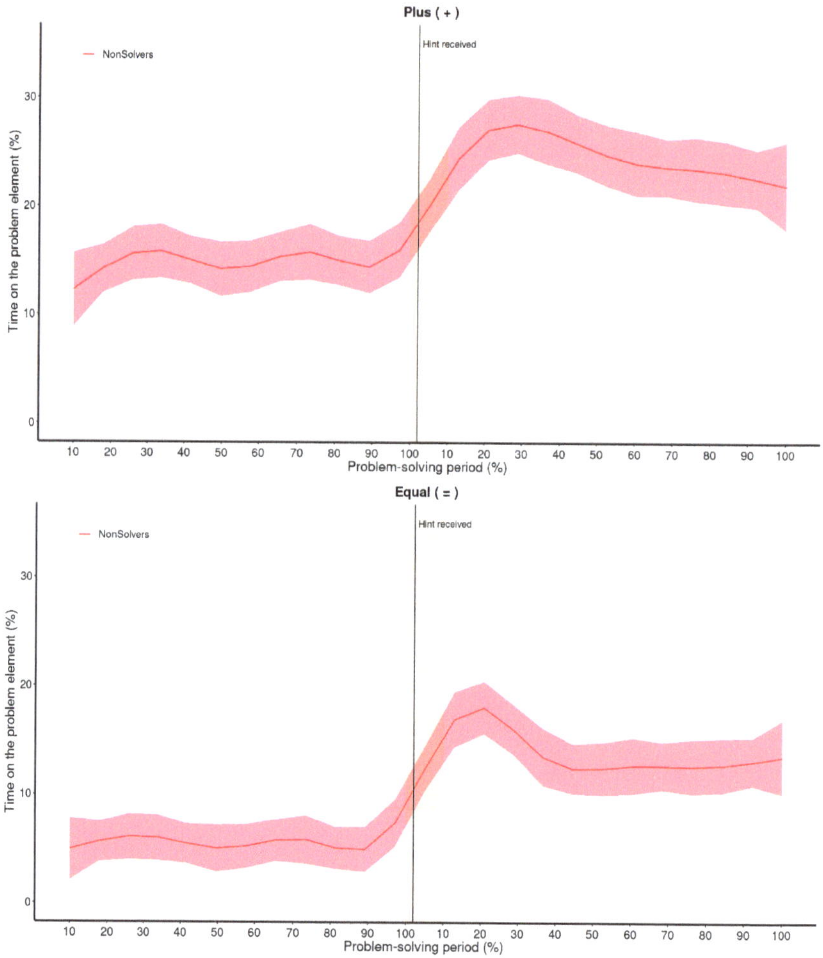

Figure 7. GAM: estimated trend line for non-solvers of the critical element (+; upper panel) and the other operator (=; lower panel). This figure illustrates that the GAM also found a nonlinear increase in the amount of time that non-solvers spend on the critical element after receiving a hint. The orange area indicates where there is a significant shift in attention.

In contrast, the switch points of the MCP analysis correctly capture where the change in attention allocation happens (see Figure 8).

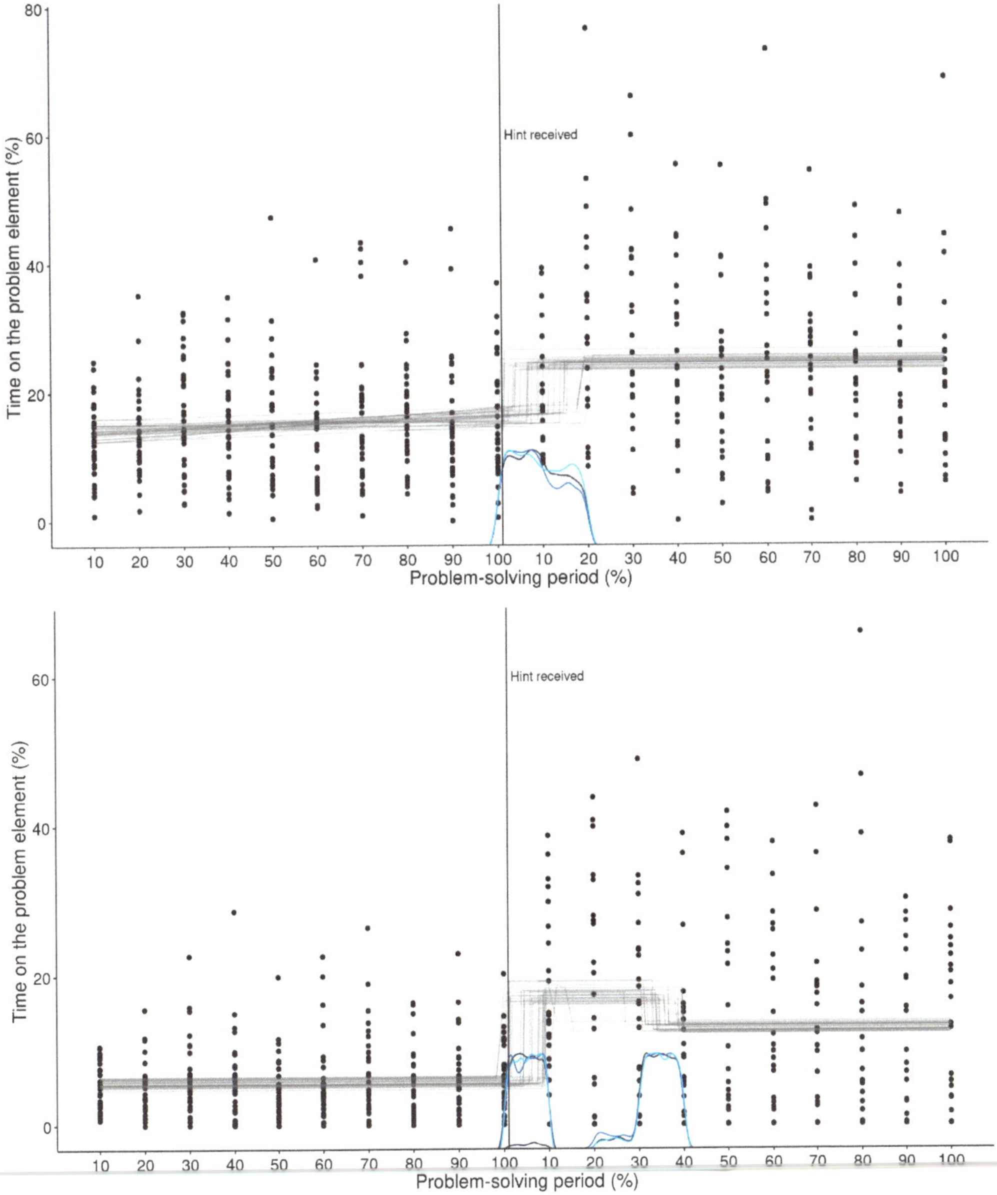

Figure 8. MCP analysis of the critical element (+; upper panel) and the other operator (=; lower panel). This figure illustrates that the switch points of the MCP analysis correctly captures where the shift in attention happens.

3.3. Does Metacognition Influence Insight Problem Solving? (Solvers vs. Non-Solvers after the Hint)

The final question we aimed to address was whether the explicit cue, and the additional knowledge about the problem associated with it, would alter the way the problem was solved. Figure 9 indicates that both solvers and non-solvers maintain the level of attention on the critical aspects throughout the problem-solving period, which is a direct consequence of the explicit cue. However, this was not sufficient for finding the solution. The eventual solvers initially shifted their attention to "=" around the 30% bin, but starting from the 50% bin, they increasingly focused on "+". This means that at this point in time, the solvers may have realized that the "+" symbol was the critical element they needed to solve the problem. Consequently, they gathered more information about the symbol by attending to it more closely.

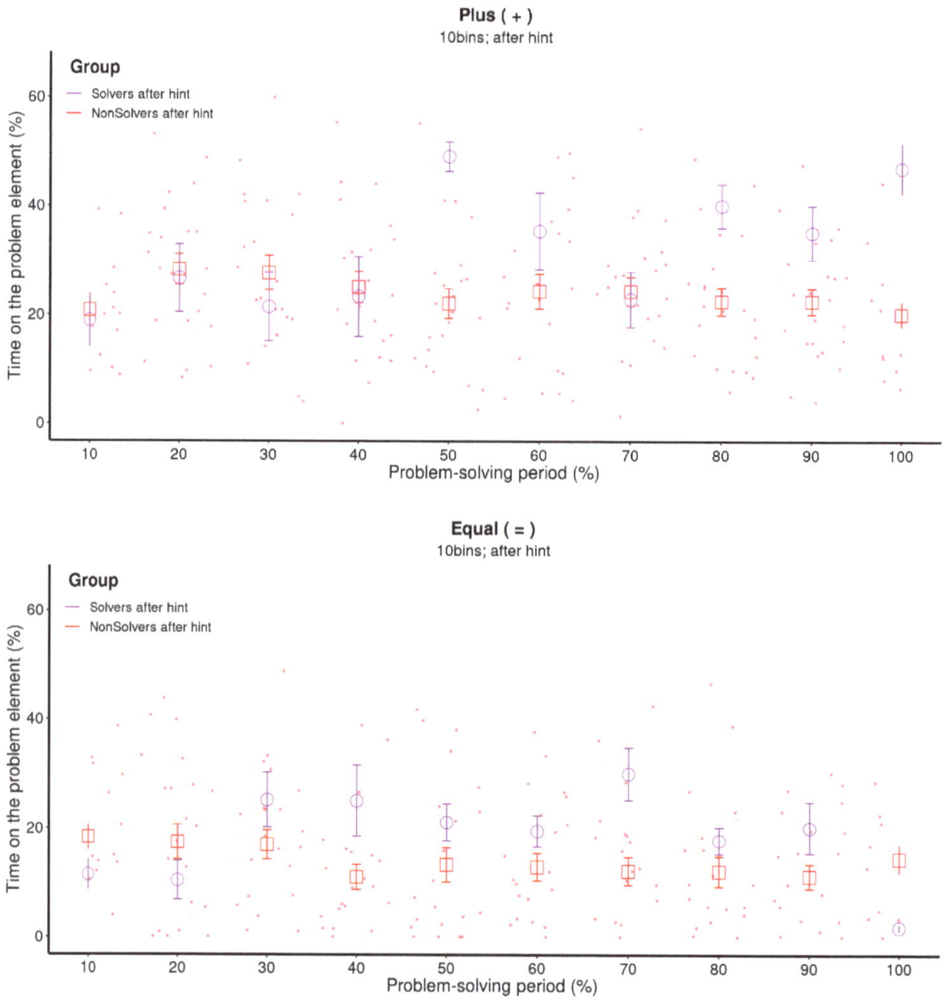

Figure 9. Raw data and means for each bin of the critical element (+; upper panel) and the other operator (=; lower panel) after the hint was provided. The raw data represent every data point of each participant over the remaining problem-solving period after the hint was given. The error bars represent the 68% confidence interval. This figure illustrates a nonlinear increase or decrease in the time solvers spend on the critical element.

This incremental pattern of solving is well captured by GAMs, as Figure 10 illustrates. While the non-solvers attended to the critical "+" operator consistently over the entire problem-solving period, but on a rather low level of 25% of their time, solvers gradually increased their attention towards it. Significant differences were found in the middle and the end of the problem-solving period. This was also the case for the other operator (=). Non-solvers attended to "=" in a consistent manner throughout the problem-solving period, while the solvers attended to "=" more in the middle of the problem-solving period and less at the very end of it, probably because they were then already focusing more on the "+" sign which needs to be changed for a solution.

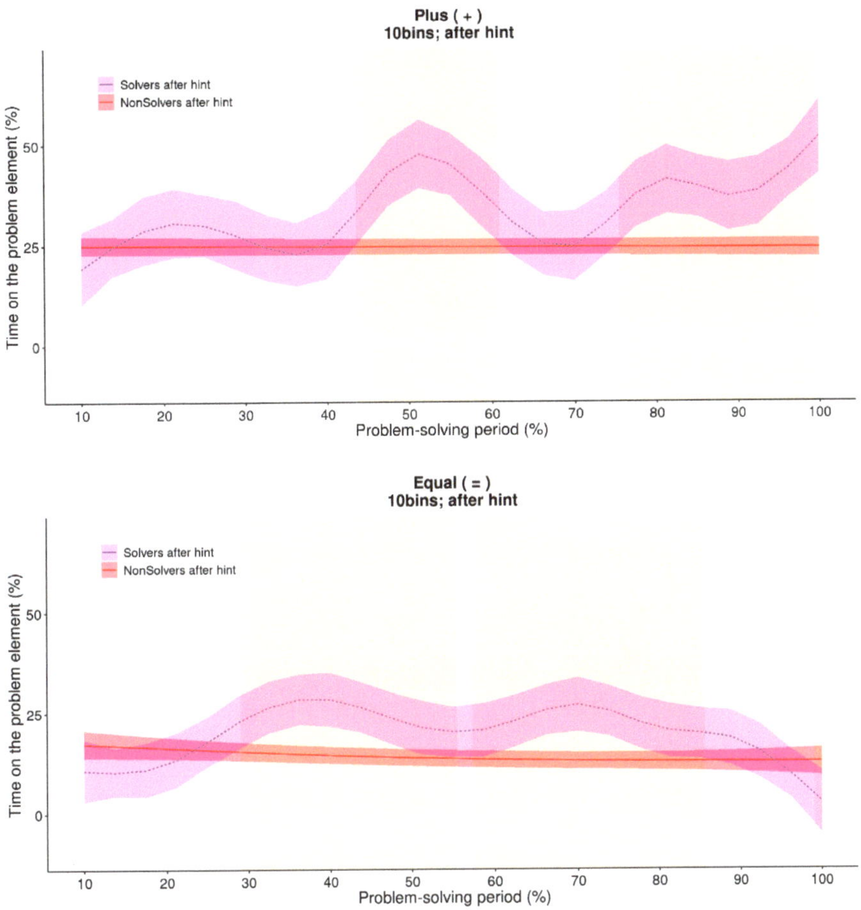

Figure 10. GAM: the difference between two estimated trend lines for solvers and non-solvers of the critical element (+; upper panel) and the other operator (=; lower panel). This figure illustrates that the GAM also found a nonlinear increase in the time solvers spend on that particular element. The orange area in the figure indicates regions where there are statistically significant differences between the attention patterns of solvers and non-solvers.

Figure 11 illustrates that the attentional shifts after receiving a hint are effectively captured by the MCP analysis. Again, non-solvers attended to the "+" operator on a consistently low level throughout the entire problem-solving process, while solvers attended to the "+" operator more and more. The same trend is observed for the "=" operator. Non-solvers attended to it less, while solvers shifted their attention to it in the middle of the problem-solving process. Towards the end of the problem-solving process, the data suggest that solvers became aware that the "=" operator was not as important for solving the problem and began to focus more on the "+" operator.

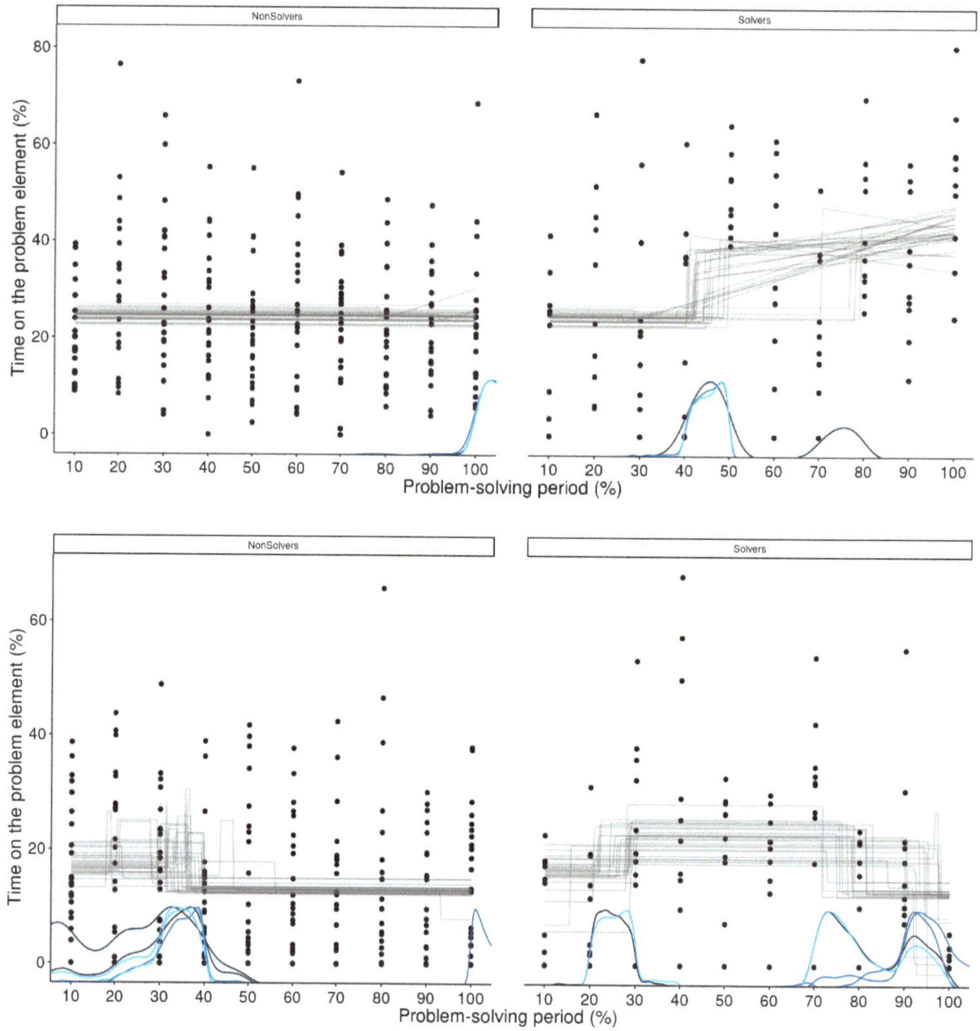

Figure 11. MCP analysis of the critical element after the hint (+; upper panel) and the other operator (=; lower panel) for non-solvers and solvers. This figure demonstrates that the switch points of the MCP analysis correctly captures the incremental pattern of solving for the critical element (+). It also demonstrates that after the hint, the non-solvers attended to the noncritical element (=) more in the beginning but not the critical element (+). As the GAMs showed already, the non-solvers attended to the critical element (+) in the same way throughout the whole problem-solving period.

4. Discussion

We have demonstrated that recording eye movements is a valuable method for gaining insight into complex cognitive processes, including mental restructuring in insight problems. It is also an adequate tool for investigating attentional shifts after receiving hints. However, it is important to use eye movement recording with appropriate analytical approaches. Our results show that it is necessary to conduct a more fine-grained analysis of the eye movement data to capture the temporal dynamics of the problem-solving process. This is particularly relevant for insight problems such as the one used here, which are believed to feature a sudden change in eye movement patterns reflecting a change in mental representation.

We were able to identify the point at which solvers and non-solvers start to differ in their attentional patterns by dividing the problem-solving period into ten equal bins. The temporal resolution of the problem-solving period is one aspect, but it is also important to choose an appropriate statistical procedure. We have demonstrated that nonlinear statistical models, such as GAM and MCP, can effectively capture the sudden change that is a hallmark of insight problem solving. The GAM analysis can effectively capture the attentional shift; however, it predicts that the change occurs prior to the correct reflection of the actual data. While GAM is considerably more flexible than regressions with polynomial terms, the same problem of interdependence of neighboring phases remains. The shift caused by the explicit cue is so drastic that the GAM needs to adjust the increase to begin earlier to account for it. In contrast, the change points of the MCP analysis correctly capture where the change in attention allocation happens. A change point is a time point where the statistical properties of a time series change abruptly. However, in contrast to GAMs, one needs a priori knowledge about the number of change points and the form of the segments in between (Lindeløv 2020). Therefore, one might decide from case to case which statistical procedure is appropriate.

Our example illustrates the importance of considering theoretical assumptions when choosing analytical and statistical procedures. The restructuring of mental representations is a key concept in theories of insight (Knoblich et al. 1999; Ohlsson 1984, 1992, 2011). It is a nonlinear process in essence, which can be operationalized as a sudden burst of attention to the relevant aspects of a problem (Bilalić et al. 2019). The shift inevitably deviates significantly from participants' previous problem solving. Seen as a part of the overall problem-solving continuum, the sudden shift is difficult to capture with linear statistical procedures. Only truly nonlinear statistical procedures can appropriately capture the sudden nature of representational change.

Providing explicit hints typically alters the dynamics of problem solving. It is obvious that the given hints were effective, as participants' patterns of attention show a drastic change, which is very well captured by both GAM and MCP. However, it is important to note that the eventual solvers, after receiving the hint, exhibit a gradual, incremental shift, with increasing attention to the main elements during the problem-solving period. In contrast, non-solvers display an immediate burst of refocusing following the hint, but subsequently, their attention to the important aspects diminishes.

Both the analytical procedure for capturing the temporal resolution and the nonlinear statistical procedures can be easily extended beyond eye movements to other tracing methods. For example, "importance-to-solution" ratings of individual problem elements that are made repeatedly during the solving process (Durso et al. 1994; Cushen and Wiley 2012; Danek et al. 2020; Danek and Wiley 2020) often reveal patterns of sudden change which could be effectively captured by GAMs and MCPs. Similarly, "Feelings-of-Warmth" that are used to assess metacognitive knowledge about solution progress (Kizilirmak et al. 2018; Hedne et al. 2016; Pétervári and Danek 2020) are another suitable candidate for nonlinear modeling with GAMs. Other tracing methods, such as mouse-tracing data (Loesche et al. 2018; van Rij et al. 2020), think-aloud protocols (Gilhooly et al. 2010; Schooler et al. 1993; Blech et al. 2020), or even self-reports (Fedor et al. 2015), are also better modeled with GAMs than with commonly applied linear methods, even if they are more appropriate than the classical ANOVA.

5. Conclusions

Our results indicate that for insight problems, the restructuring process leaves a discernible trace of suddenness. Eye movements suggest that just prior to solving the problems, participants shift their focus from elements that constituted the initial problem representation to those crucial for the solution. Our results also demonstrate that receiving hints leads to attentional shifts towards critical aspects, which in turn facilitates the generation of a correct solution. However, in order to accurately capture the sudden shift in attention, a combination of the appropriate methodological approach and statistical procedure is necessary. These nonlinear processes are best captured by nonlinear statistical procedures, such as GAMs and MCPs.

Supplementary Materials: The following supporting information can be downloaded at: https://osf.io/pwuhs/?view_only=7c52bda4e6fa481e826e5d7570b6ef34.

Author Contributions: Conceptualization, M.B. and M.G.; methodology, M.B. and M.G.; software, M.G.; validation, M.B., N.V., A.H.D., and M.G.; formal analysis, M.G. and N.V.; investigation, M.G.; data curation, M.G.; writing—original draft preparation, M.G.; writing—review and editing, M.B., A.H.D., N.V., and M.G.; visualization, M.G., N.V., and M.B.; project administration, M.G. and M.B.; funding acquisition, M.G. All authors have read and agreed to the published version of the manuscript.

Funding: This research was funded by Talent Austria der OeAD-GmbH, finanziert aus Mitteln des österreichischen Bundesministeriums für Wissenschaft, Forschung und Wirtschaft (BMWFW), grant number ICM-2017-07423 given to the first author.

Institutional Review Board Statement: Since this is a re-analysis of our paper (Bilalić et al. 2019), please refer to the original paper for the Institutional Review Board Statement.

Informed Consent Statement: Informed consent was obtained from all subjects involved in the study.

Data Availability Statement: Technical details, such as data and code for the analysis, is available at https://osf.io/pwuhs/?view_only=7c52bda4e6fa481e826e5d7570b6ef34 (accessed on 25 April 2023).

Acknowledgments: The help and cooperation from participants is greatly appreciated, as is Matthew Bladen's contribution in preparing the text.

Conflicts of Interest: The authors declare no conflict of interest.

Notes

1. The length of time taken to solve (or not solve) a problem is different from person to person, meaning that one cannot compare the eye tracking data directly between people. For example, some may need only 45 s to solve the problem, whereas others need four minutes to find a solution. In consequence, the data must be transformed in order to be able to compare the data between people properly. While the problem-solving period can be extended by adding more time phases, it is important to note that the duration should not be prolonged beyond a certain point. Utilizing too many time frames may leave too little data (e.g., a 10-second trial should not be divided into 100 bins, as each bin will have the duration of only 100 ms). This can lead to distorted eye movement patterns, masking the underlying effects present before the data were binned. On the other hand, choosing too few bins may not capture the full temporal dynamics of the problem-solving process. In either case, ANOVA is not suitable for analyzing a large number of problem-solving periods, unlike GAM and multiple change point analysis, which can easily accommodate a large number of time frames. MCP analysis is another adequate tool for this type of analysis as it can capture the shift of attention. However, in contrast to GAMs, one needs a priori knowledge about the number of change points and the form of the segments in between (Lindeløv 2020).
2. Please note that the data presented here are simulated to represent a sudden shift, which is difficult to capture by classical analyses. The original data in Bilalić et al. (2019) indicate a gradual shift.

References

Ammalainen, Artur, and Nadezhda Moroshkina. 2021. The effect of true and false unreportable hints on anagram problem solving, restructuring, and the Aha!-experience. *Journal of Cognitive Psychology* 33: 644–58. [CrossRef]

Ash, Ivan K., and Jennifer Wiley. 2006. The nature of restructuring in insight: An individual-differences approach. *Psychonomic Bulletin & Review* 13: 66–73.

Ash, Ivan K., and Jennifer Wiley. 2008. Hindsight bias in insight and mathematical problem solving: Evidence of different reconstruction mechanisms for metacognitive versus situational judgments. *Memory & Cognition* 36: 822–37. [CrossRef]

Ash, Ivan K., Benjamin Jee, and Jennifer Wiley. 2012. Investigating insight as sudden learning. *The Journal of Problem Solving* 4: 150–76. [CrossRef]

Becker, Maxi, Simone Kühn, and Tobias Sommer. 2021. Verbal insight revisited—Dissociable neurocognitive processes underlying solutions accompanied by an AHA! experience with and without prior restructuring. *Journal of Cognitive Psychology* 33: 659–84. [CrossRef]

Bilalić, Merim, and Peter McLeod. 2014. Why good thoughts block better ones. *Scientific American* 310: 74–79. [CrossRef]

Bilalić, Merim, Peter McLeod, and Fernand Gobet. 2008. Why good thoughts block better ones: The mechanism of the pernicious Einstellung (set) effect. *Cognition* 108: 652–61. [CrossRef]

Bilalić, Merim, Peter McLeod, and Fernand Gobet. 2010. The Mechanism of the Einstellung (Set) Effect: A Pervasive Source of Cognitive Bias. *Current Directions in Psychological Science* 19: 111–15. [CrossRef]

Bilalić, Merim, Mario Graf, Nemanja Vaci, and Amory H. Danek. 2019. The temporal dynamics of insight problem solving–restructuring might not always be sudden. *Thinking & Reasoning* 27: 1–37.

Blech, Christine, Robert Gaschler, and Merim Bilalić. 2020. Why do people fail to see simple solutions? Using think-aloud protocols to uncover the mechanism behind the Einstellung (mental set) effect. *Thinking & Reasoning* 26: 552–80.

Bowden, Edward M. 1997. The effect of reportable and unreportable hints on anagram solution and the Aha! Experience. *Consciousness and Cognition* 6: 545–73. [CrossRef] [PubMed]

Cushen, Patrick J., and Jennifer Wiley. 2012. Cues to solution, restructuring patterns, and reports of insight in creative problem solving. *Consciousness and Cognition* 21: 1166–75. [CrossRef] [PubMed]

Danek, Amory H., and Joshua Wiley. 2020. What causes the insight memory advantage? *Cognition* 205: 104411. [CrossRef]

Danek, Amory H., Joshua Williams, and Jennifer Wiley. 2020. Closing the gap: Connecting sudden representational change to the subjective Aha! experience in insightful problem solving. *Psychological Research* 84: 111–19. [CrossRef] [PubMed]

Duncker, Karl. 1945. On problem-solving. *Psychological Monographs* 58: 270. [CrossRef]

Durso, Francis T., Cornelia B. Rea, and Tom Dayton. 1994. Graph-theoretic confirmation of restructuring during insight. *Psychological Science* 5: 94–97. [CrossRef]

Ellis, Jessica J., and Eyal M. Reingold. 2014. The Einstellung effect in anagram problem solving: Evidence from eye movements. *Frontiers in Psychology* 5: 679. [CrossRef]

Ellis, Jessica J., Mackenzie G. Glaholt, and Eyal M. Reingold. 2011. Eye movements reveal solution knowledge prior to insight. *Consciousness and Cognition* 20: 768–76. [CrossRef] [PubMed]

Fedor, Anna, Eörs Szathmáry, and Michael Öllinger. 2015. Problem solving stages in the five square problem. *Frontiers in Psychology* 6: 1050. [CrossRef]

Fleck, Jessica I., and Robert W. Weisberg. 2013. Insight versus analysis: Evidence for diverse methods in problem solving. *Journal of Cognitive Psychology* 25: 436–63. [CrossRef]

Gilhooly, Ken J., Evie Fioratou, and N. Henretty. 2010. Verbalization and problem solving: Insight and spatial factors. *British Journal of Psychology* 101: 81–93. [CrossRef] [PubMed]

Hedne, Mikael R., Elisabeth Norman, and Janet Metcalfe. 2016. Intuitive feelings of warmth and confidence in insight and noninsight problem solving of magic tricks. *Frontiers in Psychology* 7: 1314. [CrossRef]

Holmqvist, Kenneth, Marcus Nyström, Richard Andersson, Richard Dewhurst, Jarodzka Halszka, and Joost Van de Weijer. 2011. *Eye Tracking: A Comprehensive Guide to Methods and Measures*. Oxford: OUP Oxford.

Just, Marcel A., and Patricia A. Carpenter. 1976. Eye fixations and cognitive processes. *Cognitive Psychology* 8: 441–80. [CrossRef]

Kizilirmak, Jasmin M., Violetta Serger, Judith Kehl, Michael Öllinger, Kristian Folta-Schoofs, and Alan Richardson-Klavehn. 2018. Feelings-of-warmth increase more abruptly for verbal riddles solved with in contrast to without Aha! Experience. *Frontiers in Psychology* 9: 1404. [CrossRef] [PubMed]

Knoblich, Günther, Stellan Ohlsson, Hilde Haider, and Detlef Rhenius. 1999. Constraint relaxation and chunk decomposition in insight problem solving. *Journal of Experimental Psychology: Learning, Memory, and Cognition* 25: 1534–55. [CrossRef]

Knoblich, Günther, Stellan Ohlsson, and Gary E. Raney. 2001. An eye movement study of insight problem solving. *Memory & Cognition* 29: 1000–9. [CrossRef]

Korovkin, Sergey, and Anna Savinova. 2021. The Effectiveness of Metacognitive Hints in Insight Problem Solving. In *Advances in Cognitive Research, Artificial Intelligence and Neuroinformatics*. Edited by Boris M. Velichkovsky, Pavel M. Balaban and Vadim L. Ushakov. Intercognsci 2020. Advances in Intelligent Systems and Computing. Cham: Springer, vol. 1358. [CrossRef]

Kuhn, Gustav, and Michael F. Land. 2006. There's more to magic than meets the eye. *Current Biology* 16: R950–R51. [CrossRef]

Kuhn, Gustav, Benjamin W. Tatler, and Geoff. G. Cole. 2009. You look where I look! Effect of gaze cues on overt and covert attention in misdirection. *Visual Cognition* 17: 925–44. [CrossRef]

Lindeløv, Jonas K. 2020. mcp: An R Package for Regression with Multiple Change Points. *OSF Preprints*. [CrossRef]

Loesche, Frank, Jeremy Goslin, and Guido Bugmann. 2018. Paving the way to eureka—Introducing "dira" as an experimental paradigm to observe the process of creative problem solving. *Frontiers in Psychology* 9: 1773. [CrossRef]

MacGregor, James N., and J. Barton Cunningham. 2009. The effects of number and level of restructuring in insight problem solving. *The Journal of Problem Solving* 2: 7. [CrossRef]

Metcalfe, Janet, and Arthur P. Shimamura, eds. 1994. *Metacognition: Knowing about Knowing*. Cambridge: MIT Press.

Newell, Allen, and Herbert A. Simon. 1972. *Human Problem Solving*. Englewood Cliffs: Prentice Hall.

Novick, Laura R., and Steven J. Sherman. 2003. On the nature of insight solutions: Evidence from skill differences in anagram solution. *The Quarterly Journal of Experimental Psychology Section A* 56: 351–82. [CrossRef] [PubMed]

Ohlsson, Stellan. 1984. Restructuring revisited: II. An information processing theory of restructuring and insight. *Scandinavian Journal of Psychology* 25: 117–29. [CrossRef]

Ohlsson, Stellan. 1992. Information-processing explanations of insight and related phenomena. In *Advances in the Psychology of Thinking*. Edited by M. Keane and K. J. Gilhooly. London: Harvester-Wheatsheaf, pp. 1–44.

Ohlsson, Stellan. 2011. *Deep Learning: How the Mind Overrides Experience*. New York: Cambridge University Press.
Öllinger, Michael, Gary Jones, and Günther Knoblich. 2006. Heuristics and representational change in two-move matchstick arithmetic tasks. *Advances in Cognitive Psychology* 2: 239–53. [CrossRef]
Öllinger, Michael, Gary Jones, and Günther Knoblich. 2008. Investigating the effect of mental set on insight problem solving. *Experimental Psychology* 55: 270–82. [CrossRef]
Pétervári, Judit, and Amory H. Danek. 2020. Problem solving of magic tricks: Guiding to and through an impasse with solution cues. *Thinking & Reasoning* 26: 502–33. [CrossRef]
Raftery, Adrian E., and V. E. Akman. 1986. Bayesian analysis of a Poisson process with a change-point. *Biometrika* 73: 85–89. [CrossRef]
Rayner, Keith. 1995. Eye movements and cognitive processes in reading, visual search, and scene perception. In *Studies in Visual Information Processing*. Amsterdam: Elsevier, vol. 6, pp. 3–22.
Reingold, Eyal M., Neil Charness, Richard S. Schultetus, and Dave M. Stampe. 2001. Perceptual automaticity in expert chess players: Parallel encoding of chess relations. *Psychonomic Bulletin & Review* 8: 504–10.
Schooler, Jonathan W., Stellan Ohlsson, and Kevin Brooks. 1993. Thoughts beyond words: When language overshadows insight. *Journal of Experimental Psychology: General* 122: 166. [CrossRef]
Spiridonov, Vladimir, Nikita Loginov, and Vladlen Ardislamov. 2021. Dissociation between the subjective experience of insight and performance in the CRA paradigm. *Journal of Cognitive Psychology* 33: 685–99. [CrossRef]
Takeuchi, Naoyuki, Takayuki Mori, Yoshimi Suzukamo, and Shin I. Izumi. 2019. Activity of prefrontal cortex in teachers and students during teaching of an insight problem. *Mind, Brain, and Education* 13: 167–75. [CrossRef]
Tseng, Chien-Chih, Ching-Hui Chen, Hsueh-Chih Chen, Yao-Ting Sung, and Kuo-En Chang. 2014. Verification of Dual Factors theory with eye movements during a matchstick arithmetic insight problem. *Thinking Skills and Creativity* 13: 129–40. [CrossRef]
van Rij, Jacolien, Nemanja Vaci, Lee Wurm, and Laurie Feldman. 2020. Alternative quantitative methods in psycholingistics: Implications for theory and design. In *Word Knowledge and Word Usage: A Cross-Disciplinary Guide to the Mental Lexicon*. Edited by Vito Pirrelli, Ingo Plag and Wolfgang Dressler. Berlin and Boston: De Gruyter Mouton, pp. 83–126. [CrossRef]
Wertheimer, Max. 1925. Über Schlussprozesse im produktiven Denken. In *Drei Abhandlungen zur Gestalttheorie*. Edited by Max Wertheimer. Erlangen: Verlag der Philosophischen Akademie, pp. 164–84.

Journal of
Intelligence

Article

The Role of Semantic Associations as a Metacognitive Cue in Creative Idea Generation

Yoed N. Kenett *, Noam Gooz and Rakefet Ackerman

Faculty of Data and Decision Sciences, Technion—Israel Institute of Technology, Haifa 320003, Israel
* Correspondence: yoedk@technion.ac.il

Abstract: Is my idea creative? This question directs investing in companies and choosing a research agenda. Following previous research, we focus on the originality of ideas and consider their association with self-assessments of idea generators regarding their own originality. We operationalize the originality score as the frequency (%) of each idea within a sample of participants and originality judgment as the self-assessment of this frequency. Initial evidence suggests that originality scores and originality judgments are produced by separate processes. As a result, originality judgments are prone to biases. So far, heuristic cues that lead to such biases are hardly known. We used methods from computational linguistics to examine the semantic distance as a potential heuristic cue underlying originality judgments. We examined the extent to which the semantic distance would contribute additional explanatory value in predicting originality scores and originality judgments, above and beyond cues known from previous research. In Experiment 1, we re-analyzed previous data that compared originality scores and originality judgments after adding the semantic distance of the generated ideas from the stimuli. We found that the semantic distance contributed to the gap between originality scores and originality judgments. In Experiment 2, we manipulated the examples given in task instructions to prime participants with two levels of idea originality and two levels of semantic distance. We replicated Experiment 1 in finding the semantic distance as a biasing factor for originality judgments. In addition, we found differences among the conditions in the extent of the bias. This study highlights the semantic distance as an unacknowledged metacognitive cue and demonstrates its biasing power for originality judgments.

Keywords: creativity; originality; metacognitive judgments; heuristic cues; semantic distance

1. Introduction

Metacognitive processes regulate our ongoing cognitive resources and processes. These metacognitive processes primarily consist of two levels. Object-level processes encompass basic cognitive operations, such as perceiving, remembering, classifying, and deciding; meta-level processes monitor the object-level operations (metacognitive monitoring) and allocate needed resources (metacognitive control; see Ackerman and Thompson 2017, for a review). Most metacognitive research has examined memory, reasoning, and decision making with well-defined tasks. Ill-defined tasks, such as creativity tasks, have been less studied under this framework (but see Puente-Díaz et al. 2021; Sidi et al. 2020). Thus, the exact role of metacognitive processes in creative thinking is largely unknown, especially heuristic cues that underly metacognitive judgments of originality.

A central question in metacognition is how people self-assess their confidence of success in each specific task item (e.g., an answer to a knowledge question). According to the well-established cue-utilization approach (Koriat 1997), people base their metacognitive judgments on various types of heuristic cues drawn from self-perception, information about the task and environment, and subjective experience while facing each task item (Ackerman 2019). Since metacognitive judgments rely on heuristic cues, they suffer from predictable biases stemming from differential effects of these cues on objective measures of success in the task and on the accompanied subjective self-judgments (Ackerman 2019; Bjork et al. 2013). As a result, the effectiveness of metacognitive control decisions which are based on biased judgments is expected to be compromised (Metcalfe and Finn 2008).

In the current study, we aimed to expose biases in metacognitive judgments that accompany creative thinking. In particular, we were interested in the role of the conceptual distance between concepts as a cue for originality judgments. By that, we aimed to document the contribution of conceptual (i.e., semantic) distance to the gap between measured originality scores and originality judgments of participant responses in a creativity task.

1.1. The Role of the Semantic Distance in Creative Thinking

It is common to assess creativity with tests of divergent thinking (Acar and Runco 2019), by presenting ill-defined problems and asking people to generate creative solutions. The alternative uses task (AUT) is among the most common divergent thinking tasks used in creativity research (Acar and Runco 2019). In the AUT, participants are asked to come up with as many alternative uses as they can for common objects (e.g., bucket: light shade, stool; Guilford 1967). Several studies reported moderate to large correlations between AUT performance and real-world creative achievements in the arts and sciences (Jauk et al. 2014; Runco and Jaeger 2012).

Originality is among four typical components defining creative idea generation together with fluency, flexibility, and elaboration (Guilford 1967; Torrance 1966). The process of generating original ideas has been a focus of attention in creativity research (Runco and Jaeger 2012; Sowden et al. 2014). There is a range of approaches to assess the originality of ideas. Some of these approaches rely on experts' evaluation (i.e., the Consensual Assessment Technique; Amabile 1983; Cseh and Jeffries 2019), some use standardized norms (Forthmann et al. 2020; Torrance 1972), and others involve subjective scoring by layperson judgments (Amabile 1983; Hass et al. 2018; Silvia et al. 2008). However, the need for tools allowing delving into biases in self-assessment of creativity is widely acknowledged (Kaufman 2019).

In recent years, creativity research has been increasingly moving towards the use of quantitative, objective assessment of AUT responses (Beaty and Johnson 2021; Dumas et al. 2021). These measures largely focus on the semantic distance. Based on the analysis of large textual corpora via computational linguistic models, semantic distance quantifies the conceptual dis-similarity between the AUT object, and the words in the open-ended responses participants generate when coming up with alternative uses for that object (Beaty and Johnson 2021; Dumas et al. 2021; Günther et al. 2019). Such measures correspond to the classic theory on memory, which argues that concepts in memory are organized as a semantic memory network, according to a principle of overlap of semantic features (Kumar 2021): The more semantic features two concepts share, the "closer" they are to each other in such a semantic memory network (Collins and Loftus 1975). Collins and Loftus (1975) defined the semantic distance as the 'shortest path [direct or indirect] between two nodes' (p. 412, note 3) within a person's semantic memory network. They argue for a spreading activation model: once a concept in the semantic network is activated, activation spreads from it to all its directly connected neighbor concepts; an activation which quickly decays over time and space.

The associative theory of creativity proposes that creative thinking is related to the ability to connect "remote" concepts in ones' semantic memory network and that the activation of remote concepts results in ideas with higher novelty (Kenett and Faust 2019; Mednick 1962; Volle 2018). The associative theory of creativity has been supported recently by several studies of computational modeling. These studies documented higher connectivity of the semantic memory network structure among highly creative individuals relative to low creativity ones (e.g., Kenett and Faust 2019). In addition, multiple studies have found that such quantitative measures of the semantic distance strongly correlate with self-assessments of idea originality (Beaty and Johnson 2021; Forthmann et al. 2020). Overall, computing semantic distance scores from such large textual corpora is now extensively used in creativity research, as it has been shown to strongly capture individual differences in creativity (Kenett 2019, 2018). Yet, the role of executive processes, such as metacognitive processes, in guiding creative search and novelty generation is still far from understood (Beaty et al. 2014, 2022; Benedek et al. 2023; Ovando-Tellez et al. 2022a; Volle 2018).

1.2. Metacognitive Processes Underlying Creative Thinking

We aim to advance the understanding of creativity processes from a metacognitive point of view, in line with recent calls for cross fertilization between these tightly related domains (Lebuda and Benedek 2023; Puente-Díaz 2023) and neurological evidence associating the two (Rominger et al. 2022). Recently, Lebuda and Benedek (2023) proposed a comprehensive framework of creative metacognition. This framework consists of both dynamic and static metacognitive components, involving a core cognitive process utilizing knowledge, and two metacognitive processes focused on monitoring and control. In particular, the authors review literature demonstrating how people with higher creativity and intelligence are also more accurate in their originality judgments (Guo et al. 2022; Karwowski et al. 2020; Lebuda and Benedek 2023).

In the current study, we used the AUT to expose biases when people judge their own originality. This is important because originality judgments are assumed to regulate creative efforts, similarly to other metacognitive judgments (Bjork et al. 2013; Fiedler et al. 2019; Perry-Smith and Mannucci 2017). That is, individuals are hypothesized to rely on originality judgments to guide their mental effort investment in further creative thinking (Puente-Díaz 2023). In this metacognitive causal chain, biased originality judgments, may lead people to terminate their idea generation process too early or keep investing labor in vain (Perry-Smith and Mannucci 2017). We focus on the semantic distance as a potential source for bias when people monitor their own originality. As explained above, semantic distance measures are based on large text corpora, quantifying relations between concepts that are largely shared across individuals (Günther et al. 2019; Mandera et al. 2015, 2017). This assumption underlies the large body of literature reviewed above, found to be robust and valuable in quantitatively assessing the originality of responses generated by participants in creativity tasks (Beaty and Johnson 2021; Dumas et al. 2021; Kenett 2019). We adapt this notion to the metacognitive realm.

In metacognitive research, shared semantic structure across people has been broadly used to identify the heuristic cues that underly and may bias metacognitive judgments (Ackerman 2019). In particular, Koriat (1993) defined accessibility as the amount of associations that come to mind when encountering a knowledge question. For instance, Koriat and Levy-Sadot (2001) compared people's metacognitive judgments regarding knowledge of composers (people generally know quite many names) versus choreographs (people generally know only a few). They found higher metacognitive judgments regardless of knowledge of a particular fact regarding a question when referring to a large body of knowledge (composers) relative to a small body of knowledge (choreographs). They also dissociated this number of associations from familiarity of the same terms. Ackerman and Beller (2017) applied this principle to problem solving and exposed similar biases in peoples' responses in the compound remote associates task, a commonly used task to study convergent thinking in creativity research (Bowden and Jung-Beeman 2003; Wu et al. 2020).

1.3. The Present Study

The present study is based on Sidi et al. (2020), who were the first to quantify the creative process in AUT in a manner that allowed examining metacognitive processes that underly originality. They operationalized originality as the percentage of participants in the study sample who suggested a similar idea to an object in the AUT. It is well established in the creativity literature that when people produce a list of potential ideas anchored to a single object, they start their ideation process with more mundane ideas, and get to the more original ideas later in the ideation process (Bai et al. 2021; Beaty and Silvia 2012). Sidi et al. (2020) replicated this finding. They collected originality judgments for each idea by asking participants to assess the percentage of participants that they expect will come up with the same idea as their own, in parallel to the definition of the originality score in their study. They found that originality judgments reflected the serial order properly, that is, people assessed their initial ideas as less original than their later AUT responses. Thus, the serial order was a reliable cue for originality judgments. Notably, the authors dissociated effects on originality scores and originality judgments. In their study, Sidi et al. (2020) manipulated originality scores without affecting originality judgments by providing information regarding the expected number of uses (low—2 or high—6); making the participants generate more or fewer ideas without acknowledging the changes in originality that it created. Demonstrating the other side of a double dissociation, they manipulated originality judgments but not originality scores by providing positive or negative feedback during the initial task practice. Thus, this study shows separated underlying processes for originality performance and its self-assessment. Particularly relevant for the present study is that this study demonstrates the sensitivity of both idea originality and originality judgments to instruction manipulations.

Our aim was to examine the influence of semantic distance on the discrepancy between self-assessment of originality and the originality of ideas. We hypothesized that semantic distance is a heuristic cue underlying originality judgment, but that its role in originality judgments may not correspond to its role as a predictive cue of the rarity of an idea (originality score) in a target population. This hypothesis was examined in Experiment 1, by re-analyzing Sidi et al. (2020) data while adding a semantic distance variable estimated using computational semantics tools. In Experiment 2, we manipulated task instructions as priming for systematically affecting the generated ideas. In particular, the instructions introduced examples representing a combination of the semantic distance (close/far) and originality (low/high). This 2 × 2 design allowed us to examine the unique contributions of the semantic distance separately from the effects of originality scores when predicting originality judgments. Any interaction effect between the two manipulations on cues that underlie the discrepancy between self-assessment of originality and the originality of ideas would indicate their separability. That is, although semantic distance is strongly associated with originality, they are not identical, and it may have differential effects on objective and subjective aspects of the creative process.

2. Experiment 1

2.1. Introduction

In Experiment 1, we examined the contribution of semantic distance to the explained variance of originality scores and originality judgments above and beyond other known influencing factors. We re-analyzed data from Sidi et al. (2020) Experiment 2 and Experiment 3. We refer to these experiments as 1a and 1b, respectively. In these experiments, participants performed the AUT with five objects after instructions and demonstrations with one object. For each object, participants could generate as many ideas as they wished. Immediately after generating each idea, they rated the originality of that idea on a 0–100% scale, representing the assessed rate of peers in the sample who generated the same idea. The dependent variables in both experiments were originality scores (the frequency of each idea in the sample) and originality judgments. Sidi et al. (2020) dissociated originality judgments and originality scores of responses with two manipulations, one that affected only the originality scores of the ideas (information about the number of ideas previous pseudo participants provided on average to each object being 2 or 6) and another one that affected only the judgments (positive/negative feedback).

We considered four heuristic cues that could potentially predict originality scores and originality judgments as independent variables. We examined response times (RT) and the serial order from Sidi et al. (2020) data, and added the number of words, as a proxy to idea specificity, and the semantic distance derived from the SemDis platform, for every generated idea. We hypothesized that the semantic distance would predict both originality scores and originality judgments. However, we expected to find a strong association between the semantic distance and originality scores based on the creativity literature that uses semantic distance measures as a direct measure of originality (Beaty and Johnson 2021; Dumas et al. 2021; Kenett 2019). Given that responses' SemDis scores are derived from the analysis of aggregated, large-volume text-corpora, we expected judgments of originality to be affected more by other cues and less by the semantic distance relative to predicting originality scores. This differential strength of effects for originality scores and originality judgments was expected to explain part of the discrepancy between the two, and thus expose the semantic distance as a factor impacting behavior in the AUT.

For the main data analyses of cue contribution and biasing role in originality scores and originality judgments, we used a recently suggested methodology called Bird's-Eye View of Cue Integration (BEVoCI; Ackerman, forthcoming current version is available in the OSF page of this paper). It allows to dissociate factors affecting achievements (cue validity, Koriat 1997) and judgments (cue utilization, Koriat 1997). A cue is considered misleading when it predicts achievements and judgments differently (also considered as a bias). Such a difference in prediction can be expressed by a positive, negative, or no effect on one dependent variable (e.g., performance or judgments) but not on the other (Ackerman forthcoming). Furthermore, another possibility for such cue misleading effect is related to differential predictive power (as represented by regression β) in one dependent variable compared to the other. Notably, most metacognitive experiments which demonstrate double dissociations between judgments and performance manipulated one biasing factor a time, in one experiment with one affecting judgment and not performance, and in the other affecting performance and not judgments (e.g., Metcalfe and Finn 2008; Sidi et al. 2020). The BEVoCI, in contrast, allows examining several potentially misleading cues and exposing double dissociations in the same analysis with the same sample.

The BEVoCI method is based on two hierarchical multiple regressions, one predicts an objective measure of performance by multiple considered cues, and the other one predicts a metacognitive judgment by the same cues. Then, the two sets of predictive values are compared in direction and strength of the effect for each cue. The hierarchical nature of the analyses allows to statistically compare groups of participants while taking into account the unique cue weights for each participant, and the weights' variance among the participants in each condition.

In this study, we applied the BEVoCI method for exposing the unique contribution of the semantic distance, word count as a control variable, as well as additional cues examined in metacognitive research (RT and serial order) in predicting originality scores and originality judgments. In particular, the BEVoCI method allows exposing the differential effects of each cue on the two dependent variables. Recent studies suggest that people integrate four and even five cues in their memory judgments and in confidence in problem solutions (Ackerman forthcoming; Undorf et al. 2018; Undorf and Bröder 2021). The BEVoCI method allows examining whether such multiple cue integration is reflected in originality judgments as well.

2.2. Materials and Methods

2.2.1. Participants

Sidi et al. (2020) experiments were conducted in English with an international online sample collected from Prolific.ac (71% females, Mage = 30.7 years, SD = 5.4 years, $N = 101$ and $N = 96$ participants in the two experiments, respectively) in exchange for GBP 1.3 for each participant.

2.2.2. Materials

The Alternative Uses Task (AUT)

The AUT involves presenting an object and asking for a sequence of potential uses for it (Acar and Runco 2019; Runco and Acar 2012). Sidi et al. (2020) used brick as an example. The main task's objects were: bucket, cloth hanger, wine cork, paperclip, and shoelace. Each object appeared with its name and a picture. Each trial was self-paced, and participants could generate as many AUT responses as they wanted, before moving on to the next trial.

Measures

Originality scores for each AUT response were calculated as percentages based on the following process: Two trained raters coded all of the participants' ideas according to a predefined coding scheme. Inter-reliability between the raters was strong (Cohen's K = 0.867, $p < 0.001$). The frequency (%) of each idea within the experiment's sample for each object (e.g., brick) was calculated. In this way, the higher the score, the more unoriginal (more common) the generated response was. The originality (uniqueness) score can thus be measured by taking the inverse of the frequency scores (100%- frequency).

Originality judgments were elicited for each idea on a scale of 0–100%. The higher the judgment, the more unoriginal (more common) the participant judged their response to be. Therefore, we inversed these ratings to indicate originality by calculating 100-unoriginality judgment.

The response time (RT) and the **serial order** of responses were documented by the experiment software. RT measures the time elapsed from page presentation until participants entered their ideas. The serial order reflects the location within the idea stream of each participant for each object.

The number of words in each response was counted, as it is a known confounding factor for originality in the creativity literature (Forthmann et al. 2019).

The semantic distance of AUT responses was calculated through the SemDis platform (Beaty and Johnson 2021; http://semdis.wlu.psu.edu/, accessed on 1 May 2022), which is a publicly available application for calculating the semantic distance of AUT responses. This tool operationalizes the theoretical concept of the semantic distance to a measurable dependent variable. We used the additive calculation procedure to compute the semantic distance between each participant's response relative to the presented object (e.g., Brick). This calculation is performed based on an average score over five different semantic spaces, mitigating the effects of a single semantic model and text corpus (e.g., textbooks vs. movie subtitles; Beaty and Johnson 2021; Dumas et al. 2021; Kenett 2019). Three of these spaces are built upon continuous bag of words (CBOW) prediction models (cbowukwacsubtitle, cbowsubtitle, and cbowBNCwikiukwac) and two are built upon count models (GloVe and TASA). The CBOW models use a neural network architecture (Mandera et al. 2017) that predicts a given word from surrounding context words within a given text corpus (for more details see Beaty and Johnson 2021).

Analyses

Bird's-Eye View on Cue Integration (BEVoCI). The analyses start by examining colinearity among the considered cues. Only cues with sufficient unique variance are included in the models, for exposing their unique contribution while controlling for all other cues. As a rule of thumb, Ackerman (forthcoming) suggested using $r < 0.30$ in all paired correlations as a threshold for including cues.

Two hierarchical multiple regression models were used with level 1: objects; level 2: participants. The analyses were performed by R 4.1.0 lmer package (Pinheiro et al. 2019). Prior to the analyses, we verified that all participants had variability in originality judgments and originality scores. Then, all independent and dependent variables were standardized. All cues were entered into one model predicting originality scores and another one predicting originality judgments. In these models, we controlled for participant and object variability.

2.2.3. Procedure

In both re-analyzed experiments, participants completed the AUT in two conditions to which participants were randomly assigned. The instructions used a picture of a brick to demonstrate the task and the user interface for entering each use idea. The idea elicitation phrasing was: "Please come up with as many uses as you can think of for a brick." Accompanied by a picture of a brick. Assessing originality was elicited with the following: "Please assess how many participants (%) came up with this use" with a continuous scale, with captions of "Nobody else" near the 0%, and "Everybody" near the 100%. Participants could produce as many ideas as they wished. Notably, a request to be original was not mentioned.

2.3. Results

First, for each re-analyzed experiment, we correlated all the predictors and found that in Experiment 1b RT had multicollinearity with the number of words ($r > 0.3$). We chose to exclude RT and keep the number of words for controlling for this potential confounder (Forthmann et al. 2019) and obtaining cleaner analyses of the semantic distance. Next, we applied the BEVoCI method on both re-analyzed experiments from Sidi et al. (2020) with serial order, number of words, and semantic distance as cues for both experiments, and RT for Experiment 1a only. See Table 1 the beta coefficients of the two regression analyses performed for each group in Experiment 1a and Experiment 1b.

Table 1. BEVoCI β coefficients representing cue validity in predicting originality scores and cue utilization for originality judgments in all conditions in Experiment 1 and Experiment 2.

	Originality Score	Originality Judgment	Originality Score	Originality Judgment
	Experiment 1a (Sidi et al. (2020), Experiment 2)			
Feedback manipulation:	Rare		Common	
Serial order	0.28 ***	0.32 ***	0.28 ***	0.31***
Number of words	0.09 **	0.08 **	0.10**	0.13 ***
Semantic distance	0.41 ***	0.34 ***	**0.38 *** **	**0.28 *** **
Response time	0.06 *	0.09 **	0.06 *	0.11 ***
	Experiment 1b (Sidi et al. (2020), Experiment 3)			
Anchor manipulation:	High		Low	
Serial order	0.28 ***	0.31 ***	0.46 ***	0.46 ***
Number of words	0.07 **	0.16 ***	0.06 *	0.08 **
Semantic distance	**0.49 *** **	**0.29 *** **	**0.46 *** **	**0.34 *** **
	Experiment 2—Semantic Distance Manipulation: Close			
Originality Manipulation:	High		Low	
Serial order	**0.23 *** **	**0.39 *** **	**0.31 *** **	**0.48 *** **
Number of words	**0.04 * **	**0.11 *** **	0.07 ***	0.13 ***
Semantic distance	**0.36 *** **	**0.18 *** **	**0.36 *** **	**0.20 *** **
Response time	**−0.01**	**0.03 ** **	0.03	0.07 ***
	Experiment 2—Semantic Distance Manipulation: Far			
Originality Manipulation:	High		Low	
Serial order	**0.27 *** **	**0.42 *** **	**0.29 *** **	**0.48 *** **
Number of words	0.05 *	0.12 ***	0.09 ***	0.11 ***
Semantic distance	**0.40 *** **	**0.16 *** **	**0.48 *** **	**0.16 *** **
Response time	0.04 *	0.06 **	0.03	0.03

Note. Significance of a cue as a predictor (difference from zero), *** $p \leq 0.001$; ** $p \leq 0.01$; * $p \leq 0.05$. **Bold** fonts: Significant mismatch between the association of the cue with originality scores and with originality judgments, $p < 0.05$.

Including semantic distance in BEVoCI analyses revealed that semantic distance, and no other cue, generated a significant difference in predicting originality scores and originality judgments in three out of four conditions (Table 1), while in the fourth condition the bias was trending towards significant, $t(1696) = 1.76$, $p = 0.078$. Comparing the two dependent variables when including the entire data set of Experiment 1a, without division into conditions, revealed a significant difference in the role of the semantic distance between them, $t(3404) = 3.09$, $p = 0.002$. The stronger predictive power for originality scores (higher β) than for originality judgments (Table 1) suggests that the semantic distance was underutilized for originality judgments. The found difference might either stem from different "slopes" when predicting each dependent variable—originality scores and originality judgments—by semantic distance while controlling for other cues, from the amount of "noise" around each prediction, or a combination of the two. Standard errors of the βs were 0.029 for both dependent variables, ruling out the "noise" difference as an explanation for the mismatch between the two.

2.4. Discussion

In Experiment 1, we re-analyzed data from Sidi et al. (2020). In addition to the robust serial order effect, they reported on, we considered the potential utilization of semantic distance as a predictor of originality scores and as a cue for originality judgments, while controlling for the number of words and response time. All considered cues had predictive power in all models. This finding should certainly not be taken for granted. Importantly, in most models, semantic distance was underutilized in originality judgments, unlike all other considered cues. In Experiment 2, we delved experimentally into the effects of the semantic distance on these aspects of creativity.

3. Experiment 2

3.1. Introduction

In Experiment 2, we aimed to delve into the role of semantic distance in metacognitive processes underlying originality behavior. In creativity research, semantic distance is increasingly used as a quantitative, objective measure assessing originality (e.g., Beaty and Johnson 2021; Dumas et al. 2021; Kenett 2019). The results of Experiment 1 revealed that semantic distance was a strong predictor of originality scores, and to a lesser extent of originality judgments. These findings thus lead to the hypothesis that semantic distance captures differently originality scores than originality judgments. To examine this hypothesis, we manipulated the instructions for the AUT to dissociate the priming of semantic distance from priming for originality, in a 2 × 2 between-participants design.

As part of the instructions, we provided three examples of objects, different than those appearing in the main task. For each object, we gave one example of a use for this object. Based on responses given by participants in Sidi et al. (2020) study, we chose examples of uses by two orthogonal factors, the semantic distance (close or far) and originality (low or high; see Table 2). All three examples in each condition had the same combination of the semantic distance and originality. Notably, we included the seemingly contradictory conditions of a high semantic distance with low originality and vice versa, to highlight the separability between the two.

Table 2. The examples used for instruction manipulations of originality (low, high) and distance (far, close).

Object	Hanger		Wine cork		Bucket	
Originality Semantic Distance	**Low**	**High**	**Low**	**High**	**Low**	**High**
Far	Grabber tool	Lighting fixture	Bottle drinks	Produce fire	Drinking water	Shower
Close	Hanging clothes	Plant support	Floating boat	Pin sticking	Wheelbarrow	Shovel

Our focus is on the predictive value of the examined cues. Overall, given the positive direction of the prediction by semantic distance in Experiment 1, we hypothesized that originality scores and originality judgments would rise as semantic distance increases. Moreover, we expected originality scores to be more sensitive to semantic distance manipulation than originality judgments. This is because semantic distances are calculated based on a generalized model from large textual corpora, originality scores are calculated based on co-occurrence statistics of words in the corpora, and originality judgments are idiosyncratic in nature. This explanation was indeed supported by the findings of Experiment 1 and was expected to be replicated in Experiment 2.

As for the manipulations, we expected two types of effects. The first was shifts in means of originality scores and originality judgments with both manipulations. The second was shifts in "slopes" (β values), in particular those related to semantic distance. Using BEVoCI analyses was expected to expose changes in cue validity and cue utilization given the manipulations we used. As explained above, any interaction effect of the two manipulations would indicate the unique contributions of the two manipulations to the creative process.

3.2. Materials and Methods

3.2.1. Participants

In Experiment 1, Sidi et al. (2020) had approximately 50 participants in each group, in one group, the differential effect of the semantic distance on the originality score and originality judgment was marginal, while in others it was significant. Thus, we increased the power by approximately 20–25%, in order to increase the robustness of Experiment 2. The experiment was conducted with an international online sample from Prolific.ac. In total, 262 participants took part in exchange for 2.5 GBP. They were invited for approximately 30 min. They were assigned randomly to one of four study groups (2 × 2 between subject manipulation). In each group, we had 61–65 participants after 10 participants were excluded from the data following failure to answer two or more of the following attention checks: (1) instruction understanding verification by a multiple choice question, inferring a detail from a displayed example; (2) clicking "I do not remember the use I entered" when rating originality on the following screen more than once; (3) or failing upon instruction to input a particular value to two statements in a final section of self-report questions. After exclusion, 252 participants (76% females, mean age = 34 years, *SD* = 11.6 years) were included in the analyses.

3.2.2. Materials

Alternative Uses Task (AUT)

Similarly to Experiment 1, we used the AUT. Participants were asked to generate alternative uses for nine objects: hammer, lipstick, fork, bench, apron, tie, pillow, knife, and belt. The chosen objects for the AUT reflected the nine psycholinguistic dimensions scales of the Glasgow Norms (arousal, valence, dominance, concreteness, imageability, familiarity, age of acquisition, semantic size, and gender association) to avoid linguistic biases (Scott et al. 2019). The objects were used in the same manner as in Sidi et al. (2020). In addition, we prepared three example objects with four uses each (Table 2). The uses for the example objects represented the four conditions: original or common uses which were either far or close in terms of semantic distances. The semantic distances of the examples were computed by the "fasttext" model in Python (Joulin et al. 2016) relative to the respective object word (e.g., bucket). The examples with high or low originality scores were taken from responses provided by the participants in Sidi et al. (2020; Experiment 2). Original examples were chosen from the first quartile of originality scores and common examples were chosen from the last quartile for each example object. For example, a closely semantic and original idea for a bucket was a shovel, which was one step away from bucket in the semantic network and was generated by only 1% of the participants of Sidi et al. (2020).

Notably, we deliberately phrased the use examples to be shorter in the number of words (M = 2.77 words, SD = 1.24) than the responses provided by Sidi et al.'s (2020) participants (M = 3.44 words, SD = 2.52). By that, we aimed to focus participants on their central use idea, to improve word count and semantic distance reliability relative to the previous study by minimizing peripheral words (e.g., "the" or "a"). We also expected that a shorter description of uses would encourage participants to provide more uses for each object without investing a longer time. This was expected to increase our study power and increase the explanatory value of the serial order.

Measures

The independent and dependent variables were the same as those used in Experiment 1. Unlike in Sidi et al. (2020), given the brief idea phrasings, originality scores were calculated by the frequency in the overall sample of each idea's dominant word, which was manually chosen by one of the authors (NG). The considered predictive cues were as in Experiment 1: serial order, number of words in the use idea, semantic distance of the use from the object, and RT.

Analyses

We started with examining the effects of the two manipulations on means of originality scores and originality judgments by two-way analyses of variance (ANOVA), with the semantic distance (close vs. far) and originality (low vs. high) as between-participant factors. Then, we performed the BEVoCI analyses for each condition separately, as in Experiment 1. Finally, unlike in Experiment 1, we compared cue validity and cue utilization (β coefficients) among the groups by adding the two manipulations and the interaction between them as additional factors in BEVoCI regression models.

3.2.3. Procedure

We used the AUT with nine objects accompanied by images, presented as in Experiment 1. Participants were required to generate at least one use idea, before moving to the next object. The task was self-paced and participants could generate as many use ideas as they wanted for each object. See a screenshot of one trial in Figure 1.

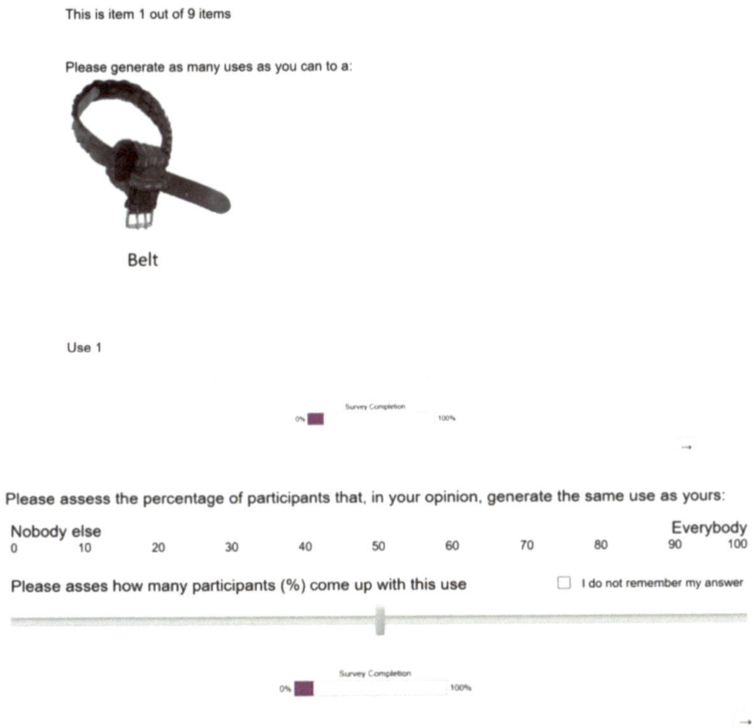

Figure 1. An example of a trial in the AUT used in Experiment 2.

3.2.4. Results

We start by examining the effects of the manipulations on the means of the dependent variables. Comparing the means of the dependent variables across the four conditions revealed that our semantic distance manipulation did not affect the means of originality scores, $F(1, 248) = 0.44$, MSE = 18.1, $p = 0.51$, $\eta^2_p = 0.002$, nor originality judgments, $F(1, 248) = 1.28$, MSE = 293.3, $p = 0.26$, $\eta^2_p = 0.005$. When the prime was close in terms of the semantic distance, uses were generated with slower RT ($M = 12.4$ sec., $SD = 7.0$) than when it was semantically far from the given object ($M = 10.9$ sec., $SD = 5.4$), $F(1, 248) = 3.95$, MSE = 154.4, $p = 0.048$, $\eta^2_p = 0.016$. Our manipulation of originality affected originality scores, $F(1, 248) = 15.76$, MSE = 640.7, $p < 0.001$, $\eta^2_p = 0.060$, with groups primed with more original examples ($M = 82.0$, $SD = 5.7$) generating more original ideas than the those who received less original examples ($M = 78.6$, $SD = 5.8$). No parallel effect of originality manipulation on means of originality judgments was found, $F(1, 248) = 0.58$, MSE = 133.5, $p = 0.45$, $\eta^2_p = 0.002$, or on RT, $F(1, 248) = 0.01$, MSE = 0.35, $p = 0.925$, $\eta^2_p = 0.000$. Finally, no interactive effects between the semantic distance and originality were found for originality scores, $F(1, 248) = 0.64$, MSE = 2.6, $p = 0.80$, $\eta^2_p = 0.000$, originality judgments, $F(1, 248) = 0.62$, MSE = 141.3, $p = 0.43$, $\eta^2_p = 0.002$, or RT, $F(1, 248) = 0.13$, MSE = 5.05, $p = 0.720$, $\eta^2_p = 0.001$. Notably, the effect of the originality manipulation on originality scores but not on originality judgments already exposes a dissociation between them.

Next, we applied the BEVoCI analysis for examining cue validity and cue utilization. There was no collinearity of above 0.30 correlations among the considered cues. The BEVoCI results for Experiment 2 are presented in Table 1. All cues were significant predictors for both originality scores and originality judgments in most models. RT was unique in being a non-significant predictor in some cases. Given that it has no unique contribution in

Experiment 1b due to collinearity, we conclude that it is a secondary, if at all, contributor to originality behavior above and beyond the other cues we considered.

Consistent with Experiment 1, in all BEVoCI models, semantic distance contributed to the mismatch between originality scores and originality judgments, as it was a stronger predictor of originality scores than originality judgments. Interestingly, and unlike Experiment 1, in all BEVoCI models, serial order contributed to this mismatch as well, with the opposite effect: a weaker predictive power for originality scores than for originality judgments. Notably, the β of the serial order for originality judgments was generally higher in Experiment 2 relative to Experiment 1. We explain this finding in the shorter responses participants provided when primed with brief examples and the production of more responses in Experiment 2 relative to Experiment 1. However, since no parallel difference in values was found regarding originality scores, it seems that the serial order affected originality judgment more than ideas' originality. Indeed, while semantic distance was consistently underutilized, serial order was consistently overutilized (see graphical representation by the β bars in Figure 2). These opposite directions support the conclusion of Sidi et al. (2020) that originality scores and originality judgment are dissociated as they are affected by different underlying processes.

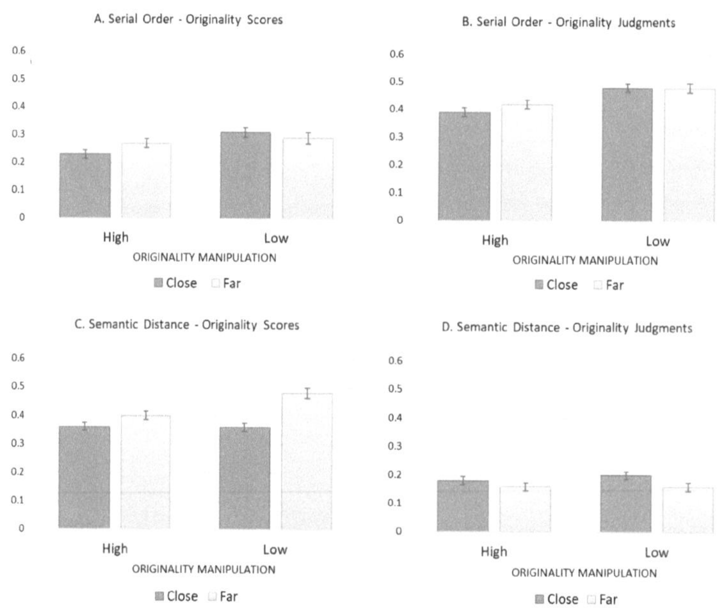

Figure 2. β of originality scores and originality judgments in the four experimental groups by the two cues that generated biases, the serial order, and the semantic distance. Error bars represent the standard error of the means for the β's.

Next, we analyzed the effects of the 2 × 2 manipulations on β for each cue separately. This was performed to examine the effect of the manipulations on the predictive power of each of the cues on originality scores and originality judgments. These regression analyses revealed several main effects and interaction effects (Table 3). As explained in our hypotheses above, the main finding is the mere existence of interactions between the two manipulations. These interactions indicate the separability and unique contribution of semantic distance manipulation on top of the effects of the originality manipulation. This finding goes beyond the effects on the means of originality scores. Interestingly, we see an effect of the originality manipulation: a higher overall serial order β (regardless of distance manipulation) of originality judgments; a higher originality score β in the far semantic distance manipulation than all others.

Table 3. Regression results of comparisons among the four groups in the β of the biasing factors (interaction effects between the group and the cue) that appear in Figure 2[1].

Manipulation / Dependent Variable (Figure 2 Panel)	Semantic Distance	Originality	Interaction
		Serial Order	
Originality Scores (Panel A)	$t(9911) = 2.29$, **$p = 0.022$**	$t(9911) = 4.52$, **$p < 0.001$**	$t(9911) = 2.11$, **$p = 0.035$**
Originality Judgments (Panel B)	$t(9911) = 1.55$, $p = 0.120$	$t(9911) = 5.94$, **$p < 0.001$**	$t(9911) = 2.33$, **$p = 0.020$**
		Semantic Distance	
Originality Scores (Panel C)	$t(9911) = 2.81$, **$p = 0.005$**	$t(9911) = 1.74$, $p = 0.08$	$t(9911) = 0.48$, $p = 0.630$
Originality Judgments (Panel D)	$t(9911) = 0.025$, $p = 0.799$	$t(9911) = 3.68$, **$p < 0.001$**	$t(9911) = 3.20$, **$p < 0.001$**

Note. **Bold** fonts indicate a significant effect.

3.3. Discussion

In Experiment 2, we replicated the main findings from Experiment 1 regarding semantic distance. We found that semantic distance was a strong predictor of both originality scores and originality judgments and was a cue that consistently predicted originality scores more strongly than originality judgments. In addition, we exposed serial order effects that were not found in Experiment 1. Interestingly, this effect was the opposite of the effect of semantic distance, as serial order was consistently a weaker predictor of originality scores than of originality judgments. Moreover, our priming manipulation demonstrated the dissociation between originality scores and originality judgments, as manipulating the originality of the examples within the instructions only affected originality scores and not originality judgments. Unlike previous studies (Heinen and Johnson 2018), our semantic distance manipulation of the examples in the AUT instructions did not lead to any significant differences in originality scores or originality judgments. However, this manipulation did affect the predictive value of the semantic distance and the serial order.

4. General Discussion

In the current study, we examined the role of semantic distance in self-evaluation of creative ideas. To do so, we re-analyzed data collected by Sidi et al. (2020) and conducted a follow-up experiment. Sidi et al. (2020) manipulated originality scores separately from originality judgments (by manipulating the expected number of generated ideas) and vice versa (by manipulating the feedback during instructions). Their findings support double dissociation between effects on performance and judgments by using a separate manipulation for each side of the dissociation. However, by considering more cues and using the BEVoCI analysis approach, our study highlights key insights for both metacognitive and creativity research.

In both experiments, we considered serial order, based on Sidi et al. (2020), and added cues from creativity literature, namely, number of words and semantic distance. We also examined RT which is often used in metacognitive research (e.g., Ackerman and Zalmanov 2012), in order to examine the unique contribution of semantic distance above and beyond well-established cues. All considered cues were found to be significant predictors of both originality scores and originality judgments in most cases. Examining the unique contribution of semantic distance led us to discover that it was a strong predictor for both originality scores and originality judgments. However, it also had a consistent biasing effect, which is a novel finding in both metacognitive and creativity research.

First, across both experiments, we found that semantic distance of AUT responses, operationalized via SemDis, is a significant predictor of originality judgments, and replicated prior work demonstrating its ability to predict originality scores (Beaty and Johnson 2021; Dumas et al. 2021; Kenett 2019). We emphasize the dissociative effect of semantic distance on originality scores versus originality judgments, highlighting a bias in the utilization of this cue. In Experiment 1 of the present study, semantic distance affected originality scores more than originality judgments. In Experiment 2, we replicated this finding and also found that serial order affected originality scores less strongly than originality judgments. These findings suggest that originality judgments, as with many other metacognitive judgments (e.g., judgment of learning, confidence; Ackerman 2019), are based on heuristic cues. Furthermore, these findings suggest that originality judgments' correspondence with behavioral performance depends heavily on the validity of the cues people utilize for inferring them (Ackerman 2019; Koriat 1997). By manipulating semantic distance and the originality of examples as two different factors in Experiment 2, our findings uniquely highlight that semantic distance and originality scores do not fully correspond to each other, although they are consistently correlated.

Second, in line with previous research with other metacognitive judgments (Ackerman forthcoming; Undorf et al. 2018; Undorf and Bröder 2021), we found cue integration of four cues in originality judgments. The fact that BEVoCI allows to directly compare the predictive role of a quantitative measure of the semantic distance relative to other cues advances metacognitive and creativity research both theoretically and methodologically (Beaty and Johnson 2021; Dumas et al. 2021; Kenett 2019). This relative comparison revealed that semantic distance was the strongest among the cues we considered to predict originality scores. However, in regard to predicting originality judgments, semantic distance was as strong as serial order in Experiment 1, and a weaker predictor compared to serial order in Experiment 2. A future direction of interest is to examine whether this trade-off between the two cues can be replicated and generalized.

Third, most metacognitive research regarding cue utilization deals with indicators of item difficulty that are accessible to the participant while performing the task (e.g., cue familiarity, text coherence; Ackerman 2019). In the present study, serial order and RT are cues of this type. In contrast, responses' SemDis scores are derived from the analysis of non-idiosyncratic, aggregated, large-volumes text-corpora (Günther et al. 2019). Indeed, as expected, originality judgments were less affected by semantic distance compared to originality scores. This is a unique situation in metacognitive research that contributes a new direction for considering other such information sources. Notably, though, although the semantic distance is defined based on information external to the participant, it is assumed to reflect knowledge structure common across people with similar cultural backgrounds (Günther et al. 2019; Mandera et al. 2015, 2017). Even more radical in this continuum of personalization is consensuality (Koriat 2008). Consensuality is the extent to which the same response is given within the experiment sample. It is not expected to be accessible to each participant but was found empirically to predict metacognitive judgments in various contexts (Ackerman et al. 2019, 2020; Bajšanski and Žauhar 2019). Consensuality is highly similar to the way we operationalized originality scores. In the present study, we highlight differences between the consensus, measured by originality scores, and the related metacognitive judgments, as reflected by originality judgments.

Finally, in Experiment 2, we conducted an instruction manipulation, manipulating different aspects of originality (low, high) and semantic distance (close, far) of the examples provided for the AUT and examined the effects of these manipulations on participants' performance. Despite previous studies demonstrating manipulations of the semantic distance of responses based on instruction manipulation (Heinen and Johnson 2018), we did not find significant effects on originality scores and originality judgment means. This finding may indicate power issues of our manipulation. However, these two manipulations (originality and semantic distance) affected participants' sensitivity to the considered metacognitive cues. This is evident in significant changes in regression slopes, as estimated via the models' β as reported in Table 3. Using state-of-the-art statistical methods, such as BEVoCi, allows unpacking more nuanced effects of such manipulations on performance and judgments in complex tasks such as the AUT.

Ill-defined tasks, such as creativity tasks, have been rarely studied under the metacognitive framework (but see Puente-Díaz et al. 2021; Sidi et al. 2020). Thus, the exact role of metacognitive processes in creative thinking is largely unknown, especially the heuristic cues underlying metacognitive judgments of originality (Lebuda and Benedek 2023). Our results highlight the complexity of the cognitive and metacognitive capacities involved in generating creative responses. More broadly, we found that each manipulation affected more strongly different cues. We suspect that this is only the tip of the iceberg and that there are more cues involved in creative processes still to be exposed. For example, Matheson and Kenett (2021) demonstrated how different motion-related strategies impact AUT responses (see also Gilhooly et al. 2007). In a recent study of theirs, Sargent et al. (2023) empirically demonstrated how environmental context and body posture impact performance on the AUT. It is yet unknown how such novel aspects affect metacognitive judgments that accompany the thinking process.

We would like to point to several methodological aspects of our study regarding how we administered the AUT. First, the way the AUT was administered in both experiments varies from common implementation in creativity research (Said-Metwaly et al. 2020). Most creativity studies administer the AUT with time constraints (Paek et al. 2021), such as two or three minutes per object, whereas we applied a self-paced AUT. Furthermore, in both experiments, originality scores were computed as the inverse of the commonality of responses in the sample, based on the approach of Sidi et al. (2020). It is possible, that asking about commonality and about rarity differ in their underlying cues, while the objective measure of originality remains unchanged. Moreover, originality scores are typically assessed subjectively, by having raters rate the originality of responses (Hass et al. 2018; Silvia et al. 2008). These judgments may share different aspects with self-judgments than with the objective operationalization of originality scores that we used, as those subjective judgments of other people, are probably heuristic based as well. Notably, such frequency-based approaches are known to suffer from reliability limitations (Forthmann et al. 2021; Silvia et al. 2008), further highlighting the need to improve assessment of idea originality (Kaufman 2019). In addition, typical creativity research using the AUT instructs participants to "be creative" in their responses (Nusbaum et al. 2014), while we refrained from any such instructions so as to not bias metacognitive processes. Finally, the instructions given in Experiment 2 used short responses as examples and this change relative to Experiment 1 seems to affect ideation fluency. This design aspect potentially limited the richness of the AUT responses, impacting their elaboration that we measured via the number of words. All these methodological aspects worth consideration in future research.

A core aspect of our current study is the computation of semantic distance scores between objects and the alternative uses generated to these objects by participants. This computation was performed via the SemDis platform (Beaty and Johnson 2021), which is based on natural language processing models that derive similarity between concepts based on co-occurrence statistics in large textual-corpora (Günther et al. 2019; Kumar 2021; Mandera et al. 2015). Semantic distance—the inverse of similarity between concepts in such a semantic space—has been consistently shown to be a quantitative, objective measure of idea originality (Beaty and Johnson 2021; Dumas et al. 2021) compared to traditional, subjective scoring (Silvia et al. 2008). Such aggregated text-based measures of the semantic distance are extremely useful in capturing commonalities across people and individual differences among them (Beaty and Johnson 2021; Dumas et al. 2021). Critically, such methods have been extremely useful in studying the role of semantic memory in creative thinking (Abraham and Bubic 2015; Benedek et al. 2023). As mentioned above, assumptions regarding common knowledge structure across people are well-established in metacognitive research as well (Ackerman and Beller 2017; Koriat 1993; Koriat and Levy-Sadot 2001). However, two potential limitations of such text-based methods are that they are based on aggregated large textual corpora that may not fully map onto each participants' semantic knowledge structure (Kenett 2019). Thus, future research should replicate and extend our current findings by applying methods to estimate individual-based semantic memory networks (Benedek et al. 2017; He et al. 2020; Ovando-Tellez et al. 2022b; Wulff et al. 2022). Such a design would allow matching individual-based originality scores (based on the semantic distance in their own semantic memory network) and originality judgments.

More generally, our findings highlight the role of metacognitive processes in creative thinking, which are still largely unknown (Puente-Díaz 2023). Our results replicate previous findings (Sidi et al. 2020) that people consistently underestimate their originality, while also judging reliably their initial ideas as more creative than their later ones (serial order effect). However, our Experiment 2 demonstrated that people sometimes rely on the serial order too strongly, on the semantic distance too little, and their creative process is affected by examples that vary in originality and the semantic distance. This collection of findings conveys insights for real-life applications. For example, encouraging higher originality judgments may better align them with the remoteness of the later responses and could also facilitate generation of more creative ideas. Furthermore, Lloyd-Cox et al. (Lloyd-Cox et al. 2022) recently showed how when people evaluate the creativity of AUT responses they over-emphasize novelty, while when evaluating the creativity of solutions to real-life problems they over-emphasize appropriateness. Similarly to Lloyd-Cox et al. (Lloyd-Cox et al. 2022), our findings demonstrate malleability of cue validity (predicting objective performance) and cue utilization (predicting judgments; Koriat 1997; Runco and Acar 2012) in creative tasks, contingent on various task demands and contexts. Finally, while we did not instruct people to "be creative" in their responses, such instructions are quite impactful on people's responses in the AUT, especially compared to "be-fluent" instructions (Forthmann et al. 2016; Niu and Liu 2009). Given the dissociation we found between originality scores and originality judgments, future studies of creativity may also consider examining a metacognitive emphasis (e.g., "be-confident" type) as part of the instructions.

5. Conclusions

The present study takes an empirical step forward in converging creativity and metacognitive research. So far, little is known about the role of metacognitive processes in creative thinking (Lebuda and Benedek 2023), and the potential dissociation between performance and judgments in creativity tasks (Benedek et al. 2016; Sidi et al. 2020). Our study uniquely shows the role of semantic distance as a metacognitive cue affecting originality judgments, its role as a biasing metacognitive cue in creativity, and a double dissociation between the opposite utilization effects of semantic distance and serial order on originality scores and originality judgments. Overall, our work highlights the need for further empirical research on the role of metacognitive processes taking place in creative thinking, research that can further elucidate the complexity of the creative process.

Author Contributions: Conceptualization, Y.N.K., N.G. and R.A.; methodology, Y.K, N.G. and R.A.; formal analysis, N.G. and R.A.; investigation, N.G.; writing—original draft preparation, Y.N.K., N.G. and R.A.; writing—review and editing, Y.N.K. and R.A.; supervision, Y.N.K. and R.A. All authors have read and agreed to the published version of the manuscript.

Funding: This research received funding a grant from the Israel Science Foundation to R.A.

Institutional Review Board Statement: This study was approved by the Institutional Review Board of the Technion—Israel Institute of Technology.

Informed Consent Statement: Informed consent was obtained from all participants involved in study two.

Data Availability Statement: Materials, data, BEVoCI paper, and analyses' R code for Experiment 2, which includes newly collected data, are available in OSF: https://osf.io/t6wby/?view_only=16049b17a58b4cf48ab0ba39df48f483 (accessed on 9 January 2023).

Acknowledgments: We thank Sidi et al. for allowing us to reanalyze their previously collected data. We thank Paul Feigin for statistical consulting, and Talia Wise for her help in editing the manuscript.

Conflicts of Interest: The authors declare no conflict of interest.

Note

1 These regressions were performed for each cue separately, without the other cues, for simplicity of the models.

References

Abraham, Anna, and Andreja Bubic. 2015. Semantic memory as the root of imagination. *Frontiers in Psychology* 6: 325. [CrossRef] [PubMed]

Acar, Selcuk, and Mark A. Runco. 2019. Divergent thinking: New methods, recent research, and extended theory. *Psychology of Aesthetics, Creativity, and the Arts* 13: 153–58. [CrossRef]

Ackerman, Rakefet. 2019. Heuristic cues for meta-reasoning judgments: Review and methodology. *Psihologijske Teme* 28: 1–20. [CrossRef]

Ackerman, Rakefet. forthcoming. Bird's-Eye View of Cue Integration: Exposing instructional and task design factors which bias problem solvers.

Ackerman, Rakefet, and Hagar Zalmanov. 2012. The persistence of the fluency–confidence association in problem solving. *Psychonomic Bulletin & Review* 19: 1187–92.

Ackerman, Rakefet, and Valerie A. Thompson. 2017. Meta-reasoning: Monitoring and control of thinking and reasoning. *Trends in Cognitive Sciences* 21: 607–17. [CrossRef]

Ackerman, Rakefet, and Yael Beller. 2017. Shared and distinct cue utilization for metacognitive judgements during reasoning and memorisation. *Thinking & Reasoning* 23: 376–408.

Ackerman, Rakefet, Avigdor Gal, Tomer Sagi, and Roee Shraga. 2019. A cognitive model of human bias in matching. Paper presented at the Pacific Rim International Conference on Artificial Intelligence, Cuvu, Yanuca Island, Fiji, August 26–30.

Ackerman, Rakefet, Elad Yom-Tov, and Ilan Torgovitsky. 2020. Using confidence and consensuality to predict time invested in problem solving and in real-life web searching. *Cognition* 199: 104248. [CrossRef]

Amabile, Teresa M. 1983. The social psychology of creativity: A componential conceptualization. *Journal of Personality and Social Psychology* 45: 357–76. [CrossRef]

Bai, Honghong, Paul P. M. Leseman, Mirjam Moerbeek, Evelyn H. Kroesbergen, and Hanna Mulder. 2021. Serial order effect in divergent thinking in five-to six-year-olds: Individual differences as related to executive functions. *Journal of Intelligence* 9: 20. [CrossRef]

Bajšanski, Igor, and Valnea Žauhar. 2019. The relationship between consistency and consensuality in syllogistic reasoning. *Psihologijske Teme* 28: 73–91. [CrossRef]

Beaty, Roger E., and Dan R. Johnson. 2021. Automating creativity assessment with SemDis: An open platform for computing semantic distance. *Behavior Research Methods* 53: 757–80. [CrossRef]

Beaty, Roger E., and Paul J. Silvia. 2012. Why do ideas get more creative over time? An executive interpretation of the serial order effect in divergent thinking tasks. *Psychology of Aesthetics, Creativity and the Arts* 6: 309–19. [CrossRef]

Beaty, Roger E., Paul J. Silvia, Emily C. Nusbaum, Emanuel Jauk, and Mathias Benedek. 2014. The roles of associative and executive processes in creative cognition. *Memory & Cognition* 42: 1186–97. [CrossRef]

Beaty, Roger E., Yoed N. Kenett, Richard W. Hass, and Daniel L. Schacter. 2022. Semantic memory and creativity: The costs and benefits of semantic memory structure in generating original ideas. *Thinking & Reasoning*, 1–35. [CrossRef]

Benedek, Mathias, Nora Nordtvedt, Emanuel Jauk, Corinna Koschmieder, Jürgen Pretsch, Georg Krammer, and Aljoscha C. Neubauer. 2016. Assessment of creativity evaluation skills: A psychometric investigation in prospective teachers. *Thinking Skills and Creativity* 21: 75–84. [CrossRef]

Benedek, Mathias, Roger E. Beaty, Daniel L. Schacter, and Yoed N. Kenett. 2023. The role of memory in creative ideation. *Nature Reviews Psychology*. [CrossRef]

Benedek, Mathias, Yoed N. Kenett, Konstantin Umdasch, David Anaki, Miriam Faust, and Aljoscha C. Neubauer. 2017. How semantic memory structure and intelligence contribute to creative thought: A network science approach. *Thinking & Reasoning* 23: 158–83.

Bjork, Robert A., John Dunlosky, and Nate Kornell. 2013. Self-regulated learning: Beliefs, techniques, and illusions. *Annual Review of Psychology* 64: 417–44. [CrossRef]

Bowden, Edward M., and Mark Jung-Beeman. 2003. One hundred forty-four compound remote associate problems: Short insight-like problems with one-word solutions. *Behavioral Research, Methods, Instruments, and Computers* 35: 634–39. [CrossRef]

Collins, Allan M., and Elizabeth F. Loftus. 1975. A spreading-activation theory of semantic processing. *Psychological Review* 82: 407–28. [CrossRef]

Cseh, Genevieve M., and Karl K. Jeffries. 2019. A scattered CAT: A critical evaluation of the consensual assessment technique for creativity research. *Psychology of Aesthetics, Creativity, and the Arts* 13: 159–66. [CrossRef]

Dumas, Denis, Peter Organisciak, and Michael Doherty. 2021. Measuring divergent thinking originality with human raters and text-mining models: A psychometric comparison of methods. *Psychology of Aesthetics, Creativity, and the Arts* 15: 645–63. [CrossRef]

Fiedler, Klaus, Rakefet Ackerman, and Chiara Scarampi. 2019. Metacognition: Monitoring and controlling one's own knowledge, reasoning and decisions. In *The Psychology of Human Thought: An Introduction*. Edited by Robert J. Sternberg and Joachim Funke. Heidelberg: Heidelberg University Publishin, pp. 89–111.

Forthmann, Boris, Anne Gerwig, Heinz Holling, Pınar Çelik, Martin Storme, and Todd Lubart. 2016. The be-creative effect in divergent thinking: The interplay of instruction and object frequency. *Intelligence* 57: 25–32. [CrossRef]

Forthmann, Boris, Dorota M. Jankowska, and Maciej Karwowski. 2021. How reliable and valid are frequency-based originality scores? Evidence from a sample of children and adolescents. *Thinking Skills and Creativity* 41: 100851. [CrossRef]

Forthmann, Boris, Oluwatosin Oyebade, Adebusola Ojo, Fritz Günther, and Heinz Holling. 2019. Application of latent semantic analysis to divergent thinking is biased by elaboration. *The Journal of Creative Behavior* 53: 55–575. [CrossRef]

Forthmann, Boris, Sue Hyeon Paek, Denis Dumas, Baptiste Barbot, and Heinz Holling. 2020. Scrutinizing the basis of originality in divergent thinking tests: On the measurement precision of response propensity estimates. *British Journal of Educational Psychology* 90: 683–99. [CrossRef]

Gilhooly, Kenneth J., Evridiki Fioratou, Susan H. Anthony, and Victor Wynn. 2007. Divergent thinking: Strategies and executive involvement in generating novel uses for familiar objects. *British Journal of Psychology* 98: 611–25. [CrossRef]

Guilford, Joy P. 1967. Creativity: Yesterday, today and tomorrow. *The Journal of Creative Behavior* 1: 3–14. [CrossRef]

Günther, Fritz, Luca Rinaldi, and Marco Marelli. 2019. Vector-space models of semantic representation from a cognitive perspective: A discussion of common misconceptions. *Perspectives on Psychological Science* 14: 1006–33. [CrossRef]

Guo, Yawei, Shengjie Lin, Selcuk Acar, Shuxian Jin, Xizheng Xu, Ye Feng, and Yuntao Zeng. 2022. Divergent Thinking and Evaluative Skill: A Meta-Analysis. *The Journal of Creative Behavior* 56: 432–48. [CrossRef]

Hass, Richard W., Marisa Rivera, and Paul J. Silvia. 2018. On the dependability and feasibility of layperson ratings of divergent thinking. *Frontiers in Psychology* 9: 1343. [CrossRef] [PubMed]

He, Li, Yoed N. Kenett, Kaixiang Zhuang, Cheng Liu, Rongcan Zeng, Tingrui Yan, Tengbin Huo, and Jiang Qiu. 2020. The relation between semantic memory structure, associative abilities, and verbal and figural creativity. *Thinking & Reasoning* 27: 268–93.

Heinen, David J. P., and Dan R. Johnson. 2018. Semantic distance: An automated measure of creativity that is novel and appropriate. *Psychology of Aesthetics, Creativity, and the Arts* 12: 144–56. [CrossRef]

Jauk, Emanuel, Mathias Benedek, and Aljoscha C. Neubauer. 2014. The road to creative achievement: A latent variable model of ability and personality predictors. *European Journal of Personality* 28: 95–105. [CrossRef]

Joulin, Armand, Edouard Grave, Piotr Bojanowski, Matthijs Douze, Hérve Jégou, and Tomas Mikolov. 2016. Fasttext. zip: Compressing text classification models. *arXiv* arXiv:1612.03651.

Karwowski, Maciej, Marta Czerwonka, and James C. Kaufman. 2020. Does intelligence strengthen creative metacognition? *Psychology of Aesthetics, Creativity, and the Arts* 14: 353–60. [CrossRef]

Kaufman, James C. 2019. Self-assessments of creativity: Not ideal, but better than you think. *Psychology of Aesthetics, Creativity, and the Arts* 13: 187–92. [CrossRef]

Kenett, Yoed N. 2018. Going the extra creative mile: The Role of Semantic distance in creativity—Theory, Research, and Measurement. In *The Cambridge Handbook of the Neuroscience of Creativity*. Edited by Rex E. Jung and Oshin Vartanian. New York: Cambridge University Press, pp. 233–48.

Kenett, Yoed N. 2019. What can quantitative measures of semantic distance tell us about creativity? *Current Opinion in Behavioral Sciences* 27: 11–16. [CrossRef]

Kenett, Yoed N., and Miriam Faust. 2019. A semantic network cartography of the creative mind. *Trends in Cognitive Sciences* 23: 271–74. [CrossRef]

Koriat, Asher. 1993. How do we know that we know? The accessibility model of the feeling of knowing. *Psychological Review* 100: 609. [CrossRef]

Koriat, Asher. 1997. Monitoring one's own knowledge during study: A cue-utilization approach to judgments of learning. *Journal of Experimental Psychology: General* 126: 349–70. [CrossRef]

Koriat, Asher. 2008. Subjective confidence in one's answers: The consensuality principle. *Journal of Experimental Psychology: Learning, Memory, and Cognition* 34: 945–59. [CrossRef]

Koriat, Asher, and Ravit Levy-Sadot. 2001. The combined contributions of the cue-familiarity and accessibility heuristics to feelings of knowing. *Journal of Experimental Psychology: Learning, Memory, and Cognition* 27: 34–53. [CrossRef]

Kumar, Abhilasha A. 2021. Semantic memory: A review of methods, models, and current challenges. *Psychonomic Bulletin & Review* 28: 40–80.

Lebuda, Isabela, and Mathias Benedek. 2023. A systematic framework of creative metacognition. *PsyArxiv*. [CrossRef]

Lloyd-Cox, James, Alan Pickering, and Joydeep Bhattacharya. 2022. Evaluating Creativity: How Idea Context and Rater Personality Affect Considerations of Novelty and Usefulness. *Creativity Research Journal* 34: 373–90. [CrossRef]

Mandera, Paweł, Emmanuel Keuleers, and Marc Brysbaert. 2015. How useful are corpus-based methods for extrapolating psycholinguistic variables? *The Quarterly Journal of Experimental Psychology* 8: 1628–42. [CrossRef] [PubMed]

Mandera, Paweł, Emmanuel Keuleers, and Marc Brysbaert. 2017. Explaining human performance in psycholinguistic tasks with models of semantic similarity based on prediction and counting: A review and empirical validation. *Journal of Memory and Language* 92: 57–78. [CrossRef]
Matheson, Heath E., and Yoed N. Kenett. 2021. A novel coding scheme for assessing responses in divergent thinking: An embodied approach. *Psychology of Aesthetics, Creativity, and the Arts* 15: 412–25. [CrossRef]
Mednick, Sarnoff A. 1962. The associative basis of the creative process. *Psychological Review* 69: 220–32. [CrossRef]
Metcalfe, Janet, and Bridgid Finn. 2008. Familiarity and retrieval processes in delayed judgments of learning. *Journal of Experimental Psychology: Learning, Memory, and Cognition* 34: 1084–97. [CrossRef]
Niu, Weihua, and Dan Liu. 2009. Enhancing creativity: A comparison between effects of an indicative instruction "to be creative" and a more elaborate heuristic instruction on Chinese student creativity. *Psychology of Aesthetics, Creativity, and the Arts* 3: 93–98. [CrossRef]
Nusbaum, Emily C., Paul J. Silvia, and Roger E. Beaty. 2014. Ready, set, create: What instructing people to "be creative" reveals about the meaning and mechanisms of divergent thinking. *Psychology of Aesthetics, Creativity, and the Arts* 8: 423–32. [CrossRef]
Ovando-Tellez, Marcela, Mathias Benedek, Yoed N. Kenett, Thomas Hills, Sarah Bouanane, Matthieu Bernard, Joan Belo, Theophile Bieth, and Emmanuelle Volle. 2022a. An investigation of the cognitive and neural correlates of semantic memory search related to creative ability. *Communications Biology* 5: 1–16. [CrossRef] [PubMed]
Ovando-Tellez, Marcela, Yoed N. Kenett, Mathias Benedek, Matthieu Bernard, Joan Belo, Benoit Beranger, Theophile Bieth, and Emmanuelle Volle. 2022b. Brain connectivity–based prediction of real-life creativity is mediated by semantic memory structure. *Science Advances* 8: eabl4294. [CrossRef] [PubMed]
Paek, Sue Hyeon, Ahmed M. Abdulla Alabbasi, Selcuk Acar, and Mark A. Runco. 2021. Is more time better for divergent thinking? A meta-analysis of the time-on-task effect on divergent thinking. *Thinking Skills and Creativity* 41: 100894. [CrossRef]
Perry-Smith, Jill E., and Pier Vittorio Mannucci. 2017. From creativity to innovation: The social network drivers of the four phases of the idea journey. *Academy of Management Review* 42: 53–79. [CrossRef]
Pinheiro, José, Douglas Bates, Saikat DebRoy, Deepayan Sarkar, and R Core Team. 2019. nlme: Linear and Nonlinear Mixed Effects Models. R Package Version 3.1–141. Available online: http://CRAN.R-project.org/package=nlme (accessed on 1 May 2022).
Puente-Díaz, Rogelio, Judith Cavazos-Arroyo, and Fernanda Vargas-Barrera. 2021. Metacognitive feelings as a source of information in the evaluation and selection of creative ideas. *Thinking Skills and Creativity* 39: 100767. [CrossRef]
Puente-Díaz, Rogelio. 2023. Metacognitive feelings as a source of information for the creative process: A conceptual exploration. *Journal of Intelligence* 11: 49. [CrossRef]
Rominger, Christian, Mathias Benedek, Izabela Lebuda, Corinna M. Perchtold-Stefan, Andreas R. Schwerdtfeger, Ilona Papousek, and Andreas Fink. 2022. Functional brain activation patterns of creative metacognitive monitoring. *Neuropsychologia* 177: 108416. [CrossRef]
Runco, Mark A., and Garrett J. Jaeger. 2012. The standard definition of creativity. *Creativity Research Journal* 24: 92–96. [CrossRef]
Runco, Mark A., and Selcuk Acar. 2012. Divergent thinking as an indicator of creative potential. *Creativity Research Journal* 24: 66–75. [CrossRef]
Said-Metwaly, Sameh, Belén Fernández-Castilla, Eva Kyndt, and Wim Van den Noortgate. 2020. Testing conditions and creative performance: Meta-analyses of the impact of time limits and instructions. *Psychology of Aesthetics, Creativity, and the Arts* 14: 15–38. [CrossRef]
Sargent, Matthew, Alex LePage, Yoed N. Kenett, and Heath E. Matheson. 2023. The effects of environmental scene and body posture on embodied strategies in creative thinking. *Creativity Research Journal: 1–15*. [CrossRef]
Scott, Graham G., Anne Keitel, Marc Becirspahic, Bo Yao, and Sara C. Sereno. 2019. The Glasgow norms: Ratings of 5500 words on nine scales. *Behavior Research Methods* 51: 1258–70. [CrossRef]
Sidi, Yael, Ilan Torgovitsky, Daniela Soibelman, Ella Miron-Spektor, and Rakefet Ackerman. 2020. You may be more original than you think: Predictable biases in self-assessment of originality. *Acta Psychologica* 203: 103002. [CrossRef]
Silvia, Paul J., Beate P. Winterstein, John T. Willse, Christopher M. Barona, Joshua T. Cram, Karl I. Hess, Jenna L. Martinez, and Crystal A. Richard. 2008. Assessing creativity with divergent thinking tasks: Exploring the reliability and validity of new subjective scoring methods. *Psychology of Aesthetics, Creativity, and the Arts* 2: 68–85. [CrossRef]
Sowden, Paul T., Andrew Pringle, and Liane Gabora. 2014. The shifting sands of creative thinking: Connections to dual-process theory. *Thinking & Reasoning* 21: 40–60. [CrossRef]
Torrance, E. Paul. 1966. Nurture of creative talents. *Theory into Practice* 5: 167–73. [CrossRef]
Torrance, E. Paul. 1972. Predictive validity of the torrance tests of creative thinking. *The Journal of Creative Behavior* 6: 236–52. [CrossRef]
Undorf, Monika, and Arndt Bröder. 2021. Metamemory for pictures of naturalistic scenes: Assessment of accuracy and cue utilization. *Memory & Cognition* 49: 1405–22.
Undorf, Monika, Anke Söllner, and Arndt Bröder. 2018. Simultaneous utilization of multiple cues in judgments of learning. *Memory & Cognition* 46: 507–19.
Volle, Emmanuelle. 2018. Associative and controlled cognition in divergent thinking: Theoretical, experimental, neuroimaging evidence, and new directions. In *The Cambridge Handbook of the Neuroscience of Creativity*. Edited by Rex E. Jung and Oshin Vartanian. New York: Cambridge University Press, pp. 333–62.

Wu, Ching-Lin, Shih-Yuan Huang, Pei-Zhen Chen, and Hsueh-Chih Chen. 2020. A systematic review of creativity-related studies applying the remote associates test from 2000 to 2019. *Frontiers in Psychology* 11: 573432. [CrossRef]
Wulff, Dirk U., Simon De Deyne, Samuel Aeschbach, and Rui Mata. 2022. Using network science to understand the aging lexicon: Linking individuals' experience, semantic networks, and cognitive performance. *Topics in Cognitive Science* 14: 93–110. [CrossRef] [PubMed]

Journal of
Intelligence

MDPI

Article

A Validated Ontology for Metareasoning in Intelligent Systems

Manuel F. Caro [1,*], Michael T. Cox [2] and Raúl E. Toscano-Miranda [1]

[1] Education, Technology & Language (EduTLan Research Group), Department of Educational Informatics, University of Córdoba, Carrera 6 No. 77-305, Montería 230002, Córdoba, Colombia
[2] Education, Department of Computer Science & Engineering, College of Engineering and Computer Science, Wright State University, Dayton, OH 45324, USA
* Correspondence: manuelcaro@correo.unicordoba.edu.co

Abstract: Metareasoning suffers from the heterogeneity problem, in which different researchers build diverse metareasoning models for intelligent systems with comparable functionality but differing contexts, ambiguous terminology, and occasionally contradicting features and descriptions. This article presents an ontology-driven knowledge representation for metareasoning in intelligent systems. The proposed ontology, called IM-Onto, provides a visual means of sharing a common understanding of the structure and relationships between terms and concepts. A rigorous research method was followed to ensure that the two main requirements of the ontology (integrity based on relevant knowledge and acceptance by researchers and practitioners) were met. The high accuracy rate indicates that most of the knowledge elements in the ontology are useful information for the integration of multiple types of metareasoning problems in intelligent systems.

Keywords: metareasoning ontology; intelligent systems; metareasoning problem; ontology validation; heterogeneity problem

1. Introduction

Metareasoning refers to the processes that monitor the progress of our reasoning and problem-solving activities and regulate the time and effort devoted to them (Ackerman and Thompson 2017; Horvitz 1987; Russell and Wefald 1991). Metareasoning is often defined as reasoning about reasoning (e.g., Hayes-Roth et al. 1983; Kuokka 1991), which implies intelligent decisions about how to think (Griffiths et al. 2019). In research on artificial intelligence, metareasoning plays a central role in the definition and design of rational agents that can operate on performance-limited hardware and interact with their environment in real time (Conitzer and Sandholm 2003; Dannenhauer et al. 2014; Svegliato and Zilberstein 2018). In the cognitive systems research community, metareasoning (or computational metacognition) is key to modeling high-level decision making, self-explanation and introspection (Cox 2011; Dannenhauer et al. 2018). Metareasoning also enables intelligent agents to optimize their own decision-making process to produce effective action in a timely manner (Svegliato et al. 2021).

However, metareasoning suffers from the heterogeneity problem, where differing metareasoning models with similar functionalities are being developed for intelligent agents in heterogeneous environments by different researchers. These models have conflicting features, descriptions, qualities, algorithms, and non-standard conventions because they come from different areas of knowledge, such as cognitive science (Ackerman and Thompson 2017), psychology (Dunlosky and Bjork 2008), education (Caro et al. 2014), computer science (Borghetti and Gini 2008), engineering (Caleiro et al. 2005), and AI (Conitzer and Sandholm 2003; Cox and Raja 2011; Svegliato and Zilberstein 2018). Due to these types of heterogeneities and the increasing application domains (e.g., online planning, anytime algorithms, brittleness problem of AI systems, introspective systems and long-duration missions), the research community now needs a common understanding of the terms and concepts related to metareasoning in intelligent systems. In this sense, there is a requirement to create a standard or global ontology that provides a common knowledge representation of the metareasoning domains.

Therefore, the main objective of this paper is to present an ontology that allows specifying a common language and a conceptualization of metareasoning in the domain of AI, which can be used to represent different metareasoning problems in intelligent systems. An ontology is defined as a formal specification of a shared conceptualization (Gruber 1995). Furthermore, ontologies allow the reuse of domain knowledge, thus making domain assumptions explicit and helping us to clarify any ambiguities (Horridge et al. 2019; Noy and McGuinness 2001).

Several studies have constructed ontologies for describing various aspects of metareasoning. Unfortunately, no comprehensive set of standardized features exists for describing the metareasoning domain. Therefore, each study has developed partial ontologies to address specific problems such as failures in AI systems (Schmill et al. 2007), the metacognitive cycle (Schmill et al. 2011) and meta-level control (Madera-Doval 2019). A general ontology for broadly describing the metareasoning domain and detailed ontologies for each metareasoning problem is still missing in the literature.

The main contributions of this work are:

- The Integrated Metareasoning Ontology, an ontology for the representation of different metareasoning problems in intelligent systems. We describe a framework for the use of the formally defined semantics of the classes, the individuals, and the properties of the ontology to construct the knowledge representation structure necessary to monitor and control the reasoning processes in intelligent agents. The proposed ontology also reuses existing ontologies that are used for partial modelling of some aspects of the metareasoning domain.

The remaining sections of the paper are as follows. Section 2 sketches the research methodology used in our work, and Section 3 details the development and validation of the ontology using this approach. It also describes a case study that applies the ontology to the problem of allocating student teams in internship programs. Section 4 then provides a discussion of the results; the summary of the study's key findings and its limitations is included in Section 5.

2. Materials and Methods

Approaches to ontology development have been evolving in recent years. Authors have proposed updated methodologies based on the review and identification of the limitations of existing ontology development methods.

The methodology employed for ontology development is a hybrid framework based on DSR (Hevner et al. 2004; Peffers et al. 2012), which leverages the "methontology" approach (Baccigalupo and Plaza 2007), the 7-step methodology for ontology development (Noy and McGuinness 2001) and the "Uschold and Gruninger" ontology building approach (Uschold and Gruninger 1996). The Uschold and Gruninger approach provides detailed information for delineating the purpose and scope, ontology formalization, evaluation, and documentation. On the other hand, the methontology approach provides a more nuanced approach toward knowledge acquisition, conceptualization, and implementation.

These phases have been identified in other studies, such as Badr et al. (2013). Figure 1 depicts this step-by-step research framework for developing the ontology and implementing it pragmatically in a case study.

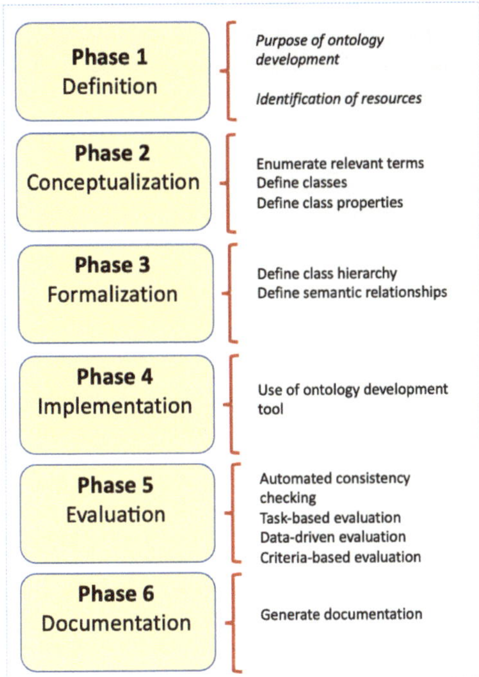

Figure 1. Research method. Source: Elaborated by the authors based on (Badr et al. 2013).

Based on this framework, it was first important to define the scope and purpose of the ontology and to identify key resources from which the ontology is derived (Phase 1). This was executed by enumerating and answering a set of ontology specification questions and by listing specific functional roles for the ontology. This phase was followed by knowledge capture and the abstraction of relevant terms and their relationships in the domain from the resources (Phase 2). The formalization process is then performed to produce meaningful models based on semantic relationships (Phase 3). This phase identifies the hierarchies of subclasses as well as the existing semantic relationships between the different classes. Beyond this stage, the ontology was formally coded using RDF/OWL (Resource Description Framework/Web Ontology Language) (Van Assem et al. 2006) to make it computer-interpretable (Phase 4). The coded ontology underwent internal logical checks and was subsequently implemented in the case study. Further validation of the results of a case study was then performed through a data-driven and criteria-based evaluation (Phase 5). The case study addressed a particular type of team composition problem in an academic unit of a higher education institution. Finally, the documentation for each class and the user manual and technical aspects of the ontology was generated (Phase 6). The next section now presents the details of these six phases.

3. Results

In summary, the methodology we follow for development includes ontology definition, conceptualization, formalization, implementation, evaluation, and documentation. Here we describe each in turn, giving specific technical elaborations.

3.1. Definition

The objective of a requirements specification is to produce a formal or informal description of the rationale behind the development of an ontology and to elucidate its potential uses (Fernández-López et al. 1997). This can be done by documenting this information in natural language, by using specification questions, and by developing intermediate representations. This study used the following set of specification questions adapted from Fernández-López et al. (1997) to help determine the ontological rationale. This is augmented by a use case diagram in response to the "intended use" question.

- What is the purpose? The objective of the ontology is to facilitate the integration of metareasoning processes into advanced intelligent systems.
- What is the scope? The ontology will include information on the processes related to metareasoning, such as allocation deliberation time, allocation evaluation effort, detection of reasoning failures, meta-explanations, introspective monitoring and metalevel control.
- Who are the intended end users? Users include research groups in cognitive science, artificial intelligence, and cognitive computing. Although the proposed ontology addresses general topics of meta-reasoning, our focus is the application in educational settings, mainly in solving academic problems in higher education institutions.
- What is the intended use? The main functional roles include the ability to model meta-reasoning problems that can occur in intelligent systems in terms of monitoring and controlling cognitive processes. The ontology is useful for reducing the discrepancies between the data structures required in different components of the meta-reasoning process. Discrepancies between data structures and language syntax make it even more difficult to exchange information between meta-reasoning models, leading to considerable information deviations when data flows are connected through the different designed components. Knowledge sharing among researchers, academics, and developers can be facilitated by having an ontology for the meta-reasoning domain. This is because the ontology reduces the ambiguity of terms and has a controlled vocabulary. The ontology provides a semantic basis for communication between designers, academics, and researchers so that designers share a common understanding of knowledge.

Existing ontologies and publications in specialized databases were the main resources identified as sources of information for the extraction of terms related to meta-reasoning. Science Direct, IEEE Xplorer, ACM, Springer and IGI Global were searched using the keywords "metareasoning", "meta-reasoning", "metareasoning problem", "meta-level control", "metacognition", "anytime algorithm" and combinations of these keywords. Similarly, a complementary search was conducted in Google Scholar and Scopus.

3.2. Conceptualization

Following the definition of the ontology requirements specification, the next step is to describe how the domain knowledge was acquired and formalized. The main steps in this phase include:

(i) Listing the relevant terms in the ontology;
(ii) Defining classes;
(iii) Defining class properties with specifications for their range and domain (Noy and McGuinness 2001).

3.2.1. Listing the Relevant Terms in the Ontology

Knowledge relevant to the ontology was initially acquired through a detailed review of existing studies on metareasoning ontologies. Two key publications proved especially valuable. Table 1 lists the main terms that appear in existing ontologies from this literature in the metareasoning domain.

Table 1. Literature sources for ontology knowledge and terms used in existing ontologies related to metareasoning.

Literature Source	Ontological Terms
(Schmill et al. 2007, 2011)	sensor, reasoning process, rebuild models, recommendation recover, reinforcement learning, replan, response ontology, result, reward, self-awareness, sensor, sensor failure, sensor malfunction, sensor not reporting, state, system, failure, time, unanticipated perturbation
(Madera-Doval 2019)	agent, metacognition, self-regulation, metamemory, introspective monitoring, meta-level control, cognitive elements, cognitive level, task, reasoning, metareasoning, reasoning task, metareasoning task, object level, cognitive function, perception, situation assessment, categorization, recognition, belief maintenance, problem solving, planning, prediction, expectation, sensor, observation

Given these terms as a base, we extended them by reviewing a select set of key papers about metareasoning and related topics. Table 2 shows the literature reviewed, including computational metacognition, introspective monitoring, and anytime planning algorithms.

Table 2. Papers reviewed for the stage of gathering information, knowledge acquisition and conceptualization.

Research Paper	Terms per Paper	Citations per Paper *
(Russell and Wefald 1991)	48	444
(Conitzer and Sandholm 2003)	67	41
(Cox 2005)	63	261
(Anderson and Oates 2007)	36	91
(Schmill et al. 2007)	77	21
(Cox and Raja 2008)	55	100
(Chen et al. 2013)	37	3
(Lin et al. 2015)	54	46
(Lieder and Griffiths 2017)	37	23
(Ackerman and Thompson 2017)	36	105
(Milli et al. 2017)	35	39
(Cserna et al. 2017)	34	9
(Karpas et al. 2018)	22	5
(Farmer 2018)	7	5
(Madera-Doval 2019)	27	1
(Houeland and Aamodt 2018)	25	5
(Parashar et al. 2018)	33	1
(Griffiths et al. 2019)	18	29
(Svegliato et al. 2018)	41	16
(Sung et al. 2021)	36	0

* Google Scholar was the source for the number of citations of the papers at the time of writing this article. Data collected in June 2021.

The concept of data saturation was taken as an indicator of when to stop the literature review process. Data saturation occurs when new information is not obtained with additional data collection effort (Fusch and Ness 2015). Figure 2 presents how data saturation was achieved in this study. In this figure, a decreasing return trend (e.g., redundant data) exists in the number of unique information attributes (used to construct concepts) identified from the reviewed papers. After article number 15, no additional unique information elements were identified.

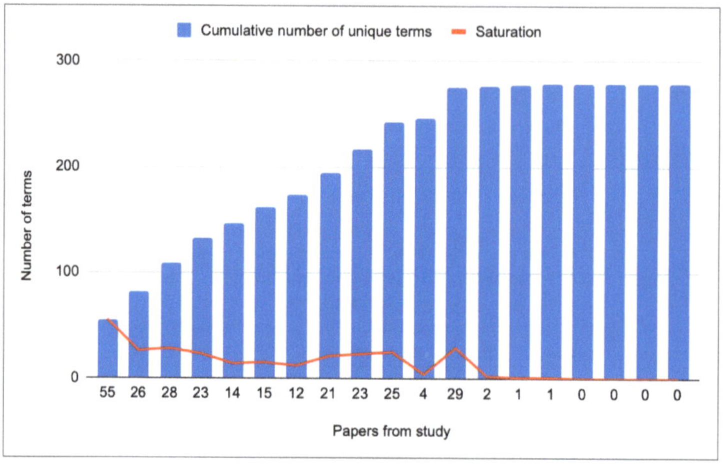

Figure 2. Saturation level as the number of novel terms added by a subsequent article. The papers were processed in the same order as in Table 2. The x-axis corresponds to the new terms provided by each paper after analyzing them in order.

3.2.2. Defining Classes

The classes of the ontology are drawn from the term list presented in Section 3.2.1. We selected terms that describe objects that have an independent existence or represent a collection of individuals or objects. The selected terms define the ontological classes and constitute the nodes in the class hierarchy (see Table 3).

Table 3. List of terms used for class definition in the ontology.

Classes of the Ontology
action, action selection, agent, algorithm, allocate deliberation time, allocating evaluation effort, answer, anytime algorithm, anytime planner, assess anomaly, bayes algorithm, best action, calculate quality of current solution, choose query to ask, cognitive level, cognitive problem, cognitive task, component, computational method, computational step, computational step result, computational time, compute expected utility, compute performance projection, compute solution quality, constraint, control, control policy, current state of the world, default action, default solution, disambiguate state, evaluation task, evaluation test, event, expectation, explain failure, explanation, failure, failure explanation, failure state, function, generate expectation, generate learning goals, get current solution, goal, ground level, ground-level story, guided response, imxp, increment time step, information gathering, information gathering policy, information gathering task, initialize performance history, initialize time step, internal state, introspection, itmxp, knowing current state of the world, knowledge test, learner, learning goal, limited time, make recommendation, mental action, meta level, metacognitive loop, metacognitive problem, metalevel control, metareasoner, metareasoning component, metareasoning problem, metareasoning task, model, model of the self, model of the word, monitor behavior, monotonicity, nag cycle, neural network, nonlinear regression, note anomaly object level, object-level process observation, online control policy, optimal information gathering policy, optimal test policy, outcome state, perception, performance history, performance predictor, performance profile, performance projection, perturbation, perturbation detection, phase, plan, planner, planning, planning task, policy, predict performance, problem, process, profile, problem space, property, question, recommendation action, reactivate learning, reasoner, reasoning, reasoning failure, reasoning problem, reasoning strategy, reasoning system, reasoning task, recommendation, recover, replan, resource, result, run test, save costs, selecting action, sensor, situation assessment, sleep, solution, solution quality, start acting, start anytime algorithm, state, state of the world, stop reasoning, stopping condition, story, story understanding task, strategy, success state, system, task, test, test policy, time, time dependent utility function, trace, unanticipated perturbation, understanding task, unit of time, unusual event, utility function, vector, violated expectation

3.2.3. Define Class Properties with Specification of the Range and Domain

This section describes the relationships and properties of some class examples, but due to the large number that has been identified, we do not report all in the ontology. In cognitive systems with a metalevel-based architecture, the function of the object level (i.e., the reasoning level) is to recognize and solve problems or situations in the environment where the system operates. In contrast, the main function of the meta-level is to monitor and control the reasoning and learning processes that take place at the object level (Cox and Raja 2011; Nelson 1990). The objective of monitoring and control is to achieve more effective results in the processes carried out at the object level.

A meta level has a problem space and a metareasoner (see Table 4). The problem space represents a collection of metareasoning problems and all possible paths to solving them. The metareasoner is the computational algorithm that executes all metareasoning tasks.

Table 4. Properties and domain range of two classes in the ontology.

Class	Property	Domain	Value Restriction
MetaLevel	rdf:subClassOf	CognitiveLevel	
	has_problemspace	ProblemSpace	Non-empty array
	has_metareasoner	Metareasoner	An algorithm
	has_current_metareasoning_loop	Integer	Positive integer
MetareasoningTask	rdf:subClassOf	MetacognitiveTask	
	has_goal	Goal	
	has_id	String	Unique value
	has_input	Array	An array of objects
	has_output	Array	Non-empty
	has_runtime	Number	Positive number
	has_preconditions	State	Non-empty array of states
	has_effects	State	Non-empty array of states
	has_name	String	Alphanumerical value

Meta-reasoning tasks are a particular type of metacognitive task whose purpose is to monitor and control the reasoning processes that take place at the object level (again, see Table 4). Some properties of meta-reasoning tasks are the input parameters, the output, the necessary conditions for its execution, the effects of its execution, as well as the event and completeness states. Table 4 shows the properties and domain range of two classes in the ontology.

3.3. Formalization

Hierarchical classifications use predefined semantic taxonomies (Silla and Freitas 2011). A taxonomy contains only one root class and defines the "Is-A" relationships between classes. An "Is-A" relationship is transitive and asymmetric. Some ontological hierarchies were generated from the explicit descriptions in the reviewed papers, as can be seen in Figures 3 and 4.

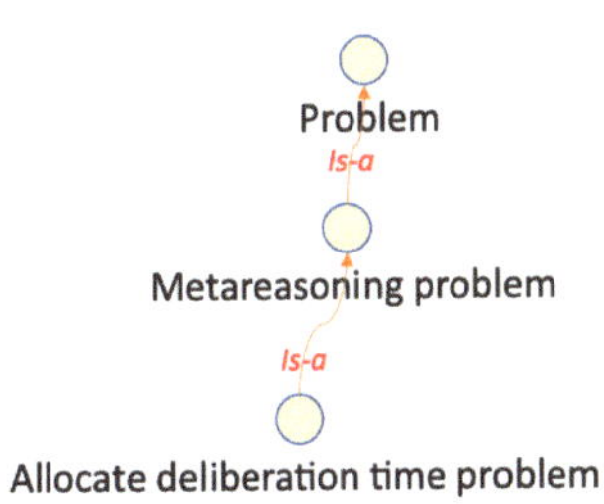

Figure 3. Class hierarchy according to (Conitzer and Sandholm 2003). Figure elaborated by the authors.

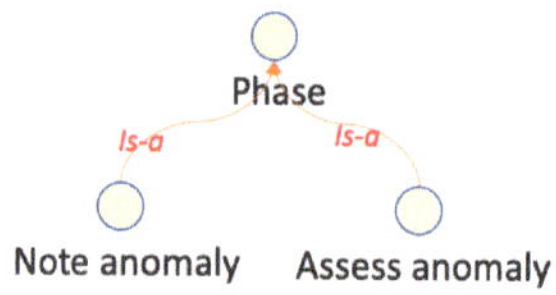

Figure 4. Class hierarchy extracted from a previous ontology in (Schmill et al. 2007). Figure elaborated by the authors.

Other class hierarchies were drawn from existing ontologies, for example:

Finally, other ontological hierarchies arose from the process of concept abstraction. For example, reasoning tasks and metareasoning tasks abstract the task class.

In this ontology, the information about the metareasoning process is modeled according to the standard vocabulary that facilitates a better semantic interpretation that can resolve ambiguities in the terms used in the different data sets, metareasoning applications, or when a little additional knowledge can lead to the discovery of new relationships between terms. In this sense, this ontology is designed to semantically represent a rich and complex knowledge about the domain of metareasoning in intelligent systems.

The semantic relationships in the ontology are given through hierarchical relationships of type Is-Un. When a relationship of this type is established between two classes, one is a superclass, and the other is a subclass. The subclass shares the structure and behavior of the superclass. In the same way, the "HAS" relationships have been implemented, which are structural and describe a set of links, which describe composition connections between the terms.

The ontology has mechanisms to verify the consistency of this knowledge and make explicit the implicit knowledge that can be generated in the monitoring and control processes of reasoning in intelligent systems.

3.4. Implementation: The Integrated Metareasoning Ontology (IM-Onto)

This section presents the technical details of the Integrated Metareasoning Ontology (IM-Onto). To make an ontology machine-readable, it is important to represent it in a formal computational language (Noy and McGuinness 2001). Hence, IM-Onto was implemented using OWL/RDF (Allemang et al. 2005) in the Protégé environment (Horridge et al. 2019).

The advantage of using OWL/RDF is that it allows for richer semantic expressions of concepts, their attributes, and the relationships between them. At the same time, the use of the OWL/RDF model supports improved interoperability or connection of information silos with the added merit of being able to generate implicit semantic inferences based on defined relationships (Horridge et al. 2019). In the context of this study, the use of this model for linking information enables a machine-readable representation (through uniform resource identifiers) that also solves a vocabulary problem in this domain. In the context of this study, the use of this model for metareasoning enables a machine-readable representation (through uniform resource identifiers) that also solves a vocabulary problem in this domain.

IM-Onto consists of seven sub-ontologies that map to key metareasoning problems. These seven are as follows: the Allocating Deliberation Time Problem (ADTP); the Allocating Evaluation Effort Problem (AEEP); the Knowledge Test Problem (KTP); the Stopping Reasoning Problem (SRP); the Gathering Computational Performance Data Problem (GCPDP); the Detection of Reasoning Failure Problem (DRFP); and finally, the Self-Explanation and Self-Understanding Problem (SE&SUP). Figure 5 shows the hierarchy of these subontologies in IM-Onto.

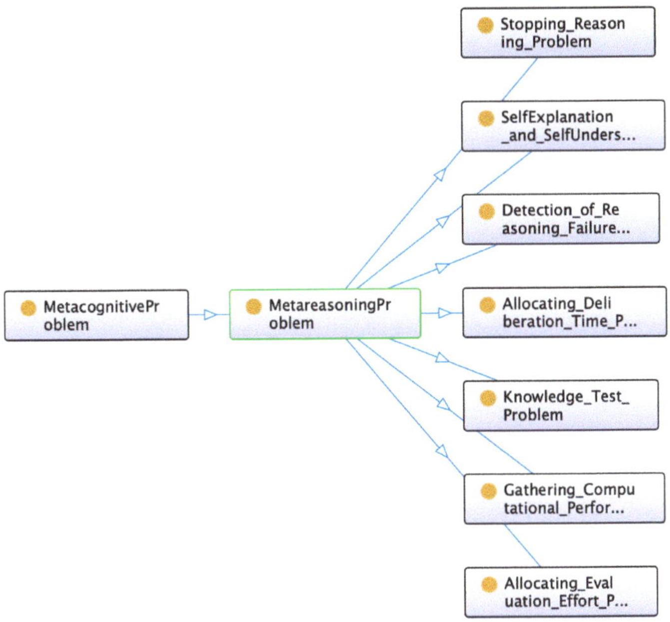

Figure 5. Subontology hierarchy.

To facilitate the ontology reading process and to differentiate an ontology concept or a concept-relation property from a regular term, the model uses the following.

- CAPITAL LETTERS denote concepts defined in the ontology. For example, ALLOCATE_DELIBERATION_TIME and OBJECT_LEVEL are important concepts and are further described in the ontology term detailed description.
- *Italics letters* refer to a property of a relation between two or more ontology concepts. For example, *has_problem_solution* is a property that relates the ontology term PROBLEM and SOLUTION.

3.4.1. The Allocating Deliberation Time Problem (ADTP) Subontology

In this section, the setting where an agent must allocate its deliberation time across different problems is represented. Figure 6 portrays the entire sub-ontology, including the ALLOCATE_DELIBERATION_TIME class with its relations. However, the concept of a METAREASONING_PROBLEM is presented first due to its key role in the overall metareasoning process.

A METAREASONING_PROBLEM is a class that represents the problem an agent has of monitoring and controlling the progress of its own reasoning and problem-solving activities and regulating the time and effort spent on them (Ackerman and Thompson 2017). This definition is a good example of the creation of inferred abstract concepts such as the concept PROBLEM. The PROBLEM class is considered part of the core of the ontology because it is used by all the sub-ontologies. In the proposed design, every PROBLEM is part of a PROBLEM_SPACE and has a SOLUTION. The meta-level analyzes the DEFAULT_SOLUTION for potential anomalies or optimizations after the object level generates it.

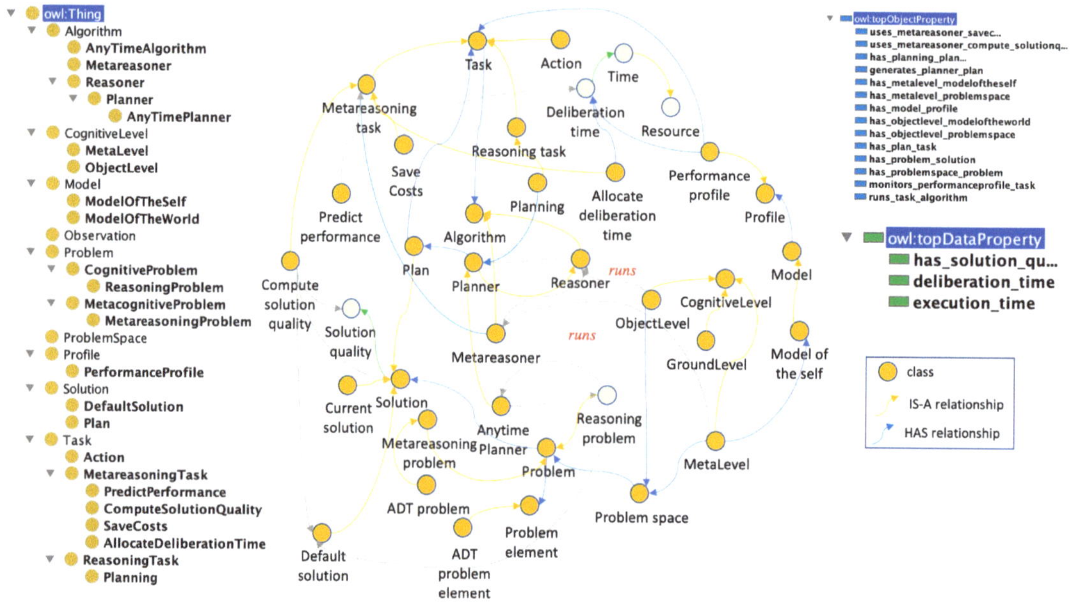

Figure 6. ADTP subontology.

The ADT_PROBLEM class is a particular type of problem for which the main elements are part of the ADT_PROBLEM_ELEMENT class and are listed below: PERFORMANCE_PROFILE, ANYTIME_ALGORITHM and METAREASONING_TASK.

- A PERFORMANCE_PROFILE generally represents a vector with the quality of the solutions of an algorithm that is monitored in reasoning time intervals.
- ANYTIME_ALGORITHM is a class that represents an anytime algorithm whose quality of results gradually improves as computation time increases and can return a valid solution to a problem even if it is interrupted before completion. This kind of algorithm offers a tradeoff between solution quality and computation time, which is expected to find better solutions the longer it keeps running. In the context of this work, an anytime algorithm works to solve a REASONING_PROBLEM.
- The METAREASONING_TASK class refers to tasks that are carried out in the META_LEVEL, and its objective is to monitor and control the reasoning processes that are carried out at the OBJECT_LEVEL. The main metareasoning tasks in this problem are COMPUTE_SOLUTION_QUALITY, PREDICT_PERFORMANCE, ALLOCATE_DELIBERATION_TIME and SAVE_COST.

In this type of problem, a system has several algorithms that can be executed in parallel. Each algorithm solves one instance of a problem, but in some circumstances where execution time is limited, the metalevel must select a subset of algorithms to execute.

For example, a shipping company may have three algorithms to calculate three routes to three different cities. Each route is optimized by an anytime algorithm, but if a time restriction is included, for example, the algorithm only has a third of the normal execution time to optimize the route and save costs, the metalevel reviews the performance profile of the three algorithms in the required time (1/3) and selects the algorithm with the best-expected performance in the required time.

3.4.2. Allocating Evaluation Effort Problem Subontology

In this section, we represent the setting where an agent is faced with multiple options (actions) from which it eventually must choose one. The agent can use deliberation or information gathering to evaluate each action, but with a given limited time, it must decide which ones to evaluate.

Figure 7 shows the entire sub-ontology, including the Dynamically Allocating Evaluation Effort Across Option class (EVALUATION_EFFORT_PROBLEM) with its relations.

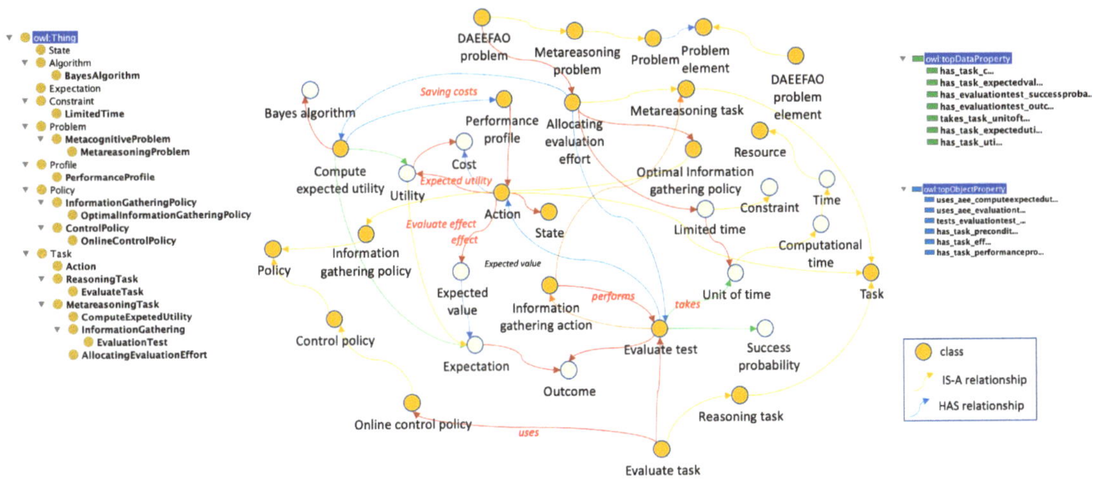

Figure 7. The AEEP subontology.

The EVALUATION_EFFORT _PROBLEM class is a particular type of problem whose main elements are part of the DAEEAO_PROBLEM_ELEMENT class and are listed below: ACTION, EVALUATION_TEST, PERFORMANCE_PROFILE and METAREASONING_TASK.

- Class ACTION represents a set of actions that an agent could evaluate.
- EVALUATION_TEST allows to verify the history of events or failures of the execution of an action. This can bring savings to the ACTION selection evaluation effort.
- The main metareasoning tasks in this problem are ALLOCATING_EVALUATION_EFFORT, COMPUTE_EXPECTED_UTILITY, and EVALUATE_TEST.

In this problem, the metareasoning process has already selected an algorithm to solve a problem, but this algorithm has several different actions to perform the same task. The meta-level must now recommend an action for the object level to perform a specific task, and the recommendation is made based on the highest EXPECTED_UTILITY. In this sense, given the current conditions of the system, the meta-level recommends that the algorithm execute the action with the best-expected utility. The EXPECTED_UTILITY is predicted based on the performance profile of an action, which is stored in the MODEL_OF_THE_SELF.

For example, an application that forms work teams may have three ways of selecting candidates to be part of the teams: (i) randomly, (ii) a person with experience in the position, and (iii) maximum score in tests of attitude. The metareasoning problem arises when the restriction is included, and there is only time to evaluate some of the available actions. Then the META_LEVEL must allocate the evaluation effort only to the actions that have the best EXPECTED_UTILITY. In this sense, the META_LEVEL monitors the PERFORMANCE_PROFILE of previous evaluations of each action in previous similar cases. The ACTION with the best EXPECTED_UTILITY will be chosen for evaluation, and the PERFORMANCE_PROFILE of the action will be modified to reflect the QUALITY of its output.

3.4.3. The Knowledge Test Problem (KTP) Subontology

In this section, we represent the problem whereby an agent has only one item to evaluate, but it must choose the order of deliberation or information-gathering actions for doing so. However, the system is not able to identify the status of the item to be evaluated and needs to run a series of tests (knowledge tests) to disambiguate the status of the item, but the system does not have enough time to run all the tests.

Figure 8 shows the entire sub-ontology, including the class STATE_DISAMBIGUATION_PROBLEM with its relations.

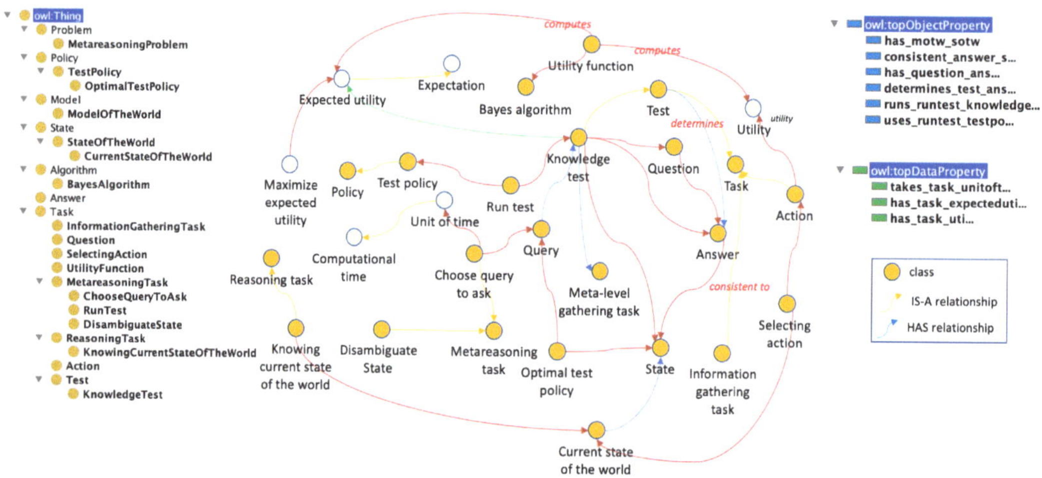

Figure 8. The KTP subontology.

The STATE_DISAMBIGUATION_PROBLEM class is a particular type of METAREASONING_PROBLEM, whose main elements are part of the STATE_DISAMBIGUATION_PROBLEM_ELEMENT class and are listed below: KNOWLEDGE_TEST, QUERY, ANSWER and METAREASONING_TASK.

- The KNOWLEDGE_TEST class determines the answers to a set of questions to determine the current state of the world.
- Class QUERY represents a set of questions to determine the current state of the world.
- Class ANSWER contains the output of the queries.

In this problem, three sub-classes exist for the METAREASONING_TASK class. They are CHOOSE_QUERY_TO_ASK, DISAMBIGUATE_STATE, and METALEVEL_GATHERING_TASK.

This type of METAREASONING_PROBLEM occurs when the OBJECT_LEVEL is not able to disambiguate the state of the world given the current observations. For example, a ship traveling at high-speed stops abruptly because it cannot identify what type of obstacle it has in front of it, whether it is a cloud, another ship, or a building. To identify the type of obstacle, the OBJECT_LEVEL must perform a series of KNOWLEDGE_TESTs, and, depending on the result of the test, it must act. If the obstacle is a cloud, then the ACTION to follow is to move forward and go through the cloud. If the obstacle is another ship, the ACTION is to decelerate and change altitude until the ship has passed, and if the obstacle is a building, the ACTION is to change course. However, if the time to perform the KNOWLEDGE_TESTs is limited and not all can be performed, then the META_LEVEL must select which KNOWLEDGE_TEST to do to change the knowledge about the state of the world. A knowledge test is based on a series of questions and answers that clarify the state of the world. The questions are predefined by the designers, but in complex systems, they can be self-generated from the agent's observations, expectations, and prior knowledge. This case of metareasoning is very particular because it requires that the systems or agents act based on their knowledge and not on the information of their internal state; most systems are not designed for this purpose.

3.4.4. The Stopping Reasoning Problem (SRP) Subontology

Stopping reasoning is the most basic decision of the metareasoning process in intelligent systems. Figure 9 shows the entire sub-ontology, including the STOPPING_REASONING_PROBLEM class with its relations. The STOPPING_REASONING_PROBLEM class is a particular type of METAREASONING_PROBLEM, whose main elements are part of the STOPPING_REASONING_PROBLEM_ELEMENT class and are listed below: TIME_DEPENDENT_UTILITY, SOLUTION_QUALITY, PERFORMANCE_PREDICTOR, PERFORMANCE_ PROFILE_HISTORY, PERFORMANCE_PROFILE_PROJECTION and METAREASONING_TASK.

- A TIME_DEPENDENT_UTILITY represents the utility of a solution computed by an anytime algorithm.
- SOLUTION_QUALITY represents the quality of the solution to a problem. In this case, a solution is generated by an algorithm to solve the current problem.
- A PERFORMANCE_PROFILE_HISTORY represents the past performance of an anytime algorithm as a vector of solution qualities.
- A PERFORMANCE_PROFILE_PROJECTION represents the future performance of an anytime algorithm as a vector of solution qualities.
- PERFORMANCE_PREDICTOR is a function that maps a PERFORMANCE_PROFILE_HISTORY to a PERFORMANCE_PROFILE_PROJECTION.

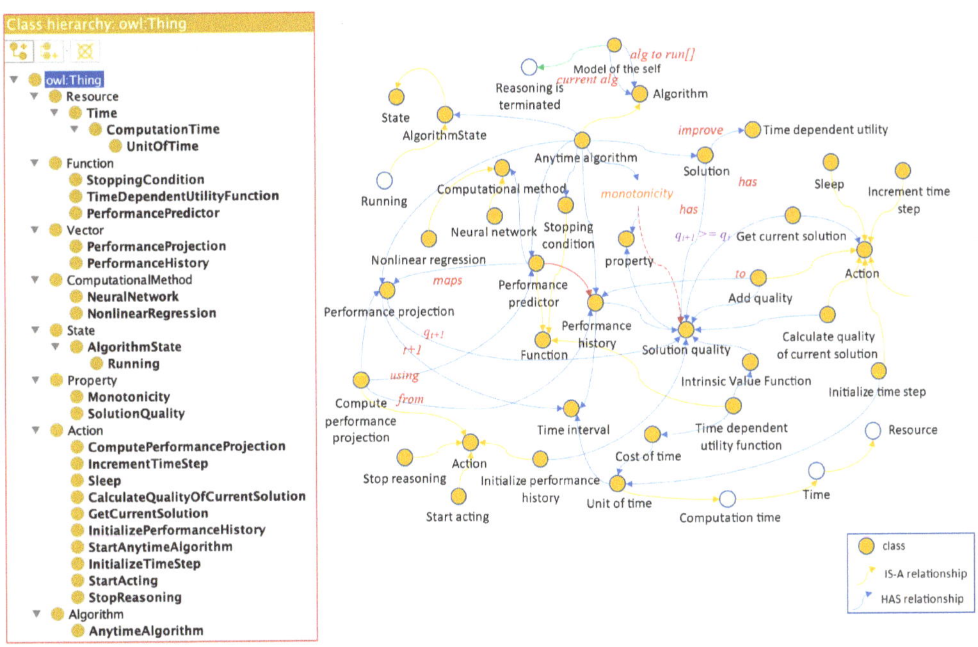

Figure 9. The SRP subontology.

In this problem, the sub-classes of METAREASONING_TASK are PREDICT_ PERFORMANCE, INTRINSIC_VALUE_FUNCTION, TIME_DEPENDENT_UTILITY_FUNCTION, COMPUTE_PERFORMANCE_PROFILE, and INIT_PERFORMANCE_PROFILE_HISTORY.

The interruption of the reasoning process is considered the most basic decision of the metareasoning process. In anytime systems that need to make decisions with limited resources, the metareasoning process is responsible for making the decision to continue reasoning or act on the current plan. For this, various techniques have been proposed in the literature, but the most common is to make the decision based on a TIME_DEPENDENT_UTILITY_FUNCTION. This is because the passage of time has a cost in itself, so a SOLUTION obtained in a reasonable time is what is desired. For example, going back to the case of the formation of teams, it is better to form a team with an EXPECTED_UTILITY of 0.7 using 5 min of reasoning than to form a team with an EXPECTED_UTILITY of 0.8 but the reasoning for nine hours (the EXPECTED_UTILITY is given by the average of the evaluation of all the skills of the team members).

The algorithm evaluates whether it is worthwhile to perform another reasoning cycle or if, on the other hand, the cycle should be stopped in any iteration, and then the CURRENT_PLAN is executed. A technique for the evaluation is to implement a predictor (PERFORMANCE_PREDICTOR) or function that predicts the EXPECTED_UTILITY of the solution in the next reasoning cycle; if the EXPECTED_UTILITY is positive, the algorithm continues, but if the utility is negative, then the META_LEVEL stops the process of reasoning.

3.4.5. The Gathering Computational Performance Data Problem (GCPDP) Subontology

Information gathering is a fundamental task for making metacognitive control decisions. Information is collected to store information in models. If the model is about the state of the world, then it is stored in the object level, and if the model is of the internal processes, then it is stored in the meta-level. Figure 10 shows the entire sub-ontology, including the INFORMATION_GATHERING_PROBLEM class with its relations.

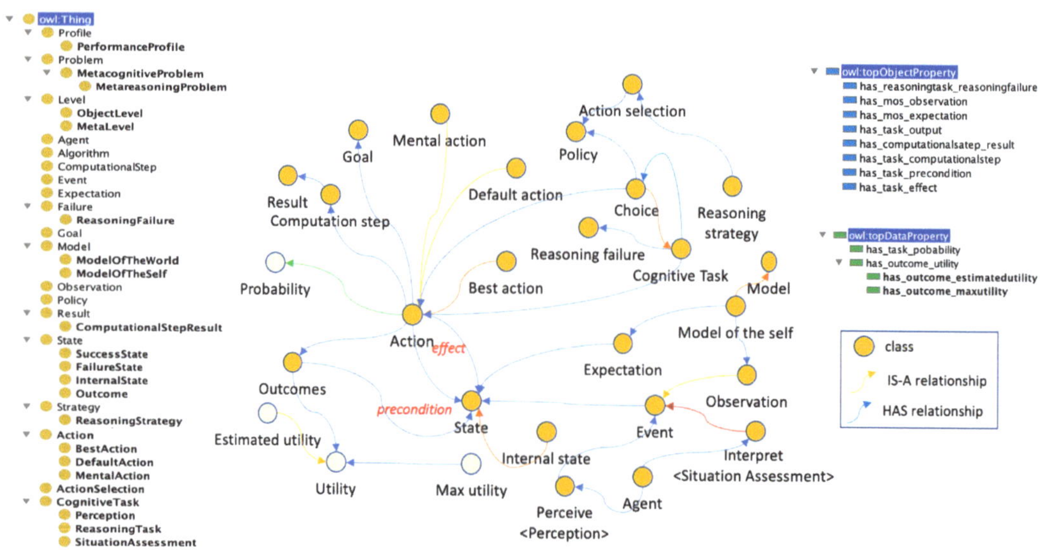

Figure 10. The GCPDP subontology.

The INFORMATION_GATHERING_PROBLEM class is a particular type of METAREASONING_PROBLEM, the main elements of which are part of the INFORMATION_GATHERING_PROBLEM_ELEMENT class and are listed below:

- MODEL_OF_THE_WORLD is an internal model that stores information related to the perception history that describes the state of the environment or world that is perceived by the system.

- MODEL_OF_THE_SELF is a dynamic model of the object level. This model is part of the meta-level and contains updated information on the current state of the reasoning processes that are carried out at the object level.

The following are the sub-classes of COGNITIVE_TASK in this problem: PERCEPTION, REASONING_TASK, and SITUATION_ASSESSMENT.

Monitoring of reasoning processes has been less studied than meta-level control. In this sense, the META_LEVEL executes a series of tasks to gather information on the state of the reasoning processes that are carried out at the OBJECT_LEVEL. The information collected is stored in performance profiles that make up the MODEL_OF_THE_SELF. The model of the self allows the META_LEVEL to have an updated view in real time of the internal states of the reasoning processes. This information is used for learning processes and for decision-making in the meta-level control.

3.4.6. Detection of Reasoning Failure Problem (DRFP) Subontology

Stopping reasoning is the most basic decision of the metareasoning process in intelligent systems. Reasoning about reasoning failures is an important research topic in metareasoning. In this sense, at some level of abstraction, the ways in which a system can fail are finite. Thus, the knowledge of how systems fail and how to recover from these failures must be represented. Figure 11 shows the full sub-ontology, including the class SOLVING_REASONING_FAILURE_PROBLEM along with its relations.

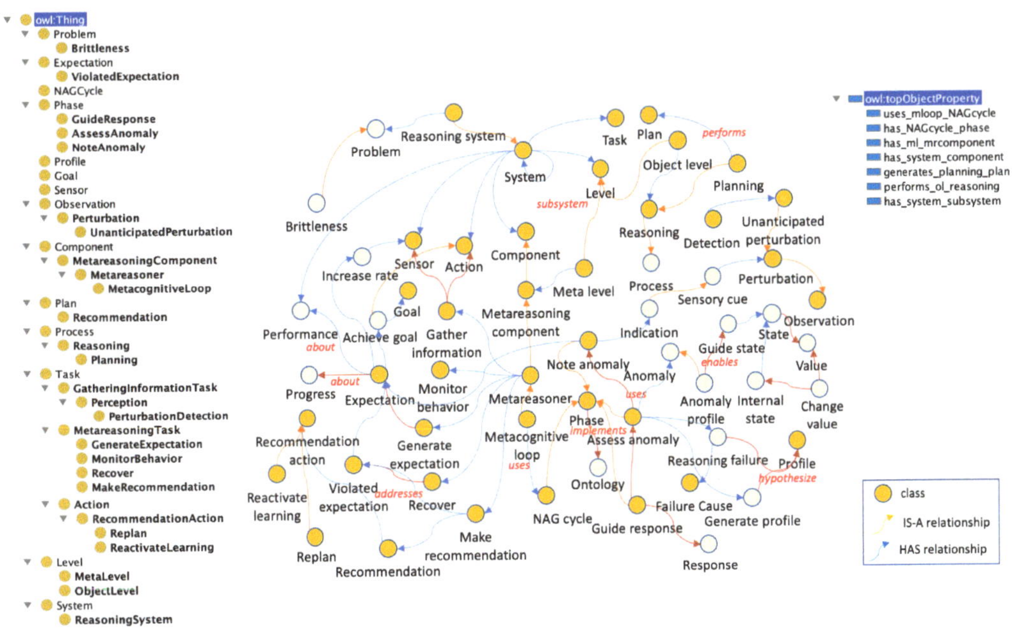

Figure 11. The DRFP subontology.

The SOLVING_REASONING_FAILURE_PROBLEM class is a particular type of METAREASONING_PROBLEM whose main elements are part of the SOLVING_REASONING_FAILURE _PROBLEM_ELEMENT class. These elements are listed below:

REASONING_FAILURE, ANOMALY, EXPECTATION, RECOMMENDATION, and GOAL.

Furthermore, in this problem, the METAREASONING_TASK class has three subclasses. They are MAKE_RECOMMENDATION, ANOMALY_DETECTION, and GENERATE_EXPECTATION.

During the reasoning process, errors such as incomplete tasks or unexpected results can occur that can affect the execution of critical tasks for an intelligent system. These errors are called reasoning failures and are represented in the class REASONING_FAILURE. An unexpected result is considered an ANOMALY in the performance of the task or process. Many designers of intelligent systems establish metareasoning points because monitoring and controlling all the reasoning processes of the system can be costly in terms of resource consumption. A METAREASONING_POINT is a TASK or process that is considered important for the operation of the system, and therefore, it is necessary to monitor its performance. The META_LEVEL keeps a record of the decisions made at the metareasoning points to analyze these traces of reasoning in the event of failures. When a REASONING_FAILURE is detected, then the META_LEVEL identifies the failure and proceeds to fix it based on the knowledge acquired from previous failures. The detection of a REASONING_FAILURE is carried out by monitoring the states of the metareasoning points. The most common failures are excessive processing time, excessive pause time waiting for a response from another task, and unexpected results that produce violations of expectations (e.g., VIOLATED_EXPECTATION). Performance expectations are generated from historical data and are contrasted against current observations of the performance of the task or process. After a REASONING_FAILURE is detected, the META_LEVEL generates a GOAL to solve the failure at the OBJECT_LEVEL. The GOAL is embodied in a RECOMMENDATION that can include any of the following options: abort the task, pause the task, reconfigure the task or resume the task. Finally, the SOLUTION is stored for future reference.

3.4.7. Self-Explanation and Self-Understanding Problem (SE&SUP) Subontology

If the reasoning that is performed at the object level (and not just its results) is represented in a declarative knowledge structure that captures the mental states and decision-making sequence, then these knowledge structures can themselves be passed to the metalevel for monitoring. Figure 12 shows the entire sub-ontology, including the SELF_METAEXPLANATION_PROBLEM class with its relations.

Figure 12. The SE&SUP subontology.

The SELF_METAEXPLANATION_PROBLEM class is a particular type of METAREASONING_PROBLEM, whose main elements are part of the SELF_METAEXPLANATION_PROBLEM_ELEMENT class and are listed below: REASONING_TRACE, IMXP, TMXP, FAILURE_EXPLANATION, and LEARNING_GOAL. In this problem, the sub-classes of

METAREASONING_TASK are STORY_UNDERSTANDING_TASK, EXPLAIN_FAILURE, GENERATE_LEARNING_GOAL, and INTROSPECTION.

The SELF_METAEXPLANATION_PROBLEM class represents the METAREASONING_PROBLEM that occurs when the OBJECT_LEVEL is unable to generate an EXPLANATION for an ANOMALY in the performance of the reasoning process when a STORY_UNDERSTANDING_TASK is running. With STORY_UNDERSTANDING_TASK a system must be able to reason introspectively about how to complete a task and what specific pieces of knowledge it needs to improve its performance at that exact moment to effectively learn the current task that is running at the object level. The META_LEVEL reads the REASONING_TRACE that contains the decisions made and the internal state of the system at the time of the decision. The META_LEVEL then runs a meta-understanding process (e.g., EXPLAIN_FAILURE) based on a REASONING_TRACE to understand the cause of the REASONING_FAILURE and why the OBJECT_LEVEL could not explain it. Once the cause is understood, the META_LEVEL generates a LEARNING_GOAL to recommend the actions that the OBJECT_LEVEL should take. The reasoning trace, the generated goals, the FAILURE_EXPLANATION and the recommendations are stored for future reference.

3.4.8. Improving the Ability of the Approach to Generalize to New (or Existing but Unstudied) Problems

The proposed ontology can be extended beyond the seven subontologies to capture new meta-reasoning problems that were not covered by this research. In this sense, it may be the case that different meta-reasoning problems induce new subontologies. To prevent new subontologies from generating counterproductive effects in terms of sharing knowledge in an easier way, a version-based approach will be adopted. In this way, when exchanging knowledge between applications or research teams, the use of different versions of the ontology can be avoided.

In relation to the creation of new subontologies and their relationship with existing ones, two different approaches can be taken to preserve the integrity of the ontology. Researchers can opt for a manual approach following the method described in this paper; the resulting subontology will be evaluated by experts to be integrated into IM-Onto. The second is a semi-automated approach based on (Althubaiti et al. 2020), in which the term extraction process for the new subontologies is carried out through an analysis of the corpus contained in documents that describe meta-reasoning problems. This approach uses a semi-automated approach to recognize the mention of IM-Onto classes in the text. The approach is based solely on labels and synonyms of the classes within the IM-Onto ontology and can be used to determine whether a word refers to an IM-Onto class; see Figure 13.

First, the IM-Onto ontology is obtained in Web Ontology Language (OWL) format, and the list of class labels and synonyms is extracted from the ontology, also using a body of text (article or description of the new meta-reasoning problem) as input to the process. Text mining tasks are performed with the Whatizit tool (Rebholz-Schuhmann et al. 2008). Word embeddings (i.e., vector space encodings of the contexts in which a word occurs) are then generated for all words in the text corpus, and a supervised machine learning model is trained to classify whether a word refers to a class in IM-Onto or not (using the ontology words and their synonyms as positive training examples and all others as negative examples). This approach broadly identifies terms that refer to classes within the domain of a meta-reasoning problem (according to the number of classes there are in the ontology). The method generates "seed" words in the text and then uses these seeds first to generate context-based features (via Word2Vec (Mikolov et al. 2013), and it uses these context-based features in a supervised machine learning classifier.

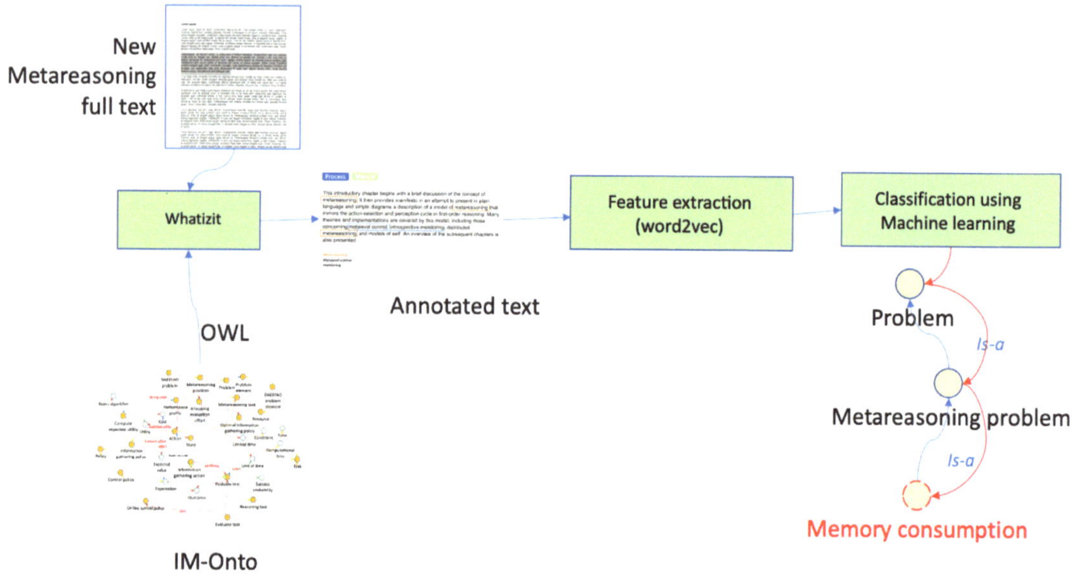

Figure 13. Semi-automated approach to creating new subontologies based on (Althubaiti et al. 2020).

3.5. Evaluation and Case Study

This section evaluates the use of IM-Onto in an application to solve a real-world problem. The resulting case study enables the use of automatic consistency checking and multiple performance evaluations (i.e., task-based, data-driven, and criteria-based evaluation). In this sense, the ontology evaluation consists of automated consistency checking (ontology verification) and task-based, data-driven, and criteria-based evaluations.

Running Example: The Team Allocation for Internship Programs (TAIP)

IM-Onto was applied to a real-world problem to demonstrate its practical use in supporting metareasoning problems. This case study has been based on the assignment of internships as a degree option in the undergraduate program of Computer Science (Licenciatura en Informática) at the University of Córdoba—Colombia. Allocating student teams to internship programs is a particular type of team composition problem (Andrejczuk et al. 2019). The problem has been selected because different studies (Georgara et al. 2020a) have used anytime algorithms to find approximate solutions to team allocation in the education context based on the Feasible Team-For-Task Allocation Problem (FTAP). FTAP considers the problem of putting together teams of students suitable for internship tasks in companies and institutions, given that it is increasingly common for students to spend some time doing internships in a company as part of their study plan. Figure 14 shows the representation of knowledge generated in the process of monitoring and control of the Team Allocation for Internship Programs Problem (TAIP).

The undergraduate program of Computer Science has several internship programs to which students can apply. An internship program is characterized by the skill requirements for students and the limitations of team size. In the internship program selected for the case study, a student is characterized by their skills and their level of mastery of each skill. Students must meet 5 skill requirements. These are (a) principles of image theory and photography, (b) STEAM (science, technology, engineering, arts, and math) skills, (c) web development, (d) fluency in the Spanish language, and (e) protégé video production, whereas the required team size is 3 members. Each student has a required mastery level in each of the skills.

Figure 14 shows a partial view of the ontology that represents the knowledge generated in the process of monitoring and control of the Team Allocation for Internship Programs Problem (TAIP) (Georgara et al. 2020b), and a knowledge base is constructed in this case study by adding instances to the classes defined in the ontology.

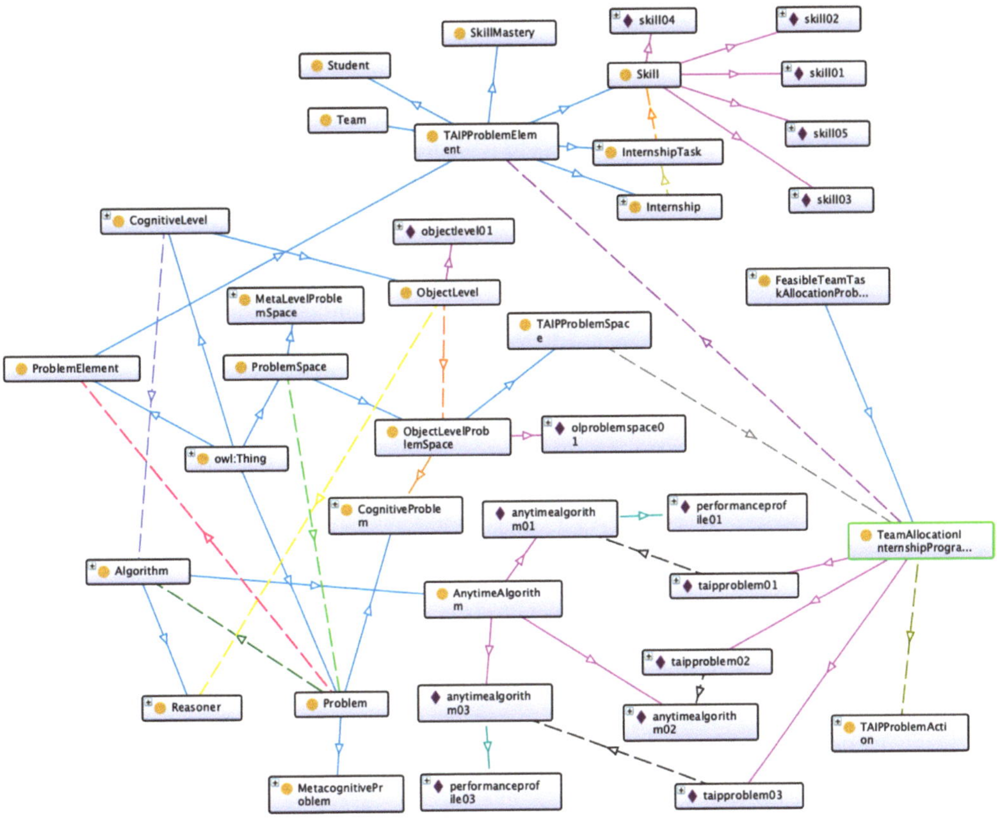

Figure 14. Representation of knowledge generated in the process of monitoring and control of the Team Allocation for Internship Programs Problem (TAIP).

The steps followed for the implementation of an intelligent agent based on the ontology specifications are presented below.

1. A Python package was developed that contains the classes common to the 7 types of problems addressed in this article. In this package, a basic cognitive system is made up of two cognitive levels, as specified in the ontology. A Beta version of the package is available at: https://github.com/dairdr/carina (login is required).
2. A polymorphic meta-reasoner was created at the meta-level, which monitors the object level by accessing the data that is updated in the model of the self (MoS); this model is designed according to the MODELOFTHESELF class. The monitoring and gathering of information are done through the MoS, which is updated in real-time from the object level. The MoS stores the profiles of the cognitive tasks that are executed at the object level; this is done automatically and does not require human intervention. Figure 15 shows a dataset based on the performance profile of the object-level reasoner and the history of a stopping reasoning problem (SRP). The meta-level uses the dataset to train SRP using a random forest algorithm; see Figure 15, section A. The MoS is stored in the working memory of the cognitive system, serving as a

bridge between the object level and the meta-level. The meta-reasoner runs in parallel with the object level and analyzes the reasoning traces of the cognitive task profile. The meta-reasoner analyzes the data using a random forest algorithm to select the method to execute according to the meta-reasoning problem detected. Profiles store data about cognitive tasks such as start time, execution time, output quality, and data that are common to any task executed at the object level. In this sense, the scaling of the meta-reasoner is facilitated, encompassing new models due to its polymorphic design.

3. A cognitive system with two cognitive levels was designed: the object level and the meta-level. The object level was configured according to the IM-Onto object level class specifications. Where the problem or cognitive task performed by the object level was defined considering the REASONINGPROBLEM class, then the elements of the problem were added according to the PROBLEMELEMENT class. In this case study, the object level is based on three anytime algorithms that are monitored and controlled by a meta-level until a suitable solution is found in cost and time. An example of the implementation of the FTAP problem is available at: https://github.com/dairdr/carina/blob/master/miscellaneous.py (login is required).
4. The system was configured with three algorithms to induce some meta-reasoning problems to observe the behavior of the meta-level. In this sense, for TAIP, one algorithm randomly selects the members of the team, another selects the most qualified members for each competition and thus assembles the team, while another algorithm receives parameters that restrict the selection; for example, if the average skill proficiency of a selected team has a student with low performance, then the algorithm replaces the student. The meta-level selects the best algorithm profile according to a set of constraints stipulated in the problem configuration, such as deliberation time and algorithm performance. Figure 15 shows some outcomes resulting from the validations of the system. Section A shows a subset of features obtained from the profiles of the cognitive tasks that are stored in the MoS and are used by the meta-level for monitoring and control; in this case, it is a training dataset for the problem of stopping the reasoning process. Section B shows the behavior of the time-dependent utility function of the algorithm that is running at the object level, which is used to predict the stopping of the reasoning process. Section C presents the profiles of the three algorithms with respect to the behavior of the time-dependent utility function.

3.6. Ontology Evaluation

This section describes four approaches for the evaluation of IM-Onto. First, the automated consistency check approach is used to evaluate the internal consistency of the ontology. Second, the task-based assessment evaluates that IM-Onto can accomplish the competency tasks defined in the specification by using completeness questions. Third, data-driven evaluation rigorously tests the integrity and conciseness of the ontology. Finally, criteria-based assessment further addresses the ontology clarity metric.

3.6.1. Automated Consistency Checking

The consistency check of an ontology is used to confirm that there are no contradictory facts according to descriptive logic (DL). IM-Onto was evaluated using the Pellet reasoner. Pellet is an OWL-DL reasoner built into the open source Protégé with reasoning support for individuals (instances), cardinality constraints, user-defined data types, sub-property axioms, reflexivity constraints, symmetric properties, and disjoint properties (Sirin et al. 2007). Pellet checks for implicit subclass relationships based on user-defined class relationships. From a development point of view, it also incorporates debugging support for the iterative process of designing and coding a DL error-free ontology. Errors in the ontology have been flagged by error messages, and inconsistent classes are marked for review. Figure 16 shows the result of several verification cycles until the ontology was free of DL errors.

A

q	t	cost_of_the_time	intrinsec_utility	reasoning_time	action_name	algorithm_name	utility_function	label
5.0	160	5.05	5.0	160	0	0	0.04000000000000036	stop_reasoning
5.0	161	5.08	5.0	161	0	0	0.009999999999999787	stop_reasoning
4.733333333333333	152	4.81	4.733333333333333	152	0	0	0.013333333333333641	stop_reasoning
4.866666666666666	154	4.87	4.866666666666666	154	0	0	0.05666666666666664	stop_reasoning
4.333333333333333	64	2.17	4.333333333333333	64	0	1	2.163333333333333	keep_reasoning
4.333333333333333	23	0.94	4.333333333333333	23	1	1	3.393333333333333	keep_reasoning
3.1333333333333333	2	0.31	3.1333333333333333	2	0	0	2.8233333333333333	keep_reasoning
4.466666666666666	26	1.03	4.466666666666666	26	0	0	3.4366666666666656	keep_reasoning
4.866666666666666	78	2.59	4.866666666666666	78	0	0	2.2766666666666664	keep_reasoning

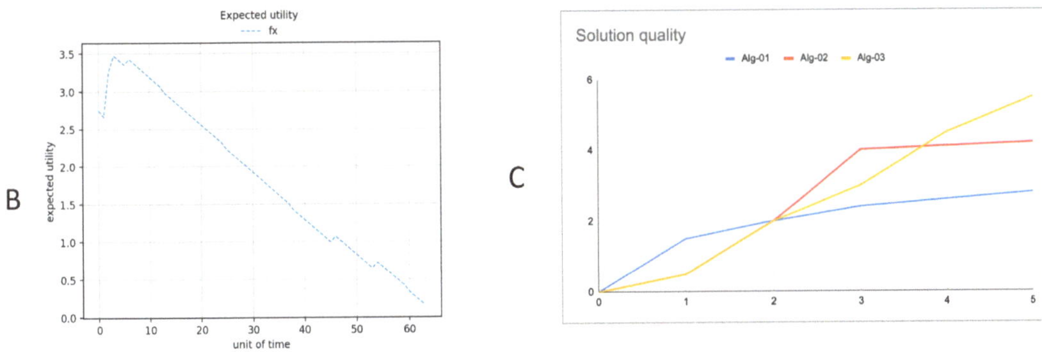

Figure 15. Data resulting from the validation process of TAIP. Section (**A**): dataset generated in a stopping reasoning problem. Section (**B**): behavior of the time-dependent utility function used to predict the stopping of the reasoning process. Section (**C**): performance profiles of three algorithms that solve TAIP, which ADTP uses to select a subset of algorithms according to the problem constraints. The x-axis represents reasoning loops.

Metrics

Axiom	161
Logical axiom count	106
Declaration axioms count	55
Class count	12
Object property count	7
Data property count	1
Individual count	35
Annotation Property count	0

Class axioms

SubClassOf	10

Figure 16. Results of verification cycle.

3.6.2. Task-Based Evaluation

This section describes the evaluation in terms of how the ontology can be used to answer questions about certain tasks (France-Mensah and O'Brien 2019), using the RDF query language (SPARQL) for information retrieval to answer sample queries related to the metareasoning problems.

When solving the FTAP problem, the object-level employs three anytime algorithms that are monitored and controlled by a meta-level until a suitable cost- and time-optimized solution is found. In the context of the Allocating Deliberation Time Problem (ADTP), the ontology can answer questions related to the information collected about the several algorithms that can be executed in parallel at the object level. Each algorithm solves one instance of FTAP, but in some circumstances where execution time is limited, the meta-level must select a subset of algorithms to further deliberation. A relevant question in the context of the ADTP is: *Which algorithm was selected by the metalevel to allocate more deliberation time to obtain a better solution that solves the problem at the object level?*

This question is written in SPARQL query in scenario 1 of Figure 17. In the response, it is observed that the meta-level has selected the anytyme_alg_01, and the ADTP was detected.

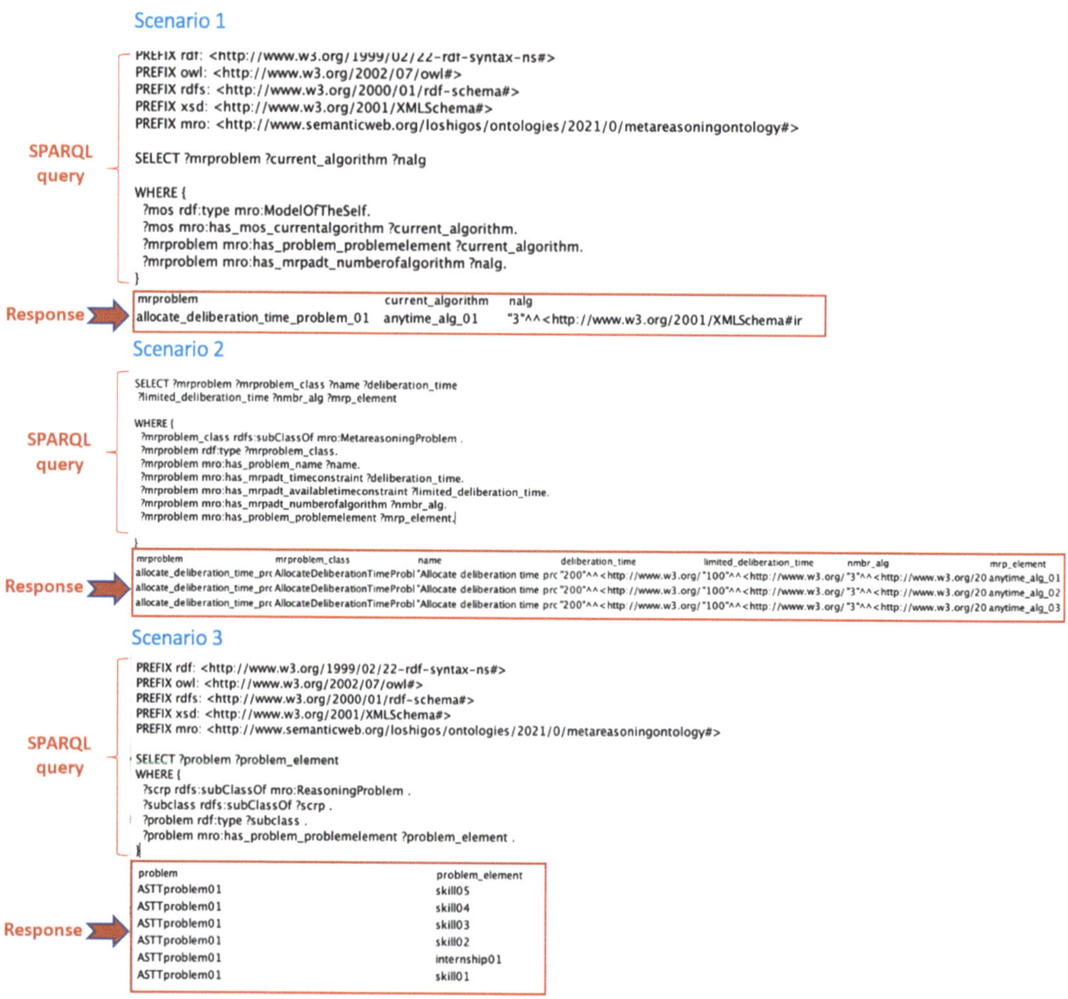

Figure 17. Questions asked to the ontology in SPARQL language.

The SPARQL query used in the second scenario corresponds to the question: *What is the information related to the configuration of the meta-reasoning problems that the meta-level can solve?* The response of the ontology provides information related to the configuration of the metareasoning problems that the meta-level can solve. In this case, the information is linked between the elements of ADTP metareasoning problem and the current state of the reasoning processes (performance, algorithms, quality of the solution, computational time).

This information provides important context for possible actions to take if a metareasoning problem is identified.

In scenario 3, the SPARQL query represents the question: *What are the elements that are part of the configuration of the problem at the object level?* The response of the ontology generates information about the context of the reasoning problems and how to address them.

3.6.3. Data-Driven Evaluation

This section describes the validation process using precision and recall as metrics for evaluating IM-Onto with respect to information retrieval. This evaluation approach makes a comparative analysis consisting of counting the related terms that appear between a predefined set of knowledge elements and the ontology (Guo and Goh 2017). Precision measures the ratio of correctly found knowledge correspondences (true positives) over the total number of returned knowledge correspondences (true positives and false positives) (Brewster et al. 2004). Recall reflects the proportion of knowledge that is accurately detected relative to all the knowledge items it should identify from the predefined set (Brewster et al. 2004). The numerator of Equations (1) and (2) describe that knowledge that is accurately detected and corresponds to the intersection of the relevant entities and the retrieved entities. In each equation, only the denominator differs. As shown in Equation (3), the F-measure we use is F_1, the harmonic means of precision and recall.

$$\text{precision} = \frac{|\{\text{relevantentities}\} \cap \{\text{retrievedentities}\}|}{|\{\text{retrievedentities}\}|} \tag{1}$$

$$\text{recall} = \frac{|\{\text{relevantentities}\} \cap \{\text{retrievedentities}\}|}{|\{\text{relevantentities}\}|} \tag{2}$$

$$F_1 = 2 \times \frac{\text{precision} \times \text{recall}}{\text{precision} + \text{recall}} \tag{3}$$

For data-driven evaluation, a corpus in the domain of metareasoning was used. The corpus is a set of answers to questions extracted from five papers that were selected as relevant in the literature review stage but that were not included for the development of the ontology due to the saturation criterion described in Section 3.2.1. For the evaluation, 35 questions were used that covered the information related to the metareasoning problems described in the literature. The questions were manually annotated to extract the main concepts needed to answer them. This is demonstrated in Figure 18. In this figure, a sample question is presented with a list of relevant and retrieved concepts.

Table 5 compares the performance (i.e., recall and precision rates) of IM-Onto with that of both the Metacognitive Loop (MCL) (Schmill et al. 2007, 2011) and Meta-level Control Ontology (MLCO) (Madera-Doval 2019). The precision and recall rate results demonstrate that the low recovery rates of MCL and MLCO reiterate the limitations of existing ontologies and further underscore the need for IM-Onto to support the design of metareasoning systems. On the other hand, the high performance of IM-Onto shows that it contains a high percentage of the relevant entities (recall = 90%) to support the design of systems with integrated support of various types of metareasoning. This reinforces the integrity of the ontology. Similarly, the accuracy rate (91%) of IM-Onto indicates that most of the knowledge elements in the ontology are useful information for the integration of various types of metareasoning in intelligent systems. This supports the conciseness of the ontology.

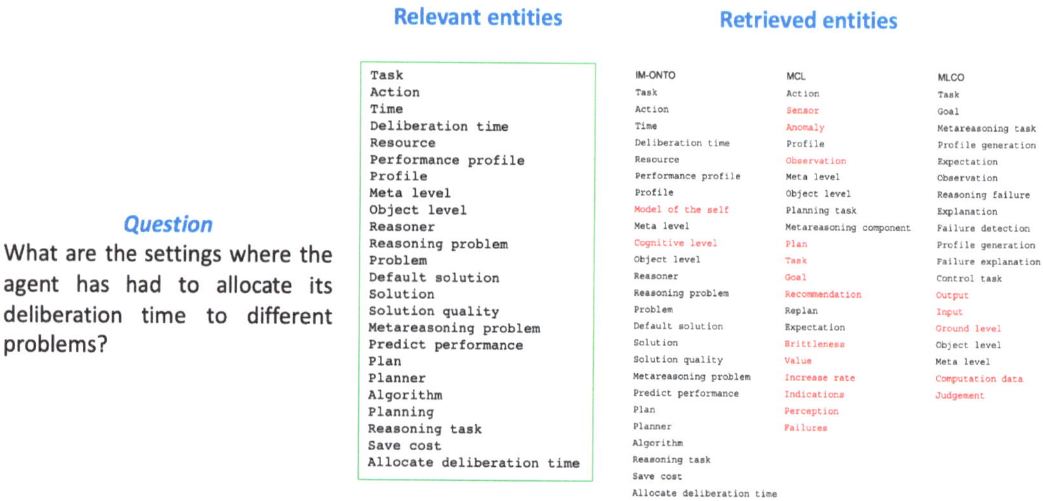

Figure 18. Sample of questions. Manually annotated questions for relevant and retrieved entities. In the response to the question formulated, the terms highlighted in red were not recovered by the corresponding ontologies.

Table 5. Precision and recall rates for comparing ontologies.

Ontologies	Precision	Recall	F-Measure
IM-Onto	92%	90%	91%
MCL (Schmill et al. 2007, 2011)	47%	37%	41%
MLCO (Madera-Doval 2019)	63%	45%	53%

3.6.4. Criteria-Based Evaluation

For this type of evaluation, the five criteria chosen include competence, consistency, integrity, clarity, and conciseness (El-Diraby 2014). The coherence and consistency criteria were primarily demonstrated by automated consistency checks and task-based assessments, respectively. The data-driven evaluation also demonstrates satisfactory performance in terms of conciseness and completeness through precision analysis and recall, respectively. In the context described, this section focuses on addressing the clarity criterion.

Clarity: The clarity of an ontology is determined when the knowledge elements of the ontology have an unambiguous meaning (El-Diraby 2014). In this study, the terms used for the ontology components were selected from the literature review. The definitions of the terms were obtained by comparing aspects common to all the ontologies studied. In this sense, it is guaranteed that the definitions are objective and independent of the social and computational context.

Three experts complemented this technical evaluation, where the experts recommended making some adjustments to the ontology due to taxonomic errors that concern the taxonomic structure and are referred to as inconsistency, incompleteness and redundancy. Some examples are presented in Figure 19.

The experts checked those three types of "inconsistency," both logical and semantic, have been highlighted: circularity errors (e.g., a concept that is a specialization of itself), partitioning errors (e.g., a concept defined as a specialization of two disjoint concepts or a concept defined as a specialization of two different classes), and semantic errors (e.g., a taxonomic relationship in contradiction with the user knowledge). In this sense, the experts found (0) errors of circularity, (1) error of partitioning and (2) semantic errors.

Incompleteness occurs when, for instance, relationships or axioms are missing. In relation to this, the experts found (2) errors. Finally, redundancy occurs when, for instance, a taxonomical relationship can be deduced from others by logical inference. In this topic, the experts found (0) errors.

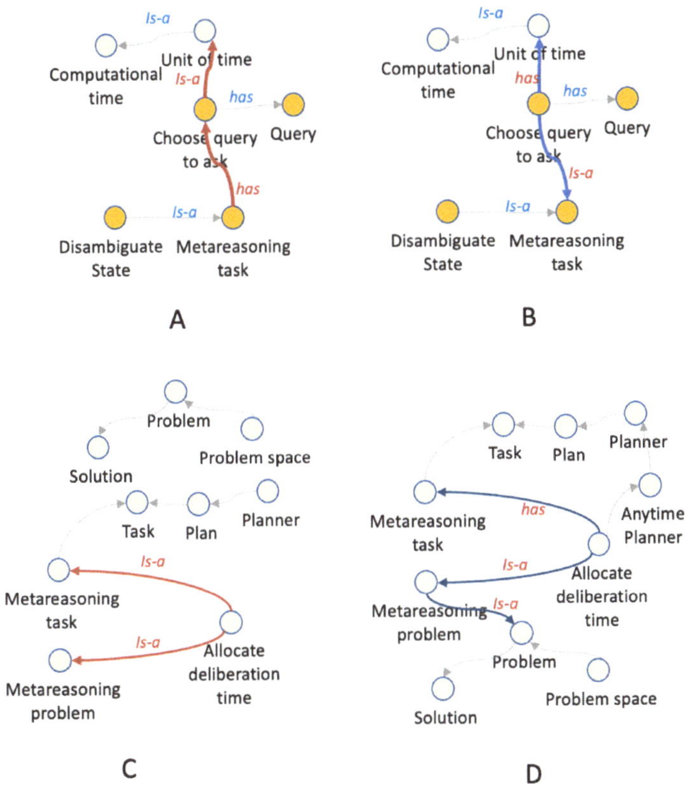

Figure 19. Examples of errors detected in expert validation. Section (**A**) shows two semantic errors in the relations colored in red, and section (**B**) shows the adjustments suggested by the experts. Section (**C**) shows an "incompleteness" error, where the experts commented that class ALLOCATE_DELIBERATION_TIME was inherited from two different classes, which was confusing. Section (**D**) presents the suggestion given by the experts.

3.7. Ontology Documentation

The IM-Onto ontology itself is available at the following link: https://tinyurl.com/2ymcrk44 (login required). In the documentation, it is possible to find a detailed description of the properties and relationships of each class that is part of the ontology. Additionally included are definitions that allow disambiguating each of the terms that the ontology contains. Similarly, instructions exist at this location for the use of the ontology in information retrieval tasks as well as reasoning tasks for intelligent systems.

4. Discussion

The main contribution of this paper was the presentation of a consistent ontology that provides a visual means of sharing a common understanding of the structure and relationships among terms and concepts related to the metareasoning domain in intelligent systems. In this paper, the application of IM-Onto to the problem of allocating student teams to internship programs has been demonstrated in three main tasks. They are (i) providing information linking the elements of a metareasoning problem and the current state of the reasoning processes; (ii) providing information related to the configuration of metareasoning problems the meta-level can solve; and (iii) generating information about the context of the reasoning problems and how to address them.

The high performance of IM-Onto shows that it contains a high percentage of the relevant entities (recall = 90%) to support the design of systems with integrated support of various types of metareasoning. On the other hand, the precision and recall rate results of the validation demonstrate that the low recovery rates of MCL (Schmill et al. 2007, 2011) and MLCO (Madera-Doval 2019) reiterate the limitations of existing ontologies.

The accuracy rate (91%) of IM-Onto indicates that most of the knowledge elements in the ontology are useful information for the integration of various types of metareasoning in intelligent systems. In this sense, MCL and MLCO demonstrate low accuracy rates. These results should be considered when considering the design of intelligent systems with metareasoning capabilities. MCL and MLCO were designed to respond to reasoning failures in intelligent systems, while IM-Onto was designed to handle a wider variety of metareasoning problems.

This study may be expanded in the future to include aspects of metareasoning in humans, metareasoning in inference engines in ontologies and monitoring and control of tasks that expire in time. The inclusion of these three topics can broaden the scope of the IM-Onto ontology, which is currently limited to the domain of metareasoning in intelligent systems. In this sense, it would be interesting to analyze the conceptual differences between metareasoning in human beings, in ontology inference engines and in the monitoring and control of computational processes.

5. Conclusions

This paper has presented an ontology called IM-Onto that captures key terms, concepts and relationships related to metareasoning and computational metacognition. The main research objective was to create a common language and conceptualization of metareasoning in the AI domain through the development of an ontology focused on the context of metareasoning problems described in published research. To achieve this, a rigorous research method was followed to guarantee that the two main requirements of the ontology were satisfied (completeness based on relevant knowledge and agreed upon by researchers and practitioners). The research method was based on (Badr et al. 2013), following the phases of definition, conceptualization, formalization, implementation, evaluation, and documentation.

IM-Onto can act as a unifying framework for data sharing across metareasoning problems in an intelligent system. This representation also solves a vocabulary problem in this domain by providing a standard semantic model for cross-functional metareasoning problems. IM-Onto consists of a sub-ontology for each metareasoning problem found on an in-depth analysis of the relevant literature as follows: Allocating Deliberation Time Problem (ADTP), Allocating Evaluation Effort Problem (AEEP), Knowledge Test Problem (KTP), Stopping Reasoning Problem (SRP), Gathering Computational Performance Data problem (GCPDP), Detection of Reasoning Failure Problem (DRFP), Self-explanation and Self-understanding Problem (SE&SUP).

Four approaches were used to evaluate IM-Onto. First, the automated consistency check approach ensures that the ontology is internally consistent. Second, the task-based assessment demonstrated that IM-Onto was able to accomplish the competency tasks defined above using completeness questions. Third, data-driven evaluation rigorously tests the integrity and conciseness of the ontology by demonstrating its comparative performance with previous ontologies. Finally, criteria-based assessment further addresses the ontology clarity metric.

Author Contributions: M.F.C., M.T.C. and R.E.T.-M. actively participated in the conceptualization, methodology, validation, in the formal writing of the first draft and the consequent revised versions. All authors have read and agreed to the published version of the manuscript.

Funding: This research was funded in part by NSF grant number 1849131 and by the Office of Naval Research grant number N00014-18-1-2009. This research was funded in part by University of Córdoba—Colombia.

Institutional Review Board Statement: The study was carried out in accordance with the Declaration of Helsinki and was approved by the Research Ethics Committee of the University of Córdoba (issue date: 15 November 2022).

Informed Consent Statement: Informed consent was obtained from all subjects involved in the study.

Data Availability Statement: The data presented in this study are available on request from the corresponding author. The data are not publicly available due to privacy issues and property rights. The IM-Onto ontology is available at the following link (account creation and login required). https://tinyurl.com/2ymcrk44.

Conflicts of Interest: The authors declare no conflict of interest.

References

Ackerman, Rakefet, and Valerie Thompson. 2017. Meta-Reasoning: Monitoring and Control of Thinking and Reasoning. *Trends in Cognitive Sciences* 21: 607–17. [CrossRef] [PubMed]

Allemang, Dean, Irene Polikoff, and Ralph Hodgson. 2005. Enterprise architecture reference modeling in OWL/RDF. In *International Semantic Web Conference (ISWC 2005)*. Edited by Yolanda Gil, Enrico Motta, V. Richard Benjamins and Mark A. Musen. Lecture Notes in Computer Science. Berlin: Springer, vol. 3729, pp. 844–57.

Althubaiti, Sara, Şenay Kafkas, Marwa Abdelhakim, and Robert Hoehndorf. 2020. Combining lexical and context features for automatic ontology extension. *Journal of Biomedical Semantics* 11: 1–13. [CrossRef]

Anderson, Michael, and Timothy Oates. 2007. A review of recent research in metareasoning and metalearning. *AI Magazine* 28: 12.

Andrejczuk, Ewa, Filippo Bistaffa, Christian Blum, Juan A. Rodríguez-Aguilar, and Carles Sierra. 2019. Synergistic team composition: A computational approach to foster diversity in teams. *Knowledge-Based Systems* 182: 104799. [CrossRef]

Baccigalupo, Claudio, and Enric Plaza. 2007. Poolcasting: A social Web radio architecture for group customisation. Paper presented at the 3rd International Conference on Automated Production of Cross Media Content for Multi-Channel Distribution, AXMEDIS 2007, Barcelona, Spain, November 28–30; pp. 115–22.

Badr, Kamal Badr Abdalla, Afaf Badr Abdalla Badr, and Mohammad Nazir Ahmad. 2013. Phases in Ontology Building Methodologies: A Recent Review. In *Ontology-Based Applications for Enterprise Systems and Knowledge Management*. Hershey: IGI Global, pp. 100–23.

Borghetti, Brett, and Maria Gini. 2008. Weighted prediction divergence for metareasoning. Metareasoning: Paper presented at the 2008 AAAI Workshop, Chicago, IL, USA, July 13–17.

Brewster, Christopher, Harith Alani, Srinandan Dasmahapatra, and Yorick Wilks. 2004. Data driven ontology evaluation. Paper presented at the 4th International Conference on Language Resources and Evaluation, LREC 2004, Lisbon, Portugal, May 26–28.

Caleiro, Carlos, Luca Vigano, and David Basin. 2005. Metareasoning about security protocols using distributed temporal logic. *Electronic Notes in Theoretical Computer Science* 125: 67–89. [CrossRef]

Caro, Manuel, Darsana Josyula, and Jovani A. Jiménez. 2014. A formal model for metacognitive reasoning in intelligent systems. *International Journal of Cognitive Informatics and Natural Intelligence* 8: 70–86. [CrossRef]

Chen, Xiaoping, Zhiqiang Sui, and Jianmin Ji. 2013. Towards metareasoning for human-robot interaction. In *Intelligent Autonomous Systems 12*. Berlin/Heidelberg: Springer, pp. 355–67. [CrossRef]

Conitzer, Vincent, and Tuomas Sandholm. 2003. Definition and complexity of some basic metareasoning problems. Paper presented at the IJCAI International Joint Conference on Artificial Intelligence, Acapulco, Mexico, August 9–15; pp. 1099–106.

Cox, Michael. 2005. Metacognition in computation: A selected research review. *Artificial Intelligence* 169: 104–41. [CrossRef]

Cox, Michael T. 2011. Metareasoning, monitoring, and self-explanation. In *Metareasoning: Thinking about Thinking*. Edited by Michael T. Cox and Anita Raja. Cambridge: MIT Press, pp. 131–49.

Cox, Michael, and Anita Raja. 2008. Metareasoning: A manifesto. In *Metareasoning: Thinking about Thinking, Proceedings of the 2008 AAAI Workshop, Chicago, IL, USA, July 13–17*. Technical Report WS-08-07. Edited by Michael T. Cox and Anita Raja. Menlo Park: AAAI Press, pp. 1–4.

Cox, Michael, and Anita Raja. 2011. Metareasoning: An introduction. In *Metareasoning: Thinking about Thinking*. Edited by Michael T. Cox and Anita Raja. Cambridge: MIT Press, pp. 3–14.

Cserna, Bence, Wheeler Ruml, and Jeremy Frank. 2017. Planning time to think: Metareasoning for on-line planning with durative actions. Paper presented at the International Conference on Automated Planning and Scheduling, Pittsburgh, PA, USA, June 18–23; vol. 27, pp. 56–60.

Dannenhauer, Dustin, Michael Cox, and Hector Munoz-Avila. 2018. Declarative metacognitive expectations for high-level cognition. *Advances in Cognitive Systems* 6: 231–50.

Dannenhauer, Dustin, Michael T. Cox, Shubham Gupta, Matt Paisner, and Don Perlis. 2014. Toward meta-level control of autonomous agents. *Procedia Computer Science* 41: 226–32. [CrossRef]

Dunlosky, John, and Robert Bjork. 2008. *Handbook of Metamemory and Memory*. New York: Psychology Press.

El-Diraby, Tamer. 2014. Validating ontologies in informatics systems: Approaches and lessons learned for AEC. *Journal of Information Technology in Construction* 19: 474–93.

Farmer, William. 2018. Incorporating quotation and evaluation into Church's type theory. *Information and Computation* 260: 9–50. [CrossRef]

Fernández-López, Mariano, Asunción Gómez-Pérez, and Natalia Juristo. 1997. Methontology: From ontological art towards ontological engineering. Paper presented at the AAAI97 Spring Symposium Series on Ontological Engineering, Palo Alto, CA, USA, March 24–25; pp. 33–40.

France-Mensah, Jojo, and William O'Brien. 2019. A shared ontology for integrated highway planning. *Advanced Engineering Informatics* 41: 100929. [CrossRef]

Fusch, Patricia, and Lawrence Ness. 2015. Are we there yet? Data saturation in qualitative research. *Qualitative Report* 20: 1408–16. [CrossRef]

Georgara, Athina, Carles Sierra, and Juan Rodríguez. 2020a. Edu2Com: An anytime algorithm to form student teams in companies. Paper presented at the AI for Social Good Workshop 2020, Virtual, July 20–21.

Georgara, Athina, Carles Sierra, and Juan Rodríguez. 2020b. TAIP: An anytime algorithm for allocating student teams to internship programs. *arXiv* arXiv:2005.09331.

Griffiths, Thomas, Frederick Callaway, Michael Chang, Erin Grant, Paul Krueger, and Falk Lieder. 2019. Doing more with less: Meta-reasoning and meta-learning in humans and machines. *Current Opinion in Behavioral Sciences* 29: 24–30. [CrossRef]

Gruber, Thomas. 1995. Toward principles for the design of ontologies used for knowledge sharing. *International Journal of Human-Computer Studies* 43: 907–28. [CrossRef]

Guo, Brian, and Yang Miang Goh. 2017. Ontology for design of active fall protection systems. *Automation in Construction* 82: 138–53. [CrossRef]

Hayes-Roth, Frederick, Donald Waterman, and Douglas Lenat. 1983. *Building Expert Systems*. Boston: Addison-Wesley Longman Publishing Co., Inc.

Hevner, Alan, Salvatore March, Jinsoo Park, and Sudha Ram. 2004. Design science in information systems research. *MIS Quarterly: Management Information Systems* 28: 75–105. [CrossRef]

Horridge, Matthew, Rafael Gonçalves, Csongor Nyulas, Tania Tudorache, and Mark Musen. 2019. WebProtégé: A cloud-based ontology editor. Paper presented at the Web Conference 2019—Companion of the World Wide Web Conference, WWW 2019, San Francisco, CA, USA, May 13–17; pp. 686–89. [CrossRef]

Horvitz, Eric. 1987. Reasoning About Beliefs and Actions Under Computational Resource Constraints. Paper presented at the Third Conference on Uncertainty in Artificial Intelligence (UAI1987), Seattle, WA, USA, July 10–12.

Houeland, Tor Gunnar, and Agnar Aamodt. 2018. A learning system based on lazy metareasoning. *Progress in Artificial Intelligence* 7: 129–46. [CrossRef]

Karpas, Erez, Oded Betzalel, Solomon Eyal Shimony, David Tolpin, and Ariel Felner. 2018. Rational deployment of multiple heuristics in optimal state-space search. *Artificial Intelligence* 256: 181–210. [CrossRef]

Kuokka, Daniel. 1991. MAX: A meta-reasoning architecture for "X". *ACM SIGART Bulletin* 2: 93–97. [CrossRef]

Lieder, Falk, and Thomas Griffiths. 2017. Strategy selection as rational metareasoning. *Psychological Review* 124: 762. [CrossRef]

Lin, Christopher, Andrey Kolobov, Ece Kamar, and Eric Horvitz. 2015. Metareasoning for Planning Under Uncertainty. Paper presented at the Twenty-Fourth International Joint Conference on Artificial Intelligence (IJCAI 2015), Buenos Aires, Argentina, July 25–31.

Madera-Doval, Dalia Patricia. 2019. A validated ontology for meta-level control domain. *Acta Scientiæ Informaticæ* 6: 26–30.

Mikolov, Tomas, Ilya Sutskever, Kai Chen, Greg Corrado, and Jeff Dean. 2013. Distributed representations of words and phrases and their compositionality. In *Advances in Neural Information Processing Systems*. Lake Tahoe: Curran Associates, Inc., vol. 26.

Milli, Smitha, Falk Lieder, and Thomas Griffiths. 2017. When does bounded-optimal metareasoning favor few cognitive systems? Paper presented at the AAAI Conference on Artificial Intelligence, San Francisco, CA, USA, February 4–9; vol. 31.

Nelson, Thomas. 1990. Metamemory: A theoretical framework and new findings. In *The Psychology of Learning and Motivation: Advances in Research and Theory*. Edited by Gordon Bower. New York: Academic Press, pp. 125–73.

Noy, Natalya, and Deborah McGuinness. 2001. Ontology Development 101: A Guide to Creating Your First Ontology. Knowledge Systems Laboratory Technical Report KSL-01-05 and Stanford Medical Informatics Technical Report SMI-2001-0880. Stanford, CA. Available online: https://protege.stanford.edu/conference/2005/slides/T1_Noy_Ontology101.pdf (accessed on 1 October 2022).

Parashar, Priyam, Ashok Goel, Bradley Sheneman, and Henrik Christensen. 2018. Towards life-long adaptive agents: Using metareasoning for combining knowledge-based planning with situated learning. *The Knowledge Engineering Review* 33: e24. [CrossRef]

Peffers, Ken, Marcus Rothenberger, Tuure Tuunanen, and Reza Vaezi. 2012. Design science research evaluation. In *International Conference on Design Science Research in Information Systems*. Berlin and Heidelberg: Springer, pp. 398–410.

Rebholz-Schuhmann, Dietrich, Miguel Arregui, Sylvain Gaudan, Harald Kirsch, and Antonio Jimeno. 2008. Text processing through Web services: Calling Whatizit. *Bioinformatics* 24: 296–98. [CrossRef]

Russell, Stuart, and Eric Wefald. 1991. Principles of metareasoning. *Artificial Intelligence* 49: 361–95. [CrossRef]

Schmill, Matt, Darsana Josyula, Michael L. Anderson, Shomir Wilson, Tim Oates, Don Perlis, and Scott Fults. 2007. Ontologies for Reasoning about Failures in AI Systems. Paper presented at the Workshop on Metareasoning in Agent Based Systems at the Sixth International Joint Conference on Autonomous Agents and Multiagent Systems, Honolulu, HI, USA, May 14–18.

Schmill, Matthew, Michael Anderson, Scott Fults, Darsana Josyula, Tim Oates, Don Perlis, Hamid Shahri, Shomir Wilson, and Dean Wright. 2011. The metacognitive loop and reasoning about anomalies. In *Metareasoning: Thinking about Thinking*. Edited by Michael Cox and Anita Raja. Cambridge: The MIT Press, pp. 183–98.

Silla, Carlos, and Alex Freitas. 2011. A survey of hierarchical classification across different application domains. *Data Mining and Knowledge Discovery* 22: 31–72. [CrossRef]

Sirin, Evren, Bijan Parsia, Bernardo Cuenca Grau, Aditya Kalyanpur, and Yarden Katz. 2007. Pellet: A practical OWL-DL reasoner. *Journal of Web Semantics* 5: 51–53. [CrossRef]

Sung, Yoonchang, Leslie Pack Kaelbling, and Tomás Lozano-Pérez. 2021. Learning When to Quit: Meta-Reasoning for Motion Planning. Paper presented at the 2021 IEEE/RSJ International Conference on Intelligent Robots and Systems (IROS), Prague, Czech Republic, September 27–October 1; pp. 4692–99.

Svegliato, Justin, and Shlomo Zilberstein. 2018. Adaptive Metareasoning for Bounded Rational Agents. Paper presented at the CAI-ECAI Workshop on Architectures and Evaluation for Generality, Autonomy and Progress in AI (AEGAP), Stockholm, Sweden, July 15.

Svegliato, Justin, Connor Basich, Sandhya Saisubramanian, and Shlomo Zilberstein. 2021. Using Metareasoning to Maintain and Restore Safety for Reliably Autonomy. Paper presented at the Submission to the IJCAI Workshop on Robust and Reliable Autonomy in the Wild (R2AW), Virtual, August 19–26.

Svegliato, Justin, Kyle Hollins Wray, and Shlomo Zilberstein. 2018. Meta-Level Control of Anytime Algorithms with Online Performance Prediction. Paper presented at the Twenty-Seventh International Joint Conference on Artificial Intelligence, Stockholm, Sweden, July 13–19.

Uschold, Mike, and Michael Gruninger. 1996. Ontologies: Principles, Methods and Applications. *The Knowledge Engineering Review* 11: 93–136. Available online: http://citeseerx.ist.psu.edu/viewdoc/download?doi=10.1.1.48.5917&rep=rep1&type=pdf (accessed on 1 October 2022). [CrossRef]

Van Assem, Mark, Aldo Gangemi, and Guus Schreiber. 2006. Conversion of WordNet to a standard RDF/OWL representation. Paper presented at the Fifth International Conference on Language Resources and Evaluation (LREC'06), Genoa, Italy, May 22–28.

Journal of
Intelligence

MDPI

Article

Progressing the Development of a Collaborative Metareasoning Framework: Prospects and Challenges

Beth H. Richardson * and Linden J. Ball

School of Psychology & Humanities, University of Central Lancashire, Preston PR1 2HE, UK; lball@uclan.ac.uk
* Correspondence: bhrichardson@uclan.ac.uk

Abstract: Metareasoning refers to processes that monitor and control ongoing thinking and reasoning. The "metareasoning framework" that was established in the literature in 2017 has been useful in explaining how monitoring processes during reasoning are sensitive to an individual's fluctuating feelings of certainty and uncertainty. The framework was developed to capture metareasoning at an *individual* level. It does not capture metareasoning during collaborative activities. We argue this is significant, given the many domains in which team-based reasoning is critical, including design, innovation, process control, defence and security. Currently, there is no conceptual framework that addresses the nature of collaborative metareasoning in these kinds of domains. We advance a framework of collaborative metareasoning that develops an understanding of how teams respond to the demands and opportunities of the task at hand, as well as to the demands and opportunities afforded by interlocuters who have different perspectives, knowledge, skills and experiences. We point to the importance of a tripartite distinction between "self-monitoring", "other monitoring" and "joint monitoring". We also highlight a parallel distinction between "self-focused control", "other-focused control" and "joint control". In elaborating upon these distinctions, we discuss the prospects for developing a comprehensive collaborative metareasoning framework with a unique focus on language as a measure of both uncertainty and misalignment.

Keywords: metareasoning; metacognitive monitoring; metacognitive control; collaboration; uncertainty; reasoning; teams; alignment; intersubjectivity; language

1. Introduction

The concept of "metareasoning" refers to the metacognitive processes that *monitor* and *control* the core cognitive operations associated with reasoning, which are often referred to as "object-level" processes (cf. Nelson and Narens 1990). These processes include those focused on interpreting and understanding available information as well as drawing inferences from it and making decisions about courses of actions to take (see Ackerman and Thompson 2017, 2018). Object-level processes can be distinguished from monitoring and control processes that function at a metalevel. At this metalevel, continual monitoring is essential to evaluate the effectiveness of object-level processes in progressing toward good outcomes, including sound inferences, and convincing decisions. Metacognitive monitoring is usually experienced subjectively in terms of one's awareness of a shifting state of certainty regarding how effectively a process is unfolding, as well as one's perception of how good a resulting outcome seems to be (Ackerman and Thompson 2017, 2018).

Like monitoring processes, control processes similarly function at a metalevel, but serve different fundamental purposes, including: (i) the allocation of essential cognitive resources, such as attention and working memory, to object-level processes; (ii) the termination of object-level processing that is failing to deliver a suitable outcome; (iii) the initiation of new object-level processing when it seems necessary to do so. It is also important to note that, during any reasoning task, there will be a continual interaction between monitoring and control processes to ensure that object-level cognition is maintained and progresses until an outcome is reached. If people feel confident about the efficacy of their object-level processing, then they will be likely to continue pursuing a particular course of action, whereas, if they reach a point where they have low confidence in what they are doing, then they might change their current strategy, ask for help or decide to give up on the task entirely (e.g., see Law et al. 2022 for an in-depth discussion of people's giving-up behaviour on reasoning tasks).

The "metareasoning framework" that has been developed by Ackerman and Thompson (2017, 2018) represents a key milestone in recent efforts to explain many findings relating to the processes that monitor and control object-level reasoning. This metareasoning framework captures the way in which monitoring processes are sensitive to fluctuating feelings of certainty and uncertainty that are experienced by a reasoner during task performance, as well as the way in which different control processes are evoked in response to heightened levels of certainty or uncertainty. It is important to emphasise, however, that this framework was developed to explain metareasoning as it arises in individual reasoners rather than in groups of reasoners who are collaborating to undertake a joint reasoning task, as might arise, for example, in team problem solving or decision making. The focus of the metareasoning framework on *individuals* clearly raises the interesting question of whether the framework—or at least some key aspects of it—can be extended to capture the nature of *collaborative* metareasoning. Addressing this question provides a key motivating rationale for the present paper, which aims to explore the prospects and challenges for progressing the development of a collaborative metareasoning framework.

We begin by highlighting the importance of understanding collaborative metareasoning for theoretical advancement, especially in relation to the monitoring and control processes that arise during team reasoning, problem solving and decision making in real-world contexts. We next overview Ackerman and Thompson's (2017, 2018) influential metareasoning framework, which is focused on individual reasoners, and we consider potential ways in which this framework can inform an understanding of collaborative metareasoning. In overviewing Ackerman and Thompson's (2017, 2018) conceptual ideas, we emphasise how their framework focuses primarily on self-report measures of fluctuating levels of certainty and uncertainty. When it comes to an analysis of *collaborative* metareasoning, however, we argue that there is an opportunity to gain more direct access to monitoring and control processes through a detailed analysis of the individual and joint language that arises in team reasoning contexts. As such, our paper progresses toward a detailed consideration of current research on the metacognitive monitoring and control processes that are critical to the occurrence of effective *dialogue* during joint action (Pickering and Garrod 2021; see also Gandolfi et al. 2023). We argue that an analysis of team dialogue has the potential to pave the way toward a rich understanding of collaborative metareasoning that has so far been starkly absent in the literature.

Throughout our paper, we continually highlight an important tripartite distinction that was articulated in the context of metareasoning by Richardson et al. (2024; see also Pickering and Garrod 2021) between "self-monitoring" (i.e., an individual's perception of their own performance), "other monitoring" (i.e., an individual's perception of the performance of others) and "joint monitoring" (i.e., the unified perception of collective performance). Furthermore, in this paper, we also introduce a parallel tripartite distinction that relates to control processes, which is captured by the notions of "self-focused control" (i.e., an individual's decisions about how to progress or terminate their own reasoning), "other-focused control" (i.e., an individual's decisions about how to control the performance of others) and "joint control" (i.e., the unified control of decisions regarding how to advance or terminate collective performance). Our detailed discussion of these distinctions culminates in a consideration of the opportunities for developing a comprehensive collaborative metareasoning framework and the associated challenges.

2. The Importance of Understanding Collaborative Metareasoning

Most existing research on metacognitive monitoring and control has been undertaken with the aim of enhancing educational outcomes (e.g., Cromley and Kunze 2020; Hacker et al. 2009; Perry et al. 2019) and has, therefore, focused on object-level cognition that is related to comprehending and remembering in learning contexts. There has, however, been a limited but nevertheless long-standing interest in metacognitive monitoring and control processes in other areas of research, particularly in situations involving reasoning (e.g., Ackerman and Beller 2017; Quayle and Ball 2000; Thompson et al. 2011, 2013) and problem solving (e.g., Ackerman 2014; Ackerman and Zalmanov 2012; Metcalfe and Wiebe 1987; Topolinski and Reber 2010). Although such research has tended to be almost exclusively laboratory-based, more recent studies have broadened their focus to investigate metareasoning outside of the laboratory. One example is a study by Pervin et al. (2015) that analysed an extensive set of tweets to reveal how the use of hashtags in Twitter appears to be driven by an individual's metacognitive experiences relating to cognitive load and confusion. Another example is a study by Roberts (2017), which used a community sample to examine the role of metacognitive monitoring and control in the acquisition of skills relating to effective web search and evaluative reasoning. Yet another example is a study by Zion et al. (2015) examining the effects of metacognitive support on metareasoning within an online discussion forum in the context of computer-supported inquiry.

Despite such exceptions, however, there remains a paucity of studies of metareasoning in relation to individuals working in real-world situations, and this dearth of research is likewise seen in relation to metareasoning in collaborating teams engaged in professional work-based practices. The relative absence of studies examining the nature of metareasoning processes during collaboration is unfortunate given that effective team-based performance is so critical in many domains. A few examples include contexts relating to design, innovation, medical diagnosis, security, defence and emergency response.

These latter domains all depend upon collaborative problem solving, reasoning and decision making operating effectively to achieve shared goals. For example, in high-risk defence and security situations, a failure to coordinate team decision making can have highly negative downstream consequences, including loss of life and damage to infrastructure. Recently, the media reported on inherent inadequacies in the response by emergency teams during the 2017 Manchester Arena attack. Three key failures were identified: (i) communication was ineffective between teams; (ii) coordination both within and between teams was weak, as reflected in a failure to identify priorities and resources; (iii) shared situational awareness was poor in that plans of action were not clearly understood. These failures were all ones relating to inadequate metareasoning. It has been argued (see Saunders 2022) that, with better team collaboration, the negative outcome of the Manchester Arena attack could have been reduced.

In emergency response situations such as those arising in the Manchester Arena attack, metacognitive monitoring processes need to be attuned to the subtleties and complexities of uncertainty at both a *personal* and *interpersonal* level. In the latter case, uncertainty will typically manifest in the language that arises in team dialogue and cross-team communication, such as the use of tentative language and hedge words (e.g., "maybe", perhaps" and "possibly"). Metacognitive control processes, on the other hand, need to affect dynamic strategy change in a coordinated manner (i.e., mutually agreed between members, either tacitly or explicitly) to ensure that successful decisions are made. Within this complex, dynamic communication context there will also be a variety of situational factors that are likely to have an impact on object-level and metalevel processing at both an *inter*personal and *intra*personal level, such as whether team members get on well with one another and whether collaborators are perceived to be competent. Recent studies of team-based emergency responding have corroborated the critical role of metacognitive monitoring and control in ensuring effective situation awareness between team members and enhanced team decision making (see Hamilton et al. 2017).

The example of decision making in the context of teams engaged in emergency responding serves to illustrate the potentially critical role played by metareasoning during real-world collaboration. The same kinds of monitoring and control processes that arise in team-based emergency responding will presumably also be involved in any jointly executed activity that involves the attainment of some common goal, whether this involves making a shared decision regarding a course of action to take or generating an agreed approach concerning how to solve a complex problem. Nevertheless, a clear conceptualisation of the metacognitive monitoring and control processes that underpin collaborative problem solving, reasoning and decision making is absent from the literature and, we contend, is in pressing need of development. To advance the formulation of such a collaborative metareasoning framework, we next review the existing literature on metareasoning at the level of the individual, before then taking this useful body of knowledge forward and extending it to contexts involving collaboration.

3. The Metareasoning Framework

The metareasoning framework, proposed by Ackerman and Thompson (2017, 2018; see Figure 1), has sparked considerable research interest in relation to metacognitive monitoring and control processes in domains that involve problem solving, reasoning and decision making. Figure 1 displays the approximate time course of object-level reasoning as well as corresponding metareasoning processes (Ackerman and Thompson 2017, 2018). The left side of the figure represents the object-level processes involved in reasoning, whilst the middle column depicts the associated monitoring processes. Monitoring processes reflect a reasoner's subjective evaluation of the probability of success or failure in relation to a given task or problem (e.g., their confidence in the unfolding process), which can occur before, during or after object-level processing.

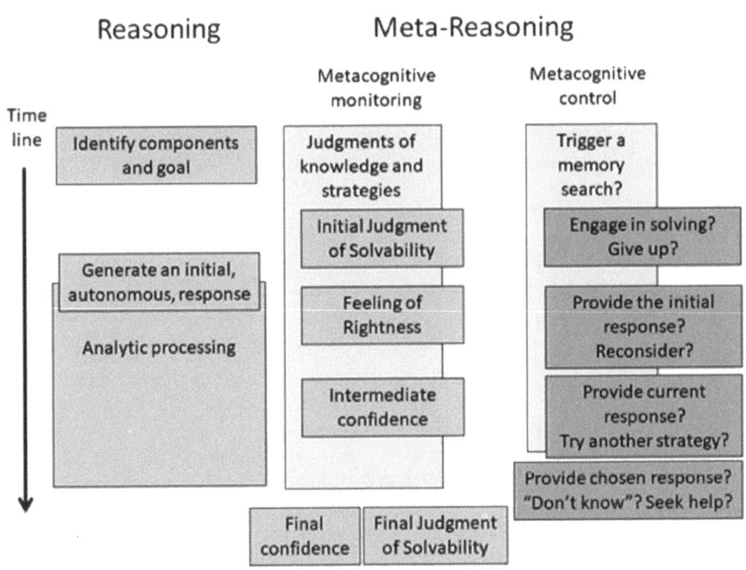

Figure 1. The approximate time course of reasoning and metareasoning processes as captured in Ackerman and Thompson's (2018) metareasoning framework. Copyright (©2018) from "Metareasoning: Shedding meta-cognitive light on reasoning research" by Ackerman and Thompson (2018). [Permission to reproduce has been sought from Taylor and Francis Group, LLC, a division of Informa plc].

Crucially, monitoring processes continually track fluctuating levels of certainty and uncertainty related to ongoing task performance or solution success. Within this metareasoning framework, monitoring processes are viewed as happening in the background and as having the capacity to "trigger" control processes. The right-hand column of Figure 1 depicts such control processes, which serve to allocate and redistribute resources in response to the outcome of ongoing monitoring. For example, if intermediate confidence is high, individuals will be likely to continue with their current course of action. In contrast, if they experience low–intermediate confidence or a feeling of uncertainty, they may activate control processes to switch their current strategy or, alternatively, they may give up.

The object-level processes that are shown in Figure 1 include ones that are associated with problem understanding and goal identification, as well as ones that generate an initial, autonomous response to the task at hand or that involve reasoning about the task analytically. Moreover, the object-level reasoning that is portrayed in Figure 1, whereby intuitive processing is followed by analytic processing, aligns with a "dual-process" architecture that is based on "default-interventionist" principles (Evans and Stanovich 2013a, 2013b). According to such an architecture, reasoning is considered to involve two qualitatively distinct types of processes, referred to as Type 1 and Type 2. Type 1 processes are viewed as being intuitive, heuristic and associative in nature, and are defined in terms of being relatively undemanding of working-memory resources as well as autonomous (running to completion whenever they are cued). Correlated features of Type 1 processes are that they tend to be high-capacity, rapid, nonconscious and capable of running in parallel. In contrast, Type 2 processes are reflective, deliberate, analytic and controlled, and are defined in terms of requiring working memory resources and having a focus on hypothetical thinking. Correlated features of Type 2 processes include their tendency to be slow, capacity-limited, serial and conscious. Furthermore, Type 2 processes are less prone to biases in comparison to Type 1 processes, although they are not invulnerable to them (e.g., see Evans 2018; Evans and Stanovich 2013a, 2013b), which may arise, for example, from the application of inadequate or inappropriate analytic operations, referred to as "defective mindware" (e.g., Stanovich 2018).

Assuming a default-interventionist version of the metareasoning framework, one key function of metacognitive monitoring is to trigger a strategic shift from default Type 1 to analytic Type 2 processing so that analytic processes can intervene to determine the accuracy of the default response. To illustrate this shift from Type 1 to Type 2 processing, consider the bat-and-ball problem that is one of the items in the Cognitive Reflection Test (Frederick 2005; Kahneman and Frederick 2002). This problem reads as follows: "A bat and a ball cost $1.10 in total. The bat costs $1.00 more than the ball. How much does the ball cost?" This problem might seem easy at first sight, yet many highly intelligent individuals generate an incorrect solution (e.g., Campitelli and Gerrans 2014; Pennycook et al. 2016; Stupple et al. 2017). Incorrect answers are attributed to a dominant, intuitive Type 1 response, whereby the answer that comes to mind rapidly and easily (i.e., 10 cents) is, in fact, wrong. When producing intuitive but incorrect responses, individuals may experience a low feeling of rightness (Thompson et al. 2011, 2013; see also Bago and De Neys 2017), which for some people will trigger a shift to Type 2 analytic reasoning, such that they might then be able to determine the correct answer.

Although Ackerman and Thompson's (2017, 2018) metareasoning framework is couched in dual-process terms that capture default-interventionist principles, they stress that the logic of monitoring processes, such as those that register a "feeling of rightness" (see Figure 1), also extends to single-process reasoning theories that do not propose two types of processing (e.g., Kruglanski and Gigerenzer's 2011, "Unified Theory of Judgment"). Moreover, the logic also pertains to theories of reasoning that propose the existence of multiple, parallel processes rather than sequential ones (e.g., see Bago and De Neys 2017). Whatever model of object-level reasoning one favours, it remains important to understand when, why and how people engage in more deliberate, reflective thinking. This, in turn, speaks to the importance of gaining insight into the way that people monitor rapid, initial answers, regardless of the type of reasoning mechanisms that are proposed to underlie their generation. The flexibility of the metareasoning framework to accommodate different theoretical perspectives on object-level reasoning is certainly a key strength.

4. Research Questions in Metareasoning Research

Ackerman and Thompson's (2017, 2018) metareasoning framework provides a rich and productive theoretical foundation for research on reasoning and metareasoning. Nevertheless, the framework is primarily just a starting point for further studies and conceptual developments in this area. The need for an extensive amount of further research was directly acknowledged by Ackerman and Thompson (2017), who presented a series of research questions that are now increasingly being addressed by several international teams. Among these research questions are ones that are concerned with whether reasoners have a degree of insight into the sources of certainty that affect their judgments, whether metareasoning processes are shaped by culture and whether reasoning performance can be enhanced through interventions that are targeted at improving metareasoning skills.

As we have also noted above, a critical area of development for the metareasoning framework relates to its extension to capture collaborative metareasoning rather than just individual metareasoning. Little research has been conducted on this topic to date, leaving many questions wide open for further research and conceptual advancement. However, before we move on to explore some of the relevant challenges and opportunities relating to the development of an understanding of collaborative metareasoning, we first describe some important ways in which Ackerman and Thompson's current framework has recently been extended. Doing this will allow us to introduce important ideas that will help inform our subsequent discussion of the complex issues that a collaborative metareasoning framework will need to tackle if it is to provide even the semblance of an account of how metareasoning operates in team contexts.

4.1. The Underpinning Basis of Metacognitive Certainty and Uncertainty

One critically important research question that needs to be addressed for an advancement of metareasoning—whether in situations that involve individual or team reasoning—concerns the underpinning basis of the fluctuating feelings of certainty and uncertainty that are experienced during task-based processing. One of the most pervasive cues that has been suggested to elicit such feelings is the perceived *ease* of task-oriented processing, which is a cue that is typically referred to as "processing fluency" (e.g., Alter and Oppenheimer 2009; Unkelbach and Greifeneder 2013). In essence, an answer or solution that comes to mind quickly and easily (i.e., with high processing fluency) gives rise to a strong feeling of rightness as well as to a heightened assessment of final confidence (Ackerman and Zalmanov 2012; Thompson and Morsanyi 2012; Thompson et al. 2013).

Processing fluency is usually indexed as the time that has elapsed from when a problem is displayed until a participant provides their response, with reasoning and problem-solving tasks typically showing a negative correlation between response time and metacognitive confidence judgments, reflecting the fact that easy items are processed more quickly than difficult items (e.g., Ackerman 2014; Ackerman and Zalmanov 2012; Baars et al. 2013). The influence of processing fluency on confidence has also been observed in studies of metacognition and memory. For example, Undorf and Erdfelder (2015) found that processing fluency—measured in terms of self-paced study time for to-be-remembered items—partially mediated the effect of item ease/difficulty on the judged probability of recalling the target (indexing "judgment of learning"). It is also important to note that the frequently observed association between processing fluency and confidence seems not merely to be correlational, but rather reflects a causal mechanism, whereby the ease of information processing directly induces a high degree of confidence in the correctness of a response (e.g., Topolinski and Reber 2010).

Other recent research on people's metacognitive judgments in relation to memory tasks has been influential in shedding further light on the sources of information that people use in addition to processing fluency when making metacognitive judgments. This research has been especially valuable in revealing how people appear to integrate *multiple* cues that stem from the same stimulus item when making judgments of learning (e.g., Undorf and Bröder 2021; Undorf et al. 2018). In such studies, Brunswik's lens model (e.g., see Kaufmann 2022) has been used to assess the overall validity of confidence judgments based on a combined set of cues. As Ackerman (2023) notes, however, this research leaves open important questions regarding the relative weights of such cues and how these weights change across task designs (e.g., different instructions) and population characteristics (e.g., different levels of prior knowledge). Ackerman (2023) presents a novel statistical approach, the "Bird's-Eye View of Cue Integration" (BEVoCI) methodology, which she argues can reveal not only the cues that determine metacognitive judgments and task success, but also the relative weights of these cues and their malleability across task designs and populations.

Ackerman's (2023) experiments using the BEVoCI method provide a wealth of findings regarding the multiplicity of weighted cues that underpin confidence judgments and solution accuracy in problem solving as well as the way in which these cues can dissociate in determining these outcome measures. Ackerman (2023) reports that people's judgments of confidence are consistently oversensitive to processing fluency (as indexed by response time), which is a phenomenon that is well-established in the metareasoning literature (e.g., Ackerman and Zalmanov 2012). We see considerable value in cue-integration methods for advancing an understanding of the information that people draw upon when providing confidence judgments regarding final solutions to reasoning problems. Nevertheless, whether such methods can also be successfully extended to predict the cues that underpin judgments of *intermediate* confidence, such as people's feeling of rightness, is yet to be determined.

4.2. Methods for Eliciting Judgments of Metacognitive Certainty and Uncertainty

A further, important research question concerning metareasoning in both individual and team contexts relates less to the issue of the cues that drive confidence judgments, and more to the question of how best to elicit confidence judgments from participants in the first place. As can be seen in Figure 1, confidence judgment can be elicited from participants at various time points, including, for example, initial judgments of solvability, intermediate feelings of rightness or confidence and final judgments of confidence or solvability. Our primary focus in this paper is on the best way to elicit *intermediate* confidence judgments at various points during the reasoning process, as this represents a critical measurement issue for understanding metacognitive monitoring in both individual and team reasoning contexts.

In problem-solving research, the traditional method that is used to obtain ephemeral, subjective judgments of certainty from a reasoner necessitates intermittently probing them for a rating of their current degree of confidence regarding the likelihood of finding a solution to a given problem. In research with *individual* problem solvers, this kind of metacognitive probe has often been described as indexing a reasoner's current "feeling of warmth"; that is, their sense that a solution is imminent (Metcalfe 1986; Metcalfe and Wiebe 1987; see also Hedne et al. 2016; Kizilirmak et al. 2018; Laukkonen and Tangen 2018). However, translating a fleeting feeling of warmth into a response on a numerical or categorical scale is a challenging requirement for participants who are midtask, and who therefore have simultaneously to maintain primary task performance while also making known their current metacognitions about likely task success.

It is also important to consider the possibility that the temporary cessation of primary task processing while generating a metacognitive judgment might have a *reactive* effect on ongoing object-level reasoning, changing its natural dynamics, trajectory, and outcome. There is only a relatively limited body of experimental evidence that has directly addressed the reactive effect of probing confidence on individual object-level reasoning, with much of the research tending to examine the reactivity of eliciting *retrospective* confidence judgments on decision time and performance accuracy in studies involving repeated trials. Intriguingly, the relevant studies that have been conducted to date provide evidence both for the idea that metacognitive confidence judgments are reactive (e.g., Baranski and Petrusic 2001; Bonder and Gopher 2019; Double and Birney 2017, 2018, 2019a; Lei et al. 2020; Petrusic and Baranski 2003; Schoenherr et al. 2010), as well as against this idea (e.g., Ackerman 2014; Ackerman and Goldsmith 2008; Ackerman et al. 2020).

We do not wish to be drawn into a detailed consideration of the potential reasons for the discrepant findings in the extant literature regarding the reactivity of confidence judgments, as the matter seems to be a long way from being resolved. Much more empirical work and conceptual development will be required to understand the conditions under which reactivity arises and the severity of its impact on performance, whether assessed in terms of processing time or outcome quality (for relevant discussion see Double et al. 2018; Double and Birney 2019b). In the absence of a solid empirical and theoretical foundation from which to make definitive predictions as to when and why some metareasoning studies show reactivity whilst others do not, we would argue that it seems judicious to conduct a parallel line of research that measures confidence using *alternative* methods. We also contend that this argument pertains as much, if not more, to team-based, rather than individual, reasoning situations. This is because reactivity arising from continually interrupting team members to elicit confidence judgments during their ongoing collaboration could impact not only the object-level processing of individual team members but also the whole team dynamic.

What, though, might alternative methods for eliciting confidence judgments entail such that they can provide a *nonreactive* measure of fluctuating certainty and uncertainty as an individual or team progresses from initial problem understanding toward final solution generation? In individual reasoning contexts, the use of think-aloud methods (e.g., Ericsson and Simon 1980), where solitary participants verbalise whatever is currently passing through their minds, may seem like a way forward to obtaining a nonreactive, dynamic index of fluctuating levels of confidence. This is because spoken language can provide an excellent source of rich information about a person's changing states of uncertainty, which can be manifest through their use of hedge words (e.g., "maybe", "perhaps" and "possibly"). Of course, the think-aloud method cannot be deployed to assess fluctuating uncertainty during team-based reasoning, as it is a technique that is only appropriate for use with individual reasoners, which immediately limits its utility. However, it may even be the case that the method is ill-suited to providing a valid, dynamic index of uncertainty in individual reasoning, given long-standing concerns about the potential for thinking aloud itself to have a reactive effect on object-level reasoning, potentially changing natural processing in profound ways (e.g., see Godfroid and Spino 2015).

The reactive effect of thinking aloud is exemplified by the phenomenon of "verbal overshadowing". This arises when participants are asked to verbalise their thoughts while attempting problems whose solution discovery benefits from restructuring processes that give rise to feelings of "insight", with these restructuring processes operating primarily at an unconscious level. Pioneering research by Schooler et al. (1993; see also Schooler and Melcher 1995) revealed that thinking aloud *hindered* the attainment of insightful solutions via unconscious restructuring processes. This effect appears to arise because the request to think aloud diverts the problem solver's attention toward strongly activated and obvious aspects of the problem that can easily be verbalised but are irrelevant to its solution (e.g., Bowden et al. 2005). In addition, more weakly activated information that *is* critical for solution success, but which resides at a level below awareness, is unable to enter consciousness because it is blocked or overshadowed by the stronger and reportable—albeit misdirected—information (cf. Ball et al. 2015; Kershaw and Ohlsson 2004; Siegler 2000). Although this verbal overshadowing effect is not always found in studies of insight problem solving, Ball et al. (2015) have suggested that occasional failures to replicate may be attributable to methodological differences between studies.

Considering the potential reactivity that can be engendered through the deployment of the think-aloud method in metareasoning research, the question arises as to whether alternative and less reactive techniques are available to detect fluctuating states of uncertainty during ongoing reasoning. Some possible options that come to mind include the use of eye-gaze tracking (e.g., Ball 2013), pupillometry (e.g., Mathôt and Vilotijević 2022) and eye-blink rate (Paprocki and Lenskiy 2017), as well as physiological measures, such as skin conductance (Figner et al. 2019) and heart-rate variability (Forte et al. 2019). The challenge with these methods, however, is that they have not traditionally been used to pinpoint states of uncertainty during task performance, instead being more closely associated with measures of fluctuating cognitive workload, arousal, stress and fatigue. It may well be that these methods can be deployed in a way that can index states of uncertainty during reasoning, but we are not aware of research that has attempted to do this to date, and it might be that the challenges to do so are simply insurmountable.

Recently, however, one interesting method that has been successfully deployed to measure continuous changes in the feeling of warmth states during problem solving involves the use of a so-called "dynamometer" (Laukkonen et al. 2021). This involves acquiring a continuous measure of fluctuating hand-grip strength from participants who are instructed to use grip intensity to indicate changing feelings of perceived progress toward a solution to a problem that can typically be solved through an unconscious restructuring process. In Laukkonen et al.'s (2021) study, participants were instructed to squeeze the dynamometer more strongly to convey greater perceived progress, and they were also asked to give the dynamometer a full-strength squeeze if they had found the solution via insight and experienced an Aha! moment, or to release their grip quickly if they had reached the solution without an Aha! moment.

The dynamometer allowed Laukkonen et al. (2021) to acquire multiple data points per second in real-time in relation to the onset of insight experiences, additionally enabling them to map such metareasoning data to other measures (i.e., solution accuracy and solution confidence) to investigate convergent validity, which was found to be high. For example, "spikes" in the dynamometer converged with participants verbally reporting after the problem-solving trial that they had experienced an Aha! moment. We note, however, that Laukkonen et al. (2021) did not directly test for an absence of reactivity on problem-solving performance arising from the deployment of the dynamometer, which is a key limitation of their study. Nevertheless, they confirm that problem-solving success rates in their study were comparable or better than those observed for similar tasks in other studies of insight problem solving (i.e., Salvi et al. 2016; Webb et al. 2016, 2018), supporting an apparent lack of reactivity arising from participants having to use the dynamometer to register feelings of progress.

Overall, Laukkonen et al. (2021) suggest that the dynamometer may be a useful tool in any context where researchers are interested in continually measuring metacognition, such as in situations that involve problem solving and insight. The nonverbal nature of the dynamometer presumably makes it far less likely that it will interfere with the demands of primary task processing, such that it should have minimal reactive impact on natural object-level reasoning. In addition, the technique could presumably be utilised with multiple, collaborating problem solvers (each using a dynamometer), although the use of a hand-held device certainly limits the feasibility of it being deployed anywhere other than in relatively simple individual or collaborative reasoning and problem-solving situations. More complex problems of the type that arise in real-world contexts would require free hand movement for activities such as gesturing, writing and collaborative interaction.

Notwithstanding the potential value of deploying dynamometers to tap into people's intermediate uncertainty in collaborative situations, we would argue that the most straightforward yet informative way to obtain a valid, nonreactive measure of fluctuating states of uncertainty during collaborative task performance involves analysing the dynamic use of language by team members. For example, tracking and analysing the dialogue arising between team members working on a joint task could provide an exciting window into moments of both individual uncertainty as well as emergent uncertainty at the level of the whole team. We will consider the value of language markers of metacognitive states in the remaining sections of this paper, as we reflect more deeply on the potential to develop a collaborative metareasoning framework that captures the nature of team-based metacognitive monitoring and control.

5. Metareasoning in Teams

As discussed above, we contend that one of the most striking and important limitations of the existing metareasoning framework relates to its sole focus on individuals. In this respect, the framework, as currently formulated, gives no consideration to the fact that much real-world reasoning takes place in situations that involve others, such as reasoning that is in the service of cooperative or competitive goals, which themselves are underpinned by many processes, including coordination, conflict resolution, deception, argumentation and persuasion. Indeed, the persuasive function of everyday reasoning is a central tenet of the theoretical approach advanced by Mercier (2016), which is focused on the critical role that is played by argumentation in social communication. Mercier (2016; see also Mercier and Sperber 2011) argues that human reasoning has evolved to enable people to devise arguments and justifications such that they can reap the benefits that come from persuading others. However, the issue of how metareasoning might function in situations that relate to persuasion does not appear to have been considered in any detail to date.

Along with situations that involve persuading others, another context in which reasoning does not operate in isolated individuals is that which is concerned with the attainment of shared or mutual goals, such as the achievement of coordinated joint action or the generation of solutions to complex problems that require collaboration between members of heterogeneous teams. Examples of such team-based reasoning and problem-solving situations are numerous in real-world professional practice. Many of these situations are focused on creative endeavours, such as design, innovation, entrepreneurship, advertising and scientific discovery, yet there are other contexts that are also important, as we have mentioned earlier, such as ones relating to defence, security and emergency response. Overall, there remains a dearth of research that has directly investigated metareasoning in these various kinds of collaborative contexts, such that our current understanding of the nature and operation of metareasoning processes in real-world collaborative activities is highly inadequate.

Notwithstanding the generally limited empirically based understanding that exists of collaborative metareasoning, we note that one of the few team-based domains that *has* been subjected to at least some degree of investigation from a metareasoning perspective is that of real-world design, where several studies have examined the monitoring and control processes that arise when designers are developing creative concepts for new products (for reviews of this literature, see Ball and Christensen 2019, and Richardson et al. 2024). A striking observation from this latter body of research relates to the way in which changes in processing are found to co-occur with the appearance in dialogue of hedge words that reflect uncertainty among team members. For example, when faced with uncertainty, designers often appear to be triggered to engage in "analogical reasoning", during which they draw upon conceptual ideas from a domain that is different to that of the problem focus and map these ideas across to the current domain (Ball and Christensen 2009; Ball et al. 2010). Likewise, uncertainty that arises in design contexts also seems to evoke "mental simulation", whereby designers enact or "run" a sequence of interdependent events in a dynamic mental model to determine cause–effect relationships (e.g., between solution components) to predict likely outcomes (Ball and Christensen 2009; Ball et al. 2010; Christensen and Schunn 2009).

Ball and Christensen (2009, 2019) have proposed that cases of analogical reasoning and mental simulation in team design reflect *strategies* that are under metacognitive control, being triggered by emerging uncertainty in the team regarding how to progress toward a design solution. When deployed, these strategies support ongoing design progress while also reducing uncertainty within the team to baseline levels. Outside of the design domain, Chan et al. (2012) have corroborated the existence of a close temporal coupling between uncertainty and the use of analogical reasoning in the context of team-based scientific problem solving. In their research, Chan et al. revealed how expressed uncertainty in team dialogue (similarly indexed through the presence of hedge words) tended to increase prior to episodes of analogical reasoning, subsequently staying at a high level during the analogising process, then returning to a baseline level just after the analogising had terminated. Heightened uncertainty in design teams has also been closely associated with strategic episodes of "problem–solution coevolution" (Wiltschnig et al. 2013). This occurs when designers *simultaneously* refine both their understanding of the design problem and their ideas relating to potential solutions, such that problem understanding informs solution development, as well as vice versa (Dorst and Cross 2001). Wiltschnig et al. (2013) demonstrated that this kind of problem–solution coevolution was more likely to arise during heightened uncertainty in the team-based dialogue and was also linked to increased analogical reasoning.

This research on collaborative metareasoning in team design highlights the critical role that dialogue plays in team interaction to enable teams to make effective control decisions and maintain progress with their design activity. Importantly, such overt dialogue (such as comments that express uncertainty) is not only discernible to interlocutors within collaborating teams but is also visible to researchers interested in investigating the monitoring and control processes that underpin both successful and unsuccessful team activity. In such situations, not only do metacognitive monitoring processes need to be alert to the shifting uncertainty that is salient in the communication of team members, but metacognitive control processes also need to affect dynamic strategy selection and strategy change in a highly coordinated manner if there is to be any hope of problem-solving success. We reiterate, however, that, to date, there has been little direct investigation of such "collaborative metareasoning" in team-based activity. We contend that this lack of research is, in large part, a consequence of the methodological challenges that arise when investigating the dynamic interplay between multiple interlocutors and their associated metareasoning processes.

6. Toward a Framework of Collaborative Metareasoning

Understanding the nature of cooperative joint activities has for a long time presented an ongoing challenge for cognitive science. Almost always, the default research approach is to analyse thought and behaviour at the level of the individual, and this is no different in the field of metareasoning. Yet, in team contexts, successful collaboration depends upon joint action and shared understanding between interlocuters. As we noted in our discussion of the metareasoning of individuals, one of the primary cues to judgments of certainty and uncertainty in relation to task-based progress is that of processing fluency; that is, the ease of generating a response. In collaborative contexts, such processing fluency has both an individual dimension, arising as part of each team member's cognitive processing, and an interpersonal dimension, arising at the level of joint processing. Interpersonal fluency is likely to rely on a variety of cues, some of which will be unique to the interpersonal context. Examples of such cues include background knowledge, familiarity with other team members, perceptions of trust and competence as well as perceived confidence in other team members. Successful team reasoning therefore requires team members to monitor both the individual and joint dimensions of fluency at various points in time, with such dynamic monitoring having the potential to have an impact on procedural decisions to continue with a current processing approach, to switch to a different strategy or potentially to terminate processing if a good solution is not forthcoming.

The importance of team members being able successfully to monitor interpersonal fluency brings to the fore a key issue for research on collaborative metareasoning, which is to investigate and understand the role of alignment and misalignment at this interpersonal level. For example, people tend to find interactions easier and more enjoyable when they like and know other team members, regardless of their individual or joint success with the problem at hand (Richardson et al. 2019). Alignment in interpersonal fluency may play a crucial role in team perceptions of progress on a given task, and hence is likely to be influential in procedural decisions to carry on without making strategy changes.

What is generally missing from existing theorising, however, is a model that refers to the interrelationships between individuals in navigating interpersonal fluency from the perspective of the whole team rather than just from the perspective of the individual members who make up the team. As noted, such a model would need to capture the way in which interpersonal fluency is monitored by team members as well as the impact of ongoing fluency dynamics on team members' control processes in determining strategic decision making.

We suggest that moving from metareasoning considerations at the individual level to considerations at the social and interpersonal level, whether reasoning is serving competitive goals (as in persuasion or deception) or cooperative goals (as in collaborative reasoning and problem solving), will require very careful and highly systematic augmentation of current models of metacognition, such as Ackerman and Thompson's (2017, 2018) metareasoning framework. Even a cursory assessment of how metareasoning might take place in team-based reasoning indicates the complexity of the theoretical issues that immediately become foregrounded. As we have noted, what is particularly fascinating, yet poorly understood, is the way in which individual members of teams invoke and coordinate both individual-level and team-level metacognitive monitoring and control processes to drive forward their joint action and strategic decision making. The metacognitive monitoring and control processes that team members need to engage will presumably have to be attuned to the subtleties and complexities of both personal and interpersonal uncertainty, with the latter most likely being made manifest in the verbal and nonverbal communication arising during team interaction.

As mentioned in our introduction, an important tripartite distinction can be drawn in relation to collaborative metareasoning between "self-monitoring" (i.e., an individual's perception of their own performance), "other monitoring" (i.e., an individual's perception of the performance of others) and "joint monitoring" (i.e., the unified perception of collective performance), as articulated by Richardson et al. (2024; see also Pickering and Garrod 2021). When considering this tripartite distinction specifically in relation to fluctuating states of certainty in reasoning contexts, it can be seen how the distinction gives rise to concerns not only with an individual's own feeling of confidence in a solution to a reasoning task but also that individual's awareness of a collaborator's feeling of confidence in this solution, as well as the collective or agreed-upon confidence in the solution. All these elements of ongoing monitoring have the potential to impact decisions about task progress in important ways and are critical in informing a shared understanding of the reasoning task at hand. As we also noted earlier, a parallel tripartite distinction can be drawn that relates to control processes, which can be captured by the notions of "self-focused control" (i.e., an individual's decisions about how to progress or terminate their own reasoning), "other-focused control" (i.e., an individual's decisions about how to control the performance of others) and "joint control" (i.e., the unified control of decisions regarding how to advance or terminate collective performance).

Table 1 presents a more detailed characterisation of key monitoring and control processes that arise across the three proposed levels of metacognition, as already discussed above and further elaborated upon in the next section. The table also summarises examples of the kinds of cues that are very likely to be detected by monitoring processes, as well as examples of the kinds of outcomes that will result from the deployment of control processes. We contend that it is possible to track many of the key monitoring and control processes associated with our tripartite distinction through an analysis of ongoing dialogue (Gonzales et al. 2010; Richardson and Nash 2022). Through such analyses, it is possible, for example, to identify periods of conflict (Tausczik and Pennebaker 2013) or uncertainty within team reasoning, problem solving and decision making (e.g., as reflected in the use of tentative language or hedge words), as well as strategic changes that arise as a result of the detection of conflict or uncertainty. We therefore propose that research on collaborative metareasoning would benefit from the development of theoretical models that capture how uncertainty, as well as both alignment and misalignment, can trigger self- and group reflection and downstream strategic decision making. Importantly, we suggest that these periods of uncertainty can be tracked dynamically by measuring social signals, specifically the linguistic, paralinguistic and proxemic features that occur during dialogue between interlocuters.

Table 1. A detailed characterisation of indicative monitoring and control processes that arise across three proposed levels of metacognition. The table also summarises examples of the cues that are likely to be detected by monitoring processes, as well as examples of the outcomes of control processes.

Levels of Monitoring	Indicative Monitoring Processes	Example Cues Detected by Monitoring Processes
Self-Monitoring An individual's perception of their own performance.	**An individual's generation of an:** initial judgment of solvability; feeling of rightness; feeling of error; feeling of warmth; intermediate confidence or uncertainty; final confidence; final judgment of solvability.	**An individual's sensitivity to:** processing fluency (ease of processing); perceived features of the presented task; perceived task complexity; study time; response time.
Other Monitoring An individual's perception of the performance of others.	**An individual's perception of someone else's:** initial judgment of solvability; feeling of rightness; feeling of error; feeling of warmth; intermediate confidence or uncertainty; final confidence; final judgment of solvability. **An individual's perception of:** alignment/misalignment.	**An individual's sensitivity to someone else's:** processing fluency (ease of processing); perceived features of the presented task; perceived task complexity; study time; response time; degree of agreement; level of understanding (potentially made manifest by language markers such as hedge words and pronoun use).
Joint Monitoring The unified perception of collective performance.	**A group's unified perception of:** initial judgment of solvability; feeling of rightness; feeling of error; feeling of warmth; intermediate confidence or uncertainty; final confidence; final judgment of solvability. **A group's perception of:** alignment/misalignment.	**A group's unified perception of:** processing fluency (ease of processing); perceived features of the presented task; perceived task complexity; study time; response time; degree of agreement; level of understanding (potentially made manifest by language markers such as hedge words and pronoun use).
Levels of Control	**Indicative Control Processes**	**Example Outcomes of Control Processes**
Self-Focused Control An individual's decisions about how to progress or terminate their own reasoning.	**An individual's procedural decision to engage in:** memory search; reasoning, problem solving or decision making; response generation; response evaluation; strategy change; giving up; help-seeking.	**An individual's generation of:** recalled information; an intermediate or final response (e.g., a solution, option or decision, including the decision to give up); an evaluation of an intermediate or final response; a new process or strategy (e.g., analogising, mental simulation); a request for help.
Other-Focused Control An individual's decisions about how to control the performance of others.	**An individual's procedural decision to engage in:** affirmation; encouragement; persuasion; argumentation; negotiation; manipulation; deception.	**An individual's generation of:** alignment (e.g., situation model alignment and linguistic alignment); intersubjectvity; shared understanding; common ground; conflict resolution; misalignment; disalignment.
Joint Control The unified control of decisions regarding how to advance or terminate collective performance.	**A group's unified procedural decision to engage in:** memory search; reasoning, problem solving or decision making; response generation; response evaluation; strategy change; giving up; help-seeking.	**A group's unified generation of:** recalled information; an intermediate or final response (e.g., a solution, option or decision, including the decision to give up); an evaluation of an intermediate or final response; a new process or strategy (e.g., analogising, mental simulation); a request for help.

7. The Role of Language in Collaborative Metareasoning

In dialogue, speakers process a great deal of information, they take and give the floor to each other as well as plan and adjust their contributions on the fly. Despite the level of cognitive effort and control that it requires, dialogue is the easiest way speakers possess to come to similar conceptualisations of the world. For example, Pickering and Garrod (2021) suggest that it is the process of alignment, a largely automatic and unconscious process whereby speakers use language in the same way to reach mutual understanding, that underpins many successful joint activities.

Research on alignment indicates that, over time, in successful communication, people tend to think in the same way (representing the world in the same way) and use similar terms of expression. For example, in problem solving, speakers align with each other by mutually controlling the flow of dialogue and by constantly monitoring their own and others' ways of representing information. Furthermore, alignment increases over time as the dialogue progresses. In a verbal description task, reported by Clark and Wilkes-Gibbs (1986), participants in the role of "director" started describing a figure to a "matcher" using long and detailed sentences (e.g., "the next one looks like a person who's ice skating, except they're sticking two arms out in front"). In subsequent turns, descriptions became simpler (e.g., "the fourth one is the person ice skating, with two arms"), until the interlocutors converged on a common description ("the ice skater") and used it effectively until the end of the task. What this means is that speakers gradually converge on a similar conceptualisation of the task (termed "situation model alignment"; Garrod and Anderson 1987) by aligning at a linguistic level (termed "linguistic alignment").

Alignment at the level of joint monitoring and control can be understood in terms of the important theoretical concept of "intersubjectivity", which refers to having a shared understanding of an object (e.g., Mori and Hayashi 2006) or an interlocuter (Gillespie and Richardson 2011). Like alignment, intersubjectivity is frequently thought of as being an implicit and automatic behavioural orientation toward others (Coelho and Figueiredo 2003; Merleau-Ponty [1945] 1962). For example, frequent, brief, paralinguistic communications such as "uh-huh" or head-nodding that punctuate a conversation serve a purely social function. They are not designed to contribute additional meaning but are instead used to provide ongoing feedback about comprehension (Schegloff 1982), thereby signalling implicit understanding to other team members. A similar role is played by the "third position repair" described by Schegloff (1992), whereby a listener will often respond to a speaker with an utterance which, instead of contributing anything new, simply displays understanding of what has been said.

Although research typically suggests that intersubjectivity and alignment are key to successful collaboration, Richardson et al. (2007) additionally propose that misalignment (or what they refer to as "disalignment") is also likely to trigger a solution to a team-based problem. Such misalignment is similar to situations in which recognition of uncertainty can trigger a strategy change that moves the task towards successful resolution (Ball and Christensen 2009, 2019). For example, misalignment can lead to teams engaging in conflict-resolution attempts that can promote team agreement on strategy change. The benefits for team-based reasoning that can derive from misalignment have been demonstrated in a study reported by Paletz et al. (2017), in which they analysed the temporal relationships between brief interpersonal disagreements (or "microconflicts") and the subsequent expression of uncertainty in conversations arising in successful and unsuccessful engineering product design teams. Paletz et al. discovered that microconflicts were followed by a relative decrease in uncertainty in successful design teams, whereas uncertainty increased after microconflicts in unsuccessful design teams. Paletz et al. (2017) interpret these findings as suggesting that the interaction between conflict and uncertainty may be critical in determining the success of design teams.

Building on the idea of the importance of misalignment and uncertainty for task progress, we further note here that Bjørndahl et al. (2015) suggest that agreement between interlocuters is not enough when working toward a solution to a problem. They identify three key interactional styles, one of which they refer to as an "integrative style", which is characterised by self–other repair via, for example, clarification requests, disagreements, questions and explicit negotiation of ideas and proposals. During a collaborative LEGO modelling task, Bjørndahl et al. (2015) found that this integrative style, which promotes explicit miscommunication, generated more innovative models as compared to an inclusive praise-based style or an instructional self-repair style. Crucially, speakers do not reach alignment in isolation, but through interaction, by manipulating each other's contributions. Furthermore, speakers are able to track this alignment by metarepresenting it (Gandolfi et al. 2023). Speakers mutually control the flow of dialogue and constantly monitor their own and their interlocutors' way of representing information.

What is missing from the existing literature on metareasoning is a framework that captures metareasoning in teams, and that offers a means of dynamically tracking key processes, such as fluency and uncertainty, between team members. Our framework (see Table 1) proposes tripartite distinctions relating to both metacognitive monitoring and control that can be meaningfully applied to understand collaborative metareasoning at the level of self-, other and joint processing, while also focusing on how collaborative metareasoning unfolds via dialogue between team members. We argue that the perception of misalignment triggers monitoring, which leads to negotiation between perspectives. During collaborative monitoring, for example, alignment can be seen as a process that requires the combination of metacognition (i.e., with respect to oneself) and social cognition (i.e., with respect to interlocutors, Gandolfi et al. 2023).

Team members not only predict and monitor the utterances of interlocutors, but they also predict and monitor their own utterances, as well as acting on joint representations by comparing their expectation of what their interlocutors might say, with what they actually say. When discrepancies arise between what is predicted and what arises, this typically leads to subsequent reformulations, expansions or clarifications that help drive the interaction forward. Specifically, when speakers metarepresent the failure of alignment they will reformulate their plans and correct their contributions to keep the dialogue on track.

Control is especially necessary for the attainment of joint reasoning activities and team decision making. We also argue that control plays out via the interaction between interlocuters, who do not reach alignment in isolation, but do so by manipulating each other's contributions in a collaborative fashion. By continuously monitoring and comparing self and others' contributions, and specifically by metarepresenting whether they believe they and their interlocutor are aligned or not, team members make decisions about whether to continue a particular course of action, terminate the task or switch strategies to find an alternative way forward.

8. Conclusions and Future Directions

A key focus in this paper has been on the importance of analysing the use of language when examining collaborative metareasoning. For example, diminishing confidence, increasing interpersonal misalignment and emerging conflict—as revealed through language change during team reasoning—appear to be linked to metacognitive control decisions to adopt new approaches or to terminate processing (cf. Paletz et al. 2017). These dynamic and objective language-based measures of unfolding metareasoning processes in teams can also be compared with more traditional self-report measures, discussed above, that relate to task performance, solution confidence and interpersonal dynamics. In addition, sophisticated approaches to language analysis in teams (Bjørndahl et al. 2015) can be invaluable for identifying important aspects of behavioural sequencing, as well as for uncovering clusters of behaviours that capture the team-based monitoring of ongoing processing fluency and disfluency.

We suggest that future research examining language use in teams as an index of metacognitive monitoring and control could explore the way in which the different interactional styles of team members modulate metareasoning processes. In this respect, we note that Gonzales et al. (2010) have shown that, in groups with a strong hierarchy, an authoritarian leadership style is often characterised by the frequent use of self over collective pronouns, and that this display of pronouns can be detrimental for team cohesion. We contend that the impact of different leadership styles is also likely to manifest in unique ways in the dialogue that relates to joint metacognitive monitoring and control during team reasoning. These modulating effects of leadership styles remain to be examined empirically, yet are clearly important to explore in high-stakes, real-world, decision-making contexts where effective team reasoning is critical.

We also propose that our collaborative metareasoning framework could be extended to take into account multimodal measures of team interaction. For example, the fine-grained analysis of real-time social signals could provide further valuable insights into fluctuating levels of confidence and cohesion in teams in relation to ongoing monitoring and control processes (cf. Casakin et al. 2015). Indeed, although we have emphasised the way in which an analysis of language can provide valuable information about ongoing monitoring and control processes in team-based reasoning, this is not to dismiss the insights into collaborative metareasoning that might also be gained from other forms of behavioural analysis. In this respect, we acknowledge the existence of substantial bodies of behavioural research relating to the complex interplay that arises between individuals engaged in the achievement of common goals, including studies concerning the nature of joint action (e.g., Sebanz and Knoblich 2009), collective intelligence (e.g., Krause et al. 2010), synchrony (e.g., Lakens and Stel 2011; Miles et al. 2011), perspective taking (Gillespie and Richardson 2011) and distributed cognition (e.g., Hutchins 1995). A detailed review of this literature might well pinpoint important findings that could be informative about collaborative metareasoning. What seems more likely, however, is that research in these areas—apart from studies of synchrony and perspective taking—has typically been more concerned with the nature of people's task-oriented, object-level processing (e.g., coordinated action or team decision making) than with people's metareasoning processes. Nevertheless, until these other research areas are carefully examined through a metareasoning lens, it remains possible that important knowledge is being missed.

What also seems clear is the need for our proposed collaborative metareasoning framework to be evaluated and extended through extensive empirical research. Indeed, the tripartite distinction that we articulate relating to self-, other and joint metareasoning only presents a starting point for further conceptual development regarding the nature of these metacognitive processes and their interconnections. In this respect, we concede that the literature on collaborative metareasoning is simply not sufficiently well advanced to enable much in the way of detail and precision concerning the highly complex interrelationships that are certain to exist between individual metareasoning and collaborative metareasoning processes. More positively, it does seem that major conceptual advancements are increasingly likely as interest in metareasoning continues to grow (e.g., see De Neys 2023).

What is also encouraging for the ongoing development of the field of collaborative metareasoning is the potential for a multimethod investigative approach to be able to elicit highly rich and informative data regarding the nature and interplay between individual and collaborative metareasoning processes. Such a multimethod approach can, for example, lend itself to the identification of key subjective metrics (e.g., self-report measures that index trust and confidence in oneself and in others) as well as objective metrics (e.g., language-based markers of uncertainty and eye-gaze measures to pinpoint challenges arising in joint cognition). Empirical research additionally needs to track processes of alignment and misalignment via dialogue (Richardson and Nash 2022) to identify periods of conflict or uncertainty within reasoning (e.g., as reflected in the use of tentative language, hedge words or pronouns), and compare these indices with measurements of traditional metareasoning concepts (e.g., fluency, intermediate confidence and strategy change).

We contend that the conceptual progress that can derive from our proposals for a collaborative metareasoning framework can additionally provide valuable opportunities to develop interventions that might support enhanced reasoning, problem solving and decision making in real-world contexts, including those that we identified throughout our paper (e.g., ones relating to design, innovation, defence, security, surveillance and emergency response). Interventions targeted at supporting metareasoning might assist, for example, with the identification of periods of uncertainty, points of impasse or instances of misunderstanding, which can then prompt teams to reflect on more effective strategic choices. Such metareasoning support tools would represent an original and radical approach to facilitating successful collaboration in applied contexts, augmenting natural reasoning and metareasoning processes in useful ways that teams could capitalise upon.

We have attempted here to advance the metareasoning literature by commencing the development of an understanding of the metareasoning processes that arise in team-based contexts. We propose a collaborative metareasoning framework that emphasises the importance of a tripartite distinction between self-, other and joint metareasoning for capturing the different levels of monitoring and control that arise during task-based processing. We also suggest that the interplay that occurs between different levels of metacognitive monitoring and control is critical for a team's procedural decision making, such as collective decisions to accept an initial solution, to switch strategy or to give up as well as to a team's potential achievement of a successful task outcome. We additionally argue for the utility of analysing language as a means of tracking the fluctuating states of uncertainty and misalignment that arise, with such states seemingly having the capacity to act as metacognitive triggers for strategy change. In conclusion, our framework begins to address a significant gap in the literature surrounding the monitoring and control processes that play out during collaborative reasoning, whereby complex, interpersonal interactions occur as people work together to achieve shared goals.

Author Contributions: Conceptualization, B.H.R. and L.J.B.; methodology, B.H.R. and L.J.B.; writing—original draft preparation, B.H.R. and L.J.B.; writing—review and editing, B.H.R. and L.J.B.; project administration, B.H.R. and L.J.B. All authors have read and agreed to the published version of the manuscript.

Funding: This research received no external funding.

Institutional Review Board Statement: Not applicable.

Informed Consent Statement: Not applicable.

Data Availability Statement: No new data were created or analyzed in this study. Data sharing is not applicable to this article.

Conflicts of Interest: The authors declare no conflict of interest.

References

Ackerman, Rakefet. 2014. The diminishing criterion model for metacognitive regulation of time investment. *Journal of Experimental Psychology: General* 143: 1349–68. [CrossRef]

Ackerman, Rakefet. 2023. Bird's-eye view of cue integration: Exposing instructional and task design factors which bias problem solvers. *Educational Psychology Review* 35: 55. [CrossRef]

Ackerman, Rakefet, and Hagar Zalmanov. 2012. The persistence of the fluency–confidence association in problem solving. *Psychonomic Bulletin & Review* 19: 1187–92.

Ackerman, Rakefet, and Morris Goldsmith. 2008. Control over grain size in memory reporting: With and without satisficing knowledge. *Journal of Experimental Psychology: Learning, Memory, & Cognition* 34: 1224–45.

Ackerman, Rakefet, and Valerie A. Thompson. 2017. Meta-reasoning: Monitoring and control of thinking and reasoning. *Trends in Cognitive Sciences* 21: 607–17. [CrossRef] [PubMed]

Ackerman, Rakefet, and Valerie A. Thompson. 2018. Meta-reasoning: Shedding meta-cognitive light on reasoning research. In *The Routledge International Handbook of Thinking and Reasoning*. Edited by Linden J. Ball and Valerie A. Thompson. Abingdon: Routledge, pp. 164–82.

Ackerman, Rakefet, and Yael Beller. 2017. Shared and distinct cue utilization for metacognitive judgements during reasoning and memorisation. *Thinking & Reasoning* 23: 376–408.

Ackerman, Rakefet, Elad Yom-Tov, and Ilan Torgovitsky. 2020. Using confidence and consensuality to predict time invested in problem solving and in real-life web searching. *Cognition* 199: 104248. [CrossRef] [PubMed]

Alter, Adam L., and Daniel M. Oppenheimer. 2009. Uniting the tribes of fluency to form a metacognitive nation. *Personality & Social Psychology Review* 13: 219–35.

Baars, Martine, Sandra Visser, Tamara Van Gog, Anique de Bruin, and Fred Paas. 2013. Completion of partially worked-out examples as a generation strategy for improving monitoring accuracy. *Contemporary Educational Psychology* 38: 395–406. [CrossRef]

Bago, Bence, and Wim De Neys. 2017. Fast logic? Examining the time course assumption of dual process theory. *Cognition* 158: 90–109. [CrossRef]

Ball, Linden J. 2013. Eye-tracking and reasoning: What your eyes tell about your inferences. In *New Approaches in Reasoning Research*. Edited by Wim De Neys and Magda Osman. Hove: Psychology Press, pp. 51–69.

Ball, Linden J., and Bo T. Christensen. 2009. Analogical reasoning and mental simulation in design: Two strategies linked to uncertainty resolution. *Design Studies* 30: 169–86. [CrossRef]

Ball, Linden J., and Bo T. Christensen. 2019. Advancing an understanding of design cognition and design metacognition: Progress and prospects. *Design Studies* 65: 35–59. [CrossRef]

Ball, Linden J., Balder Onarheim, and Bo T. Christensen. 2010. Design requirements, epistemic uncertainty and solution development strategies in software design. *Design Studies* 31: 567–89. [CrossRef]

Ball, Linden J., John E. Marsh, Damien Litchfield, Rebecca L. Cook, and Natalie Booth. 2015. When distraction helps: Evidence that concurrent articulation and irrelevant speech can facilitate insight problem solving. *Thinking & Reasoning* 21: 76–96.

Baranski, Joseph V., and William M. Petrusic. 2001. Testing architectures of the decision–confidence relation. *Canadian Journal of Experimental Psychology/Revue Canadienne de Psychologie Expérimentale* 55: 195–206. [CrossRef]

Bjørndahl, Johanne S., Riccardo Fusaroli, Svend Østergaard, and Kristian Tylén. 2015. Agreeing is not enough: The constructive role of miscommunication. *Interaction Studies* 16: 495–525. [CrossRef]

Bonder, Taly, and Daniel Gopher. 2019. The effect of confidence rating on a primary visual task. *Frontiers in Psychology* 10: 2674. [CrossRef]

Bowden, Edward M., Mark Jung-Beeman, Jessica Fleck, and John Kounios. 2005. New approaches to demystifying insight. *Trends in Cognitive Sciences* 9: 322–28. [CrossRef]

Campitelli, Guillermo, and Paul Gerrans. 2014. Does the cognitive reflection test measure cognitive reflection? A mathematical modeling approach. *Memory & Cognition* 42: 434–47.

Casakin, Hernan, Linden J. Ball, Bo T. Christensen, and Petra Badke-Schaub. 2015. How do analogizing and mental simulation influence team dynamics in innovative product design? *AI EDAM* 29: 173–83. [CrossRef]

Chan, Joel, Susannah B. Paletz, and Christian D. Schunn. 2012. Analogy as a strategy for supporting complex problem solving under uncertainty. *Memory & Cognition* 40: 1352–65.

Christensen, Bo T., and Christian D. Schunn. 2009. The role and impact of mental simulation in design. *Applied Cognitive Psychology* 23: 327–44. [CrossRef]

Clark, Herbert H., and Deanna Wilkes-Gibbs. 1986. Referring as a collaborative process. *Cognition* 22: 1–39. [CrossRef]

Coelho, Nelson E., Jr., and Luis C. Figueiredo. 2003. Patterns of intersubjectivity in the constitution of subjectivity: Dimensions of otherness. *Culture & Psychology* 9: 193–208.

Cromley, Jennifer G., and Andrea J. Kunze. 2020. Metacognition in education: Translational research. *Translational Issues in Psychological Science* 6: 15–20. [CrossRef]

De Neys, Wim. 2023. Advancing theorizing about fast-and-slow thinking. *Behavioral and Brain Sciences* 46: e111. [CrossRef] [PubMed]

Dorst, Kees, and Nigel Cross. 2001. Creativity in the design process: Co-evolution of problem–solution. *Design Studies* 22: 425–37. [CrossRef]

Double, Kit S., and Damian P. Birney. 2017. Are you sure about that? Eliciting confidence ratings may influence performance on Raven's progressive matrices. *Thinking & Reasoning* 23: 190–206.

Double, Kit S., and Damian P. Birney. 2018. Reactivity to confidence ratings in older individuals performing the Latin square task. *Metacognition & Learning* 13: 309–26.

Double, Kit S., and Damian P. Birney. 2019a. Do confidence ratings prime confidence? *Psychonomic Bulletin & Review* 26: 1035–42.

Double, Kit S., and Damian P. Birney. 2019b. Reactivity to measures of metacognition. *Frontiers in Psychology* 10: 2755. [CrossRef]

Double, Kit S., Damian P. Birney, and Sarah A. Walker. 2018. A meta-analysis and systematic review of reactivity to judgements of learning. *Memory* 26: 741–50. [CrossRef] [PubMed]

Ericsson, K. Anders, and Herbert A. Simon. 1980. Verbal reports as data. *Psychological Review* 87: 215–51. [CrossRef]

Evans, Jonathan St. B. T. 2018. Dual-process theories. In *The Routledge International Handbook of Thinking and Reasoning*. Edited by Linden J. Ball and Valerie A. Thompson. Abingdon: Routledge, pp. 151–66.

Evans, Jonathan St. B. T., and Keith E. Stanovich. 2013a. Dual-process theories of higher cognition: Advancing the debate. *Perspectives on Psychological Science* 8: 223–41. [CrossRef] [PubMed]

Evans, Jonathan St. B. T., and Keith E. Stanovich. 2013b. Theory and metatheory in the study of dual processing: Reply to comments. *Perspectives on Psychological Science* 8: 263–71. [CrossRef] [PubMed]

Figner, Bernd, Ryan O. Murphy, and Paul Siegel. 2019. Measuring electrodermal activity and its applications in judgment and decision-making research. In *A Handbook of Process Tracing Methods*, 2nd ed. Edited by Michael Schulte-Mecklenbeck, Anton Kuehberger and Joseph G. Johnson. Abingdon: Routledge, pp. 163–84.

Forte, Giuseppe, Francesca Favieri, and Maria Casagrande. 2019. Heart rate variability and cognitive function: A systematic review. *Frontiers in Neuroscience* 13: 710. [CrossRef] [PubMed]

Frederick, Shane. 2005. Cognitive reflection and decision making. *Journal of Economic Perspectives* 19: 25–42. [CrossRef]

Gandolfi, Greta, Martin J. Pickering, and Simon Garrod. 2023. Mechanisms of alignment: Shared control, social cognition and metacognition. *Philosophical Transactions of the Royal Society B* 378: 20210362. [CrossRef]

Garrod, Simon, and Anthony Anderson. 1987. Saying what you mean in dialogue: A study in conceptual and semantic co-ordination. *Cognition* 27: 181–218. [CrossRef]

Gillespie, Alex, and Beth Richardson. 2011. Exchanging social positions: Enhancing perspective taking within a cooperative problem-solving task. *European Journal of Social Psychology* 41: 608–16. [CrossRef]

Godfroid, Aline, and Lee Anne Spino. 2015. Reconceptualizing reactivity of think-alouds and eye tracking: Absence of evidence is not evidence of absence. *Language Learning* 65: 896–928. [CrossRef]

Gonzales, Amy L., Jeffrey T. Hancock, and James W. Pennebaker. 2010. Language style matching as a predictor of social dynamics in small groups. *Communication Research* 37: 3–19. [CrossRef]

Hacker, Douglas J., John Dunlosky, and Arthur C. Graesser, eds. 2009. *Handbook of Metacognition in Education*. Abingdon: Routledge.

Hamilton, Katherine, Vincent Mancuso, Susan Mohammed, Rachel Tesler, and Michael McNeese. 2017. Skilled and unaware: The interactive effects of team cognition, team metacognition, and task confidence on team performance. *Journal of Cognitive Engineering and Decision Making* 11: 382–95. [CrossRef]

Hedne, Mikael R., Elisabeth Norman, and Janet Metcalfe. 2016. Intuitive feelings of warmth and confidence in insight and non-insight problem solving of magic tricks. *Frontiers in Psychology* 7: 1314. [CrossRef] [PubMed]

Hutchins, Edwin. 1995. *Cognition in the Wild*. Cambridge, MA: MIT Press.

Kahneman, Daniel, and Shane Frederick. 2002. Representativeness revisited: Attribute substitution in intuitive judgment. In *Heuristics and Biases: The Psychology of Intuitive Judgment*. Edited by Thomas Gilovich, Dale Griffin and Daniel Kahneman. Cambridge, MA: Cambridge University Press, pp. 49–81.

Kaufmann, Esther. 2022. Lens model studies: Revealing teachers' judgements for teacher education. *Journal of Education for Teaching* 49: 236–51. [CrossRef]

Kershaw, Trina C., and Stellan Ohlsson. 2004. Multiple causes of difficulty in insight: The case of the nine-dot problem. *Journal of Experimental Psychology: Learning, Memory, and Cognition* 30: 3–13. [CrossRef] [PubMed]

Kizilirmak, Jasmin M., Violetta Serger, Judith Kehl, Michael Öllinger, Kristian Folta-Schoofs, and Alan Richardson-Klavehn. 2018. Feelings-of-warmth increase more abruptly for verbal riddles solved with in contrast to without Aha! experience. *Frontiers in Psychology* 9: 1404. [CrossRef] [PubMed]

Krause, Jens, Graeme D. Ruxton, and Stefan Krause. 2010. Swarm intelligence in animals and humans. *Trends in Ecology & Evolution* 25: 28–34.

Kruglanski, Arie W., and Gerd Gigerenzer. 2011. Intuitive and deliberate judgments are based on common principles. *Psychological Review* 118: 97–109. [CrossRef]

Lakens, Daniël, and Mariëlle Stel. 2011. If they move in sync, they must feel in sync: Movement synchrony leads to attributions of rapport and entitativity. *Social Cognition* 29: 1–14. [CrossRef]

Laukkonen, Ruben E., and Jason M. Tangen. 2018. How to detect insight moments in problem solving experiments. *Frontiers in Psychology* 9: 282. [CrossRef]

Laukkonen, Ruben E., Daniel J. Ingledew, Hilary J. Grimmer, Jonathan W. Schooler, and Jason M. Tangen. 2021. Getting a grip on insight: Real-time and embodied Aha experiences predict correct solutions. *Cognition and Emotion* 35: 918–35. [CrossRef] [PubMed]

Law, Marvin K., Lazar Stankov, and Sabina Kleitman. 2022. I choose to opt-out of answering: Individual differences in giving up behaviour on cognitive tests. *Journal of Intelligence* 10: 86. [CrossRef] [PubMed]

Lei, Wei, Jing Chen, Chunliang Yang, Yiqun Guo, Pan Feng, Tingyong Feng, and Hong Li. 2020. Metacognition-related regions modulate the reactivity effect of confidence ratings on perceptual decision-making. *Neuropsychologia* 144: 107502. [CrossRef] [PubMed]

Mathôt, Sebastiaan, and Ana Vilotijević. 2022. Methods in cognitive pupillometry: Design, preprocessing, and statistical analysis. *Behavior Research Methods* 55: 3055–77. [CrossRef] [PubMed]

Mercier, Hugo. 2016. The argumentative theory: Predictions and empirical evidence. *Trends in Cognitive Sciences* 20: 689–700. [CrossRef] [PubMed]

Mercier, Hugo, and Dan Sperber. 2011. Why do humans reason? Arguments for an argumentative theory. *Behavioral & Brain Sciences* 34: 57–74.

Merleau-Ponty, Maurice. 1962. *Phenomenology of Perception*. Translated by Colin Smith. London: Routledge & Kegan Paul. First published 1945.

Metcalfe, Janet. 1986. Premonitions of insight predict impending error. *Journal of Experimental Psychology: Learning, Memory, and Cognition* 12: 623–34. [CrossRef]

Metcalfe, Janet, and David Wiebe. 1987. Intuition in insight and noninsight problem solving. *Memory & Cognition* 15: 238–46.

Miles, Lynden K., Joanne Lumsden, Michael J. Richardson, and C. Neil Macrae. 2011. Do birds of a feather move together? Group membership and behavioral synchrony. *Experimental Brain Research* 211: 495–503. [CrossRef]

Mori, Junko, and Makoto Hayashi. 2006. The achievement of intersubjectivity through embodied completions: A study of interactions between first and second language speakers. *Applied Linguistics* 27: 195–219. [CrossRef]

Nelson, Thomas O., and Louis Narens. 1990. Metamemory: A theoretical framework and new findings. In *The Psychology of Learning and Motivation: Advances in Research and Theory*. Edited by Gordon Bower. Cambridge, MA: Academic Press, pp. 125–73.

Paletz, Susannah B., Joel Chan, and Christian D. Schunn. 2017. The dynamics of micro-conflicts and uncertainty in successful and unsuccessful design teams. *Design Studies* 50: 39–69. [CrossRef]

Paprocki, Rafal, and Artem Lenskiy. 2017. What does eye-blink rate variability dynamics tell us about cognitive performance? *Frontiers in Human Neuroscience* 11: 620. [CrossRef] [PubMed]

Pennycook, Gordon, James A. Cheyne, Derek J. Koehler, and Jonathan A. Fugelsang. 2016. Is the cognitive reflection test a measure of both reflection and intuition? *Behavior Research Methods* 48: 341–48. [CrossRef] [PubMed]

Perry, John, David Lundie, and Gill Golder. 2019. Metacognition in schools: What does the literature suggest about the effectiveness of teaching metacognition in schools? *Educational Review* 71: 483–500. [CrossRef]

Pervin, Nargis, Tuan Q. Phan, Anindya Datta, Hideaki Takeda, and Fujio Toriumi. 2015. Hashtag popularity on twitter: Analyzing co-occurrence of multiple hashtags. In *Proceedings of Social Computing and Social Media: 7th International Conference, SCSM 2015*. Berlin/Heidelberg: Springer International Publishing, pp. 169–82.

Petrusic, William M., and Joseph V. Baranski. 2003. Judging confidence influences decision processing in comparative judgments. *Psychonomic Bulletin & Review* 10: 177–83.

Pickering, Martin J., and Simon Garrod. 2021. *Understanding Dialogue: Language Use and Social Interaction*. Cambridge, MA: Cambridge University Press.

Quayle, Jeremy D., and Linden J. Ball. 2000. Working memory, metacognitive uncertainty and belief bias in syllogistic reasoning. *Quarterly Journal of Experimental Psychology* 53: 1202–23. [CrossRef] [PubMed]

Richardson, Beth H., and Robert A. Nash. 2022. 'Rapport myopia' in investigative interviews: Evidence from linguistic and subjective indicators of rapport. *Legal & Criminological Psychology* 27: 32–47.

Richardson, Beth H., Kathleen C. McCulloch, Paul J. Taylor, and Helen J. Wall. 2019. The cooperation link: Power and context moderate verbal mimicry. *Journal of Experimental Psychology: Applied* 25: 62–76. [CrossRef]

Richardson, Beth H., Linden J. Ball, Bo T. Christensen, and John E. Marsh. 2024. Collaborative meta-reasoning in creative contexts: Advancing an understanding of collaborative monitoring and control in creative teams. In *The Routledge International Handbook of Creative Cognition*. Edited by Linden J. Ball and Frédéric Vallée-Tourangeau. Abingdon: Routledge, pp. 709–27.

Richardson, Daniel C., Rick Dale, and Natasha Z. Kirkham. 2007. The art of conversation is coordination. *Psychological Science* 18: 407–13. [CrossRef]

Roberts, Linds W. 2017. Research in the real world: Improving adult learners web search and evaluation skills through motivational design and problem-based learning. *College & Research Libraries* 78: 527–51.

Salvi, Carola, Emanuela Bricolo, John Kounios, Edward Bowden, and Mark Beeman. 2016. Insight solutions are correct more often than analytic solutions. *Thinking & Reasoning* 22: 443–60.

Saunders, John. 2022. *Manchester Arena Inquiry, Volume 2: Emergency Response*. London: His Majesty's Stationery Office.

Schegloff, Emanuel A. 1982. Discourse as an interactional achievement: Some uses of 'uh huh' and other things that come between sentences. In *Analyzing Discourse: Text and Talk*. Edited by Deborah Tannen. Washington, DC: Georgetown University Press, pp. 71–93.

Schegloff, Emanuel A. 1992. Repair after next turn: The last structurally provided defense of intersubjectivity in conversation. *American Journal of Sociology* 97: 1295–345. [CrossRef]

Schoenherr, Jordan R., Craig Leth-Steensen, and William M. Petrusic. 2010. Selective attention and subjective confidence calibration. *Attention, Perception, and Psychophysics* 72: 353–68. [CrossRef]

Schooler, Jonathan W., and Joseph Melcher. 1995. The ineffability of insight. In *The Creative Cognition Approach*. Edited by Steven Smith, Thomas Ward and Ronald Finke. Cambridge, MA: MIT Press, pp. 97–133.

Schooler, Jonathan W., Stellan Ohlsson, and Kevin Brooks. 1993. Thoughts beyond words: When language overshadows insight. *Journal of Experimental Psychology: General* 122: 166–83. [CrossRef]

Sebanz, Natalie, and Guenther Knoblich. 2009. Prediction in joint action: What, when, and where. *Topics in Cognitive Science* 1: 353–67. [CrossRef] [PubMed]

Siegler, Robert S. 2000. Unconscious insights. *Current Directions in Psychological Science* 9: 79–83. [CrossRef]

Stanovich, Keith E. 2018. Miserliness in human cognition: The interaction of detection, override and mindware. *Thinking & Reasoning* 24: 423–44.

Stupple, Edward J. N., Melanie Pitchford, Linden J. Ball, Thomas E. Hunt, and Richard Steel. 2017. Slower is not always better: Response-time evidence clarifies the limited role of miserly information processing in the Cognitive Reflection Test. *PLoS ONE* 12: e0186404. [CrossRef]

Tausczik, Yla R., and James W. Pennebaker. 2013. Improving teamwork using real-time language feedback. Paper presented at the SIGCHI Conference on Human Factors in Computing Systems, Paris, France, April 27–May 2; pp. 459–68.

Thompson, Valerie A., Jamie A. P. Turner, and Gordon Pennycook. 2011. Intuition, reason, and metacognition. *Cognitive Psychology* 63: 107–40. [CrossRef]

Thompson, Valerie A., Jamie A. P. Turner, Gordon Pennycook, Linden J. Ball, Hannah Brack, Yael Ophir, and Rakefet Ackerman. 2013. The role of answer fluency and perceptual fluency as metacognitive cues for initiating analytic thinking. *Cognition* 128: 237–51. [CrossRef]

Thompson, Valerie, and Kinga Morsanyi. 2012. Analytic thinking: Do you feel like it? *Mind & Society* 11: 93–105.

Topolinski, Sascha, and Rolf Reber. 2010. Immediate truth: Temporal contiguity between a cognitive problem and its solution determines experienced veracity of the solution. *Cognition* 114: 117–22. [CrossRef] [PubMed]

Undorf, Monika, and Arndt Bröder. 2021. Metamemory for pictures of naturalistic scenes: Assessment of accuracy and cue utilization. *Memory & Cognition* 49: 1405–22.

Undorf, Monika, and Edgar Erdfelder. 2015. The relatedness effect on judgments of learning: A closer look at the contribution of processing fluency. *Memory & Cognition* 43: 647–58.

Undorf, Monika, Anke Söllner, and Arndt Bröder. 2018. Simultaneous utilization of multiple cues in judgments of learning. *Memory & Cognition* 46: 507–19.

Unkelbach, Christian, and Rainer Greifeneder. 2013. A general model of fluency effects in judgment and decision making. In *The Experience of Thinking*. Edited by Christian Unkelbach and Rainer Greifeneder. Hove: Psychology Press, pp. 21–42.

Webb, Margaret E., Daniel R. Little, and Simon J. Cropper. 2016. Insight is not in the problem: Investigating insight in problem solving across task types. *Frontiers in Psychology* 7: 1424. [CrossRef]

Webb, M. E., Daniel R. Little, and Simon J. Cropper. 2018. Once more with feeling: Normative data for the aha experience in insight and noninsight problems. *Behavior Research Methods* 50: 2035–56. [CrossRef]

Wiltschnig, Stefan, Bo T. Christensen, and Linden J. Ball. 2013. Collaborative problem–solution co-evolution in creative design. *Design Studies* 34: 515–42. [CrossRef]

Zion, Michal, Idit Adler, and Zemira Mevarech. 2015. The effect of individual and social metacognitive support on students' metacognitive performances in an online discussion. *Journal of Educational Computing Research* 52: 50–87. [CrossRef]

MDPI AG
Grosspeteranlage 5
4052 Basel
Switzerland
Tel.: +41 61 683 77 34

Journal of Intelligence Editorial Office
E-mail: jintelligence@mdpi.com
www.mdpi.com/journal/jintelligence

www.ingramcontent.com/pod-product-compliance
Lightning Source LLC
Chambersburg PA
CBHW041418210926
51850CB00013B/3308
9783725840526